Topics in Applied Physics
Volume 97

Available  at
SpringerLink.com

Topics in Applied Physics is part of the SpringerLink service. For all customers with standing orders for Topics in Applied Physics we offer the full text in electronic form via SpringerLink free of charge. Please contact your librarian who can receive a password for free access to the full articles by registration at:

springerlink.com → Orders

If you do not have a standing order you can nevertheless browse through the table of contents of the volumes and the abstracts of each article at:

springerlink.com → Browse Publications

Topics in Applied Physics

Topics in Applied Physics is a well-established series of review books, each of which presents a comprehensive survey of a selected topic within the broad area of applied physics. Edited and written by leading research scientists in the field concerned, each volume contains review contributions covering the various aspects of the topic. Together these provide an overview of the state of the art in the respective field, extending from an introduction to the subject right up to the frontiers of contemporary research.

Topics in Applied Physics is addressed to all scientists at universities and in industry who wish to obtain an overview and to keep abreast of advances in applied physics. The series also provides easy but comprehensive access to the fields for newcomers starting research.

Contributions are specially commissioned. The Managing Editors are open to any suggestions for topics coming from the community of applied physicists no matter what the field and encourage prospective editors to approach them with ideas.

Managing Editors

Dr. Claus E. Ascheron

Springer-Verlag GmbH
Tiergartenstr. 17
69121 Heidelberg
Germany
Email: claus.ascheron@springer-sbm.com

Dr. Hans J. Koelsch

Springer-Verlag New York, LLC
233, Spring Street
New York, NY 10013
USA
Email: hans.koelsch@springer-sbm.com

Assistant Editor

Dr. Werner Skolaut

Springer-Verlag GmbH
Tiergartenstr. 17
69121 Heidelberg
Germany
Email: werner.skolaut@springer-sbm.com

Kiyomi Sakai (Ed.)

Terahertz Optoelectronics

With 270 Figures

 Springer  National Institute of Information and Communications Technology

Dr. Kiyomi Sakai
Kansai Advanced Research Center
National Institute of Information
and Communications Technology
588-2 Iwaoka, Nishi-ku
Kobe-shi, Hyogo 651-2492, Japan
sakai@crl.go.jp

Library of Congress Control Number: 2004114339

Physics and Astronomy Classification Scheme (PACS):
41.20.Jb, 42.25.Bs, 42.50.Nn, 42.65.Re, 42.72.Ai

ISSN print edition: 0303-4216
ISSN electronic edition: 1437-0859
ISBN-10 3-540-20013-4 Springer Berlin Heidelberg New York
ISBN-13 978-3-540-20013-0 Springer Berlin Heidelberg New York

Springer is a part of Springer Science+Business Media

springeronline.com

© Springer-Verlag Berlin Heidelberg 2005
Printed in Germany

Typesetting: DA-TEX · Gerd Blumenstein · www.da-tex.de
Production: LE-TEX Jelonek, Schmidt & Vöckler GbR, Leipzig
Cover design: design & production GmbH, Heidelberg

Printed on acid-free paper   57/3141/YL   5 4 3 2 1 0

Preface

Situated in the gap between electronics and optics, the terahertz frequency range of the electromagnetic spectrum has long been of interest to only a limited number of scientists and engineers due to a lack of efficient and affordable terahertz sources and detectors. But the advent of femtosecond lasers in the 1980s and photoconductive switches in 1984 have made the terahertz gap accessible, while at the same time advances in electronics and optics have made it narrower. Research activities in terahertz frequencies have risen dramatically since that time and funding has increased remarkably in the last decade. In spite of the enormous recent interest in the terahertz range, there are as yet very few books describing selected efforts in this field. In particular, Japanese work has not been published coherently at all, although much of the progress in the terahertz field originates from this country. In this book we will present recent and important developments in the field of terahertz radiation with emphasis on terahertz pulse radiation and on work carried out in Japan.

The first contribution by *Sakai* and *Tani* provides an historical overview of this modern and rapidly developing field and its bases. The next Chapter by *Kono* et al. describes the latest advances in emission and detection of terahertz radiation. The third Chapter by *Gu* and *Tani* deals with the emission of terahertz radiation from semiconductor surfaces followed by the contribution by *Ohtake* et al. that describes the enhancement of the surface radiation with an external magnetic field. The fifth Chapter by *Kadoya* and *Hirakawa* describes the microscopic ultrafast carrier transport in bulk and quantum structures in relation to the terahertz radiation. The Chapter by *Matsuura* and *Ito* deals with the continuous-wave terahertz radiation by means of photomixing that shows promising novel photomixers. The following Chapter by *Nishzawa* et al. includes various aspects of terahertz time-domain spectroscopy and shows advantages over traditional spectroscopy, indicating the results by applying the method to various phases of materials. The eighth Chapter by *Kida* et al. focuses on the terahertz radiation and strongly correlated electron systems. The last Chapter by *Herrmann* et al. deals with terahertz imaging, in which some important results are shown and discussed.

Throughout this book basic ideas are explained briefly but effectively. The book has been compiled in such a way that each chapter can stand on its own while relations to other chapters are observed. So it is believed that the book will be useful not only for readers intending to enter this area of science and technology but also for those already working in this area.

Kobe, December 2004   *Kiyomi Sakai*

Contents

Terahertz Time-Domain Spectroscopy

Seizi Nishizawa, Kiyomi Sakai, Masanoi Hangyo, Takeshi Nagashima,
Mitsuo Wada Takeda, Keisuke Tominaga, Asako Oka,
Koichiro Tanaka, Osamu Morikawa

Terahertz Optics in Strongly Correlated Electron Systems

Noriaki Kida, Hironaru Murakami, Masayoshi Tonouchi

Introduction to Terahertz Pulses

Kiyomi Sakai[1] and Masahiko Tani[1,2]

[1] National Institute of Information and Communications Technology
588-2 Iwaoka, Nishi-Ku, Kobe 651-2492, Japan
sakai@nict.go.jp
[2] Institute of Laser Engineering, Osaka University
2-6 Yamadaoka, Suita, Osaka 565-0871, Japan

Abstract. An historical review of the optoelectronic terahertz technologies, initiated by *Auston* and *Lee* and developed remarkably, keeping in step with the rapid progress of short pulse lasers is given first, showing the variety of activities until recently. Then the principles of some terahertz-pulse generation and detection are explained showing experimental results, as the basis of the current booming technology.

1 Historical Introduction

The technology of generating short optical pulses has advanced remarkably in recent years (Fig. 1). The availability of short optical pulses, especially those of femtosecond laser pulses, and the development of semiconductor technology, including the ultrafast photoconductive (PC) thin films and semiconductor quantum structures as a fraction of its products, have fostered an innovative field calle terahertz (THz) optoelectronics. It is in essence based on the PC switching pioneered by *Auston* [1] and *Lee* [2].

In the mid-1970s, optical pulses from a mode-locked Nd:glass laser and high-resistivity Si [1] or Cr-doped semi-insulating GaAs (SI-GaAs) [2] were used for switching. This optoelectronic means of switching with the use of PC material, i.e., PC switching is often called Auston switching. Most of the early works are summarized by *Auston* and *Lee* in the book edited by *Lee* [3].

In 1980, *Auston* et al. [4] demonstrated a sampling technique by using transmission-line structures and amorphous Si (a-Si) film on fused silica substrate as a photoconductor. *Smith* et al. [5], in 1981, developed an excellent PC film, radiation-damaged Si-on-sapphire (RD-SOS), of which carrier lifetime is controlled by ion implantation. The RD-SOS was then constantly used as a fast response photoconductor for about 10 years.

Mourou et al. [6] in 1981 and *Heidemann* et al. [7] in 1983 used the PC switching to drive their antennas and emitted picosecond microwave transients in free space. Then, 0.5 ps optical pulses from continuous-wave (CW) mode-locked dye laser and Cr-doped SI-GaAs were used and 1 ps to 3 ps microwave transients emitted from each antenna were detected with respective standard microwave detectors. It is to be noted that the PC switch played

K. Sakai (Ed.): Terahertz Optoelectronics, Topics Appl. Phys. **97**, 1–30 (2005)

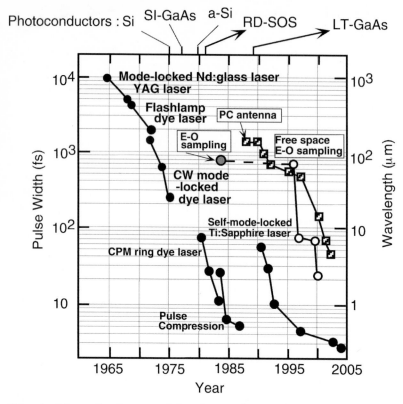

Fig. 1. Schematic diagram of the shortest laser pulses reported vs year and the detected highest frequencies (shortest wavelengths) vs. year. Improvements of available ultrafast photoconductors are shown on the *top*

a subsidiary role in these experiments and the radiation source was not the DC-biased PC switch itself.

Auston et al. [8], in 1983, refined their sampling technique by using the RD-SOS with lifetimes shorter than 2.6 ps and with 0.3 ps optical pulses from the dye laser. They showed coherent detection of a short burst of millimeter-wave radiation without the need for additional rectifying elements.

In 1984, *Auston, Cheung* and *Smith* [9] were able to emit approximately 1.6 ps electromagnetic (EM) pulses, let them propagate freely and detect them coherently using such a configuration as the PC switch arranged symmetrically on opposite sides of a dielectric slab. The radiation source was the ultrafast transient current in the PC switch and the detection was performed coherently by sampling the repetitively incident EM pulses. They used 100 fs optical pulses from colliding-pulse passively mode-locked (CPM) ring dye laser and RD-SOS for the experiment. This experiment is substantially the origin of the current booming optoelectronic THz technologies referred to as THz optoelectronics.

In the next step, *DeFonzo* et al. [10, 11] emitted EM pulses in air, let them propagate in air and then coherently detected them both by using identical broadband tapered slot antennas monolithically integrated on RD-SOS substrate in 1987.

In 1988, *Smith*, *Auston* and *Nuss* [12] showed a refined antenna structure with the PC dipole on RD-SOS. They used the antenna both as an emitter (or transmitter) and a detector (or receiver), and tested by pumping the center of each dipole with 120 fs optical pulses from a CPM dye laser for coherent emission and detection. It was revealed that such dipoles have frequency spectra that extend from ≈ 100 GHz to over 1 THz and this paper is regarded as the first paper that mentioned that the EM transient emitted from the PC antenna involves THz frequency components.

In 1989, *van Exter*, *Fattinger* and *Grischkowsky* [13] showed a refined dipole antenna with a coplanar transmission line, introduced by *Ketchen* et al. [14], to match the pulse generated in the gap with the propagating modes, and showed an EM beam system including GHz and THz frequency components with off-axis paraboloidal mirrors and with a lens on each PC dipole. Furthermore, they characterized their EM beam system precisely and showed the system is useful as a spectroscopic tool (terahertz time-domain spectroscopy, THz-TDS) [15].

Since the discovery of self-mode-locked operation in Ti:sapphire oscillators [16], there has been rapid development in the generation of sub-100 fs optical pulses. The laser oscillation by the solid gain media is more stable compared with liquid such as dye, though the shortest pulse width of dye lasers already attained 6 fs in 1987. Early in the 1990s, Ti:sapphire femtosecond lasers became commercially available and THz optoelectronics, which was conducted by those engaged in the ultrafast optoelectronics such as those at AT & T Bell Laboratories and IBM's T. J. Watson Research Center, began to spread all over the world and the word "THz-pulse" began to be used to express ultrafast EM transients, gradually.

At almost the same time as the Ti:sapphire sub-100 fs laser was brought to market, the molecular-beam-epitaxy (MBE) grown low temperature GaAs (LT-GaAs) thin film began to be used. This film was first developed by *Smith* et al. [17] as a buffer layer in GaAs MESFETs and later its high optical and electrical quality as a PC material was pointed out by *Gupta* et al. [18]. The carrier lifetime of LT-GaAs is controlled by the substrate temperature in epitaxial growth by MBE. It has the shortest carrier lifetime compared to any other materials, while keeping relatively high mobility. The carrier dynamics of LT-GaAs has been discussed in many papers (e.g., [19]). Since then, infinitesimal dipoles, or Hertzian dipoles, fabricated on LT-GaAs substrate have been used frequently as an emitter or detector driven by PC shortening the dipole gap pumping with a sub-100 fs Ti:sapphire laser.

Pedersen et al. [20] and *Jepsen* et al. [21] at Keiding's group in Denmark investigated carrier transport and the screening effect of semiconductor antenna. *Ludwig* and *Kuhl* [22], or *Tani* et al. [23] and *Piao* et al. [24] in Sakai's group in Japan examined emission characteristics of PC antennas. *Jepsen* et al. [25] and *Rudd* et al. [26] reported radiation patterns from lens-coupled THz antennas. *Cai* with *Brener* et al. [27] showed the design and performance of various PC dipole antennas. Recently, *Duvillaret* et al. [28] in Coutaz' group in France described a sophisticated analysis of PC antennas including the width of short optical pulses.

Around 1990, extended PC antennas, i.e., large-aperture PC antennas having a simple form or large-aperture Si p-i-n diodes were shown by *Darrow* et al. [29], *Hu* et al. [30] and *Xu* et al. [31] in Auston's group. More precise theoretical analyses compared with experimental results were carried out by *Darrow* et al. [32] or *Benicewicz* et al. [33], *Rodriguez* [34, 35] at Taylor's group. In 1993, *You* et al. [36] generated the most powerful half-cycle pulses with pulse energies as high as $0.8\,\mu J$ and pulse length less than $500\,fs$, with a repetition rate of $10\,Hz$.

Here, it is worth noting that the PC antenna fabricated on the high-speed LT-GaAs has been used as photomixers to generate CW THz radiation initiated by E. Brown et al. in 1993. This technique has progressed so much and has formed an active research field. The precise description of CW THz radiation will be shown, with its history, in the Chapter by *Matsuura* et al. Expect in that chapter, the word "THz radiation" is used for the pulsed THz radiation throughout this book.

In the beginning of the 1970s, two groups, *Yajima* and *Takeuchi* [37, 38] and *Yang, Richards* and *Shen* [39] reported the far-infrared (FIR, roughly overlaps with the current definition of the THz frequency range) generation, independently, due to the optical rectification with the use of a second-order nonlinear (NL) crystal and picosecond pulses from mode-locked Nd:glass laser. In 1983, *Auston* [40] predicted that subpicosecond electro-optic (EO) shock waves are radiated in a Cherenkov cone from a moving polarization, due to the optical rectification, when an ultrashort optical pulse is focused in a NL material. This Cherenkov radiation with a broad bandwidth was soon realized using femtosecond laser pulses and a $LiTaO_3$ crystal [41], and the radiation was utilized for coherent TDS in succession [42]. Recently, such a technique to emit broadband radiation has become even more important as the detection bandwidth expanded unexpectedly, as is stated in the Chapter by *Kono* et al. Because of this, optical rectification attracts much attention as it emits an extremely broadband radiation limited only by the duration of the optical pulse [43].

Returning to the shock wave, *Grischkowsky* et al. [44] have pointed out that it is radiated from transmission lines. This radiation is, however, considered to be a loss for an emitter having a transmission line in its configuration.

In 1990, *Zhang* et al. [45] generated subpicosecond EM pulses from a semiconductor surface. They explained the pulses are emitted due to the surface current that flows normal to the semiconductor surface. *Hu* et al. [46] investigated some semiconductors by changing sample temperatures. *Chuang* et al. [47] suggested that FIR pulse generation at semiconductor surfaces includes that due to electric-field-induced optical rectification in the depletion field, which is sensitive to the crystal orientation. *Green* et al. [48] and *Zhang* et al. [49] observed a THz-pulse emission dependence on the crystal orientations.

Apart from the radiation from the bulk materials, emission from semiconductor quantum structures due to charge oscillations was observed early in the 1990s. THz-pulse emission from asymmetric coupled quantum wells was studied first by *Leo* et al. [50] in 1991 using time-resolved degenerate four-wave-mixing and by *Roskos* et al. [51] in 1992 using a PC dipole antenna, as well as the observation of emission from single quantum wells by *Planken* et al. [52] in 1992 or that from Bloch oscillations in a superlattice by *Waschke* et al. in Kurz's group [53] in 1993.

In 1995, *Dekorsy* et al. [54] in the same group reported the first observation of THz-pulse emission arising from coherent infrared (IR)-active phonons in a semiconductor. *Kersting* et al. [55] then in Gornik's group observed THz-pulse emission from cold plasma oscillations in a semiconductor. *Tani* et al. [56] and *Gu* et al. [57] also observed emission due to coherent LO phonons and LO phonon–plasmon coupling mode, respectively.

In 1995, *Hangyo* et al. [58] observed THz-pulse emission for the first time from a high-T_c superconducting thin-film bridge with bias voltage and in 1996, *Tonouchi* et al. [59] observed THz-pulse emission from the same devices under external magnetic field, both by exciting the bridge with femtosecond optical pulses.

In contrast to the above-mentioned hybrid optical-electronic approach, an all-electronic system has been developed by *Bloom* and his coworkers at Stanford University since 1987 [60,61,62]. An all-electronic system uses monolithic nonlinear transmission lines (NLTLs) as ultrafast voltage-step generators for both generating picosecond pulses and driving monolithically integrated diode samplers for the detection of these pulses. The advantages of this approach lie in its relative simplicity and its robustness, on the other hand, the disadvantage is its limited available frequency range, less than about 1 THz.

As early as 1984, *Auston* et al. [41] used the EO effect for the detection of Cherenkov shock waves using the same EO crystal (a NL crystal is also called an EO crystal when its EO effect is used). In 1995, *Wu* and *Zhang* [63] applied this technique for free-space THz-pulse sensing. This sensing is based on the Pockels effect, and it is particularly attractive because it offers a flat frequency response over an ultrabroad bandwidth [64] reaching nowadays over 100 THz [65] and because it offers a large detection area that is useful for imaging.

On the other hand, until recently it was believed that the bandwidth of the PC antenna is limited to below about 7 THz, but very recently *Kono* et al. [66, 67, 68] have broken the limitation and extended the upper limit as high as 60 THz with the use of 15 fs light pulses. The extension of the detection bandwidth is added in Fig. 1, which shows the development of short laser pulses. The figure also shows various photoconductors with their beginnings of utilization.

In writing this section, the authors referred to the second thesis written by *Keiding* [69], especially the history before 1990. The articles relating to THz optoelectronics are found in the books [70, 71, 72, 73, 74, 75].

2 Principles of Terahertz-Pulse Generation

2.1 Terahertz-Pulse Emission from Photoconductive Antennas

The THz pulses have been generated by different methods, such as irradiation of PC antennas, semiconductor surfaces, or quantum structures with femtosecond optical pulses. Among these methods, THz-pulse emission from the PC antenna is crucial and fundamental. Figure 2 shows a standard THz-pulse generation scheme, which has been developed by many researchers as stated in Sect. 1.

For the construction of the PC antenna, a PC substrate with such excellent properties as short carrier lifetime, high mobility and high breakdown

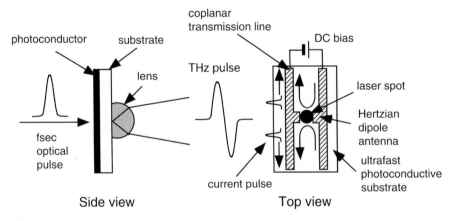

Fig. 2. Standard THz-pulse generation from a PC antenna pumped with a femtosecond optical pulse. Propagation of current pulses along the coplanar transmission line are shown in the *top view*

Table 1. Characteristics of ultrafast photoconductive materials

Photoconductive materials	Carrier lifetime (ps)	Mobility $(cm^2/(V \cdot s))$	Resistivity $(\Omega \cdot cm)$ (Breakdown field, V/cm)	Band gap (eV at R.T.)
Cr:doped SI-GaAs	50–100.0	$\approx 1\,000$	10^7	1.43
LT-GaAs	0.3	150– 200	10^6 (5×10^5)	1.43
SI-InP	50–100.0	$\approx 1\,000$	4×10^7	1.34
Ion-Implanted InP	2– 4.0	200	$> 10^6$	1.34
RD-SOS	0.6	30		1.10
Amorphous Si	0.8– 20.0	1	10^7	1.10
MOCVD CdTe	0.5	180		1.49
LT-In$_{0.52}$Al$_{0.48}$As	0.4	5		1.45
Ion-implanted Ge	0.6	100		0.66

voltage are needed. Historically, such PC materials shown in Table 1 have been used.

Among them, RD-SOS and LT-GaAs have been used most frequently. The RD-SOS is prepared by implanting argon, silicon, or oxygen ions into SOS samples consisting of, e.g., $1\,\mu m$ thick $50\,\Omega \cdot cm$ (100) silicon films on 0.32 mm sapphire substrates. By implanting these ions, dislocations are formed that cause shortening of the carrier lifetime. The carrier lifetime of RD-SOS depends strongly on the dose of ion implantation. Figure 3a shows an example of reflectivity change for a RD-SOS measured by means of the pump-and-probe method and Fig. 3b shows a relation between the carrier lifetime and dose of ion implantation [76].

Since the advent of LT-GaAs layers much attention has been paid and they have been extensively studied for ultrafast optoelectronic applications due to their improved subpicosecond carrier lifetime, relatively high carrier mobility, and high breakdown fields. The properties of the LT-GaAs depend on both the growth conditions during the MBE process, and on the post-growth annealing. It is known that during the growth step, excess As is incorporated, which leads to the formation of a high density ($10^{17}\,cm^{-3}$ to $10^{20}\,cm^{-3}$) of As$_{Ga}$-related point defects. These defects form a midgap band, and act as nonradiative recombination centers that decrease the carrier lifetimes. When the growth temperature is lowered, more defects are incorporated and the carrier lifetime is reduced. In order to increase the resistivity the sample is *in situ* annealed. By annealing above 600 °C in As overpressure a large number of As-rich precipitates are formed. Although as-grown LT-GaAs is relatively conducting ($\rho \approx 10\,\Omega \cdot cm$) at room temperature, annealed LT-GaAs is semi-insulating ($\rho \approx 10^7\,\Omega \cdot cm$). Despite the high density of point defects and As precipitates, the Hall mobility at room temperature in annealed LT-GaAs is relatively high ($\mu = 1000\,cm^2/(V \cdot s)$) preserving short carrier lifetime. Thus the LT-GaAs films with, e.g., $1.5\,\mu m$ thickness were grown on the, e.g., 0.4 mm semi-insulating GaAs (SI-GaAs). Figure 4a shows

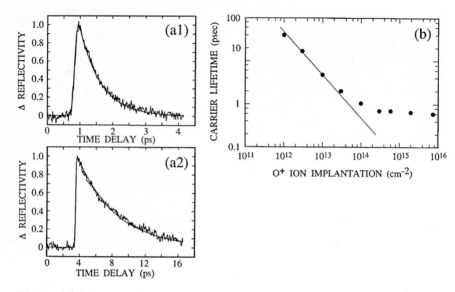

Fig. 3. (a) Measured change in reflectivity for RD-SOS samples ion implanted at doses of $2 \times 10^{15}\,\mathrm{cm}^{-2}$ (**a1**) and $1 \times 10^{13}\,\mathrm{cm}^{-2}$ (**a2**). *Smooth curves* are fits obtained using a 140 fs instrument response convolution with 0.65 ps (**a1**) and 4.8 ps (**a2**) exponential decays [76], (**b**) Carrier lifetimes vs. ion-implantation dose. Lifetimes are derived from individual reflectivity data at each dose. The *solid line* shown has a slope of unity [76]

an example of the time-resolved reflectance change for LT-GaAs grown at various substrate temperatures and *in situ* annealed, and Fig. 4b shows the growth-temperature dependence of the carrier lifetime [77, 78].

The PC antennas were fabricated on the LT-GaAs substrate having the shortest carrier lifetime. The structure of the antenna is shown in Fig. 2 right (top view). A coplanar transmission line and a dipole antenna with a small gap at the center made of AuGe/Ni/Au metal layer, was fabricated on the substrate.

The physical properties of a THz emitter system is described based on the PC antenna shown in Fig. 2 right, which behaves as a Hertzian dipole. When the PC gap is pumped with femtosecond optical pulses with an energy greater than the bandgap of the semiconductor ($E_\mathrm{g} = 1.43\,\mathrm{eV}$ for GaAs at room temperature), free electrons (holes) are generated in the conduction (valence) band. The carriers are then accelerated in phase by the bias field and decay with a time constant determined by the carrier lifetime, resulting in a pulsed photocurrent (or a step-function-like photocurrent in long-carrier-lifetime semiconductors) in the PC antenna. Current modulation occurs in the subpicosecond regime and thus emits a subpicosecond EM transient, i.e., THz pulse. For an elementary Hertzian dipole antenna in free space, the radiated electric field $E(r, t)$ at a distance r (much greater than the wavelength

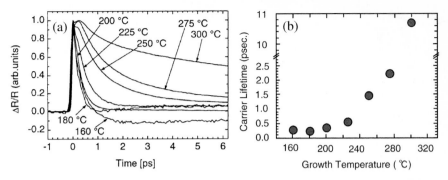

Fig. 4. (a) Time-resolved reflectance changes for LT-GaAs grown at temperatures 160 °C, 180 °C, 200 °C, 225 °C, 250 °C, 275 °C and 300 °C measured at $\lambda = 835$ nm. The peak values are normalized to unity. The inversion between the 160 °C and 180 °C curves is attributed to the crystallinity of the film. Each film above 180 °C is single-crystalline-like and the film of 160 °C is polycrystalline like, (b) Growth-temperature dependence of the carrier lifetime for LT-GaAs layers [78]

of the radiation) and time t are described as

$$E(r,t) = \frac{l_e}{4\pi\varepsilon_0 c^2 r}\frac{\partial J(t)}{\partial t}\sin\theta \propto \frac{\partial J(t)}{\partial t}, \tag{1}$$

where $J(t)$ is the current in the dipole, l_e the effective length of the dipole, ε_0 the dielectric constant of a vacuum, c the velocity of light in a vacuum, and θ the angle from the direction of the dipole.

Equation (1) indicates that the radiation amplitude is proportional to the time derivative of the transient photocurrent $\partial J(t)/\partial t$ and the effective antenna length l_e. The photocurrent density is described as [28]

$$j(t) \propto I(t) \otimes [n(t)qv(t)], \tag{2}$$

where $\otimes$ denotes the convolution product, $I(t)$ is the optical intensity profile, and $q, n(t)$ and $v(t)$ are, respectively, the charge, the density, and the velocity of photocarriers. The dynamics of photogenerated free carriers in a semiconductor is well described by the classical Drude model. According to this model, the average velocity of free carriers obey the differential equation

$$\frac{dv(t)}{dt} = -\frac{v(t)}{\tau} + \frac{q}{m}E(t), \tag{3}$$

where τ is the momentum relaxation time and m is the effective mass of the carrier. The current density $n(t)qv(t)$ represents the impulse response of the PC antenna, i.e., the response to a delta-function-like optical pumping. Figure 5 shows a typical temporal behavior of the photocurrent density in

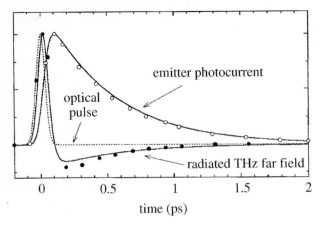

Fig. 5. Calculated photocurrent in the emitter and amplitude of the radiated field vs. time. The temporal shape of the optical pulse is drawn as a *dotted line* [28]

the emitter and of the associated radiated THz far field, and the temporal shape of the optical pulse is assumed to be Gaussian.

It is illustrated in Fig. 2 left (side view) that the generated THz pulse is emitted from the substrate side. This is based on the antenna theory that shows that a dipole antenna on the surface of dielectric material emits roughly $\varepsilon^{3/2}/2$ times more power to the dielectric material than to the air, where ε is the relative dielectric constant of the substrate [79].

2.2 Terahertz-Pulse Emission from Extended Photoconductive Sources

In order to generate high-power and large-aperture THz pulse, extended PC source, i.e., large-aperture photoconductors biased with a high DC voltage are used by pumping them with amplified femtosecond laser pulses. Large-aperture PC antennas are defined to have an optically illuminated area of dimensions much greater than the center wavelength of the emitted THz radiation. The construction of the large-aperture PC antenna is simple, as can be seen in Fig. 6.

The electrodes are formed on photoconductors such as InP, GaAs, RD-SOS and so on [29, 32]. The separation between the electrodes ranges from a few millimeter to a few centimeter. A model of the radiated THz field near the emitting large-aperture antenna can be constructed from the boundary conditions on the electric and magnetic fields at the surface of the emitter.

The analysis [32] shows that the inward and outward propagating electric fields are comparable but the inward propagating magnetic field is $\sqrt{\varepsilon}$ larger than the outward propagating field. Hence, the intensity of the inward and outward waves will be approximately in the ratio of $\sqrt{\varepsilon} : 1$. This behavior is

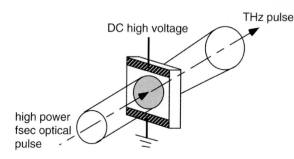

Fig. 6. Typical experimental arrangement for the generation of THz pulses from an extended source (large-aperture PC antenna)

in sharp contrast to the radiation from an elementary Hertzian dipole on the surface of a dielectric material. The maximum THz energy is obtained from a 3.5 cm × 3.5 cm GaAs wafer biased with an external field of 10.7 kV/cm, which is 0.8 μJ, with a 450 fs pulse duration and with a repetition rate of 10 Hz [36].

2.3 Terahertz-Pulse Emission other than Photoconductive Antennas

In addition to the PC antenna, the generation of THz pulses has been achieved by other techniques, such as optical rectification in NL media, surge current drive by the surface depletion field of semiconductors, charge oscillations in semiconductor quantum structures, coherent excitation of polar optical phonons, modulation of supercurrent in a biased high-T_c bridge, and NLTLs. These mechanisms are illustrated in Fig. 7. Some of them are described here precisely and the rest are described in the Chapters by *Gu* et al., *Kadoya* et al., and *Kida* et al.

2.3.1 Optical Rectification in Nonlinear Media

The THz-pulse emission due to the optical rectification is caused by a second-order NL process (Fig. 7a). Typically, a train of visible or near-IR (NIR) short pulses is focused on a second-order NL materials including 1. dielectric crystals, 2. semiconductors or 3. organic materials. The optical rectification is nothing other than all the possible difference-frequency generation in the broad frequency spectra of the short optical pulse.

The NL polarization $P(\omega)$ created by the pump fields $E(\omega)$ is written as

$$P(\omega) = \varepsilon_0 \chi^{(2)}(\omega = \omega_1 - \omega_2) E(\omega_1) E^*(\omega_2). \tag{4}$$

The Fourier transformation of $P(\omega)$ yields the polarization in the time domain $P(t)$ and the emitted THz field is written as

$$E_{\mathrm{THz}}(t) \propto \frac{\partial^2 P(t)}{\partial t^2}. \tag{5}$$

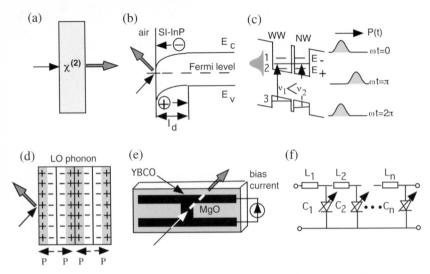

Fig. 7. THz-pulse emission mechanisms other than PC antenna, (**a**) optical rectification in NL media, (**b**) surge current and NL process, (**c**) semiconductor quantum structure, (**d**) coherent LO phonons, (**e**) high-T_c superconducting bridge, (**f**) NLTL

The generation of a THz pulse through optical rectification has been demonstrated in various materials, such as 1. LiNbO$_3$, LiTaO$_3$ [37, 80, 81], 2. ZnTe, InP, GaAs, GaSe, etc. [37, 43, 47, 48, 49, 82] and 3. DAST, MA1:MMA [83, 84].

In spite of the lower emission power, compared with PC antenna, optical rectification attracts much attention, recently with the extension of the detection bandwidth to higher frequencies, for its simplicity and its spectral broadness arising from the instantaneous occurrence of the phenomena. In order to improve the emission efficiency, the phase-matching condition has been taken into consideration, recently. This is described in the following Chapter by *Kono* et al.

2.3.2 Surge Current at the Semiconductor Surface

The THz-pulse emission from most wide-bandgap semiconductors can be explained by the effect of the surface-depletion field. A clean semiconductor surface is covered with a monolayer composed of atoms with covalent-bond-free orbitals in the surface known as dangling bonds. These bonds form a large number of surface states and produce influences on electrical properties of the semiconductors surfaces.

Donors and acceptors inside the crystal are trapped by the surface state and tend to pin the Fermi level, which forms space charge layers on the surface, and bend the energy bands near the surface of semiconductor/air interfaces, forming a charge-depletion region and thus a built-in surface electric field. The depletion field is directional normal to the surface, and the strength of the field is a function of the Schottky-barrier potential and the doping concentration. In general, the energy band is bent upward in an n-type semiconductor and is bent downward in a p-type semiconductor.

Figure 7b schematically shows a band diagram of a p-type semiconductor. When an ultrafast optical pulse pumps a semiconductor surface with a photon energy greater than the bandgap, photons are absorbed and create electron–hole pairs. The built-in static field drives the two kinds of carriers in opposite directions, i.e., in the case of Fig. 7b, the electrons move to the surface and the holes to the wafer. The free carriers are swept across the depletion layer, with the width of l_d, the photocurrent flows and a dipole layer builds up. This current flow (called surge current) emits THz pulses according to (1).

For narrow-bandgap semiconductors with bandgap energy less than approximately 0.4 eV, the surge current is induced by different diffusion velocities between electrons and holes, and subsequent relaxation of charge distribution, called the photo-Dember effect [54].

The rise of the photocurrent, which plays an important role in THz-pulse emission, is on the order of the laser-pulse duration in both cases. The direction of the emitted THz beam into the free space and into the wafer satisfy the generalized Fresnel's law. The beam into the free space is collinear with the reflected pump beams. It is obvious that the emitted THz-pulse radiation caused by the surge current can not be observed from the direction normal to the surface, as the dipole is formed normal to the surface. The THz radiation caused by the surge current is usually overlapped with the radiation arising from the optical rectification due to the $\chi^{(2)}$ inherent in the semiconductor or induced by the symmetry-breaking surface-depletion field in systems with inversion symmetry [47]. The most striking evidence for the $\chi^{(2)}$ process is a strong modulation of the detected field of THz radiation as the sample is rotated about its surface normal. The crystal-orientation dependence of the emitted THz radiation is shown in Fig. 8 [85]. It is seen that THz radiation from both origins are overlapping.

The discussion of this section is extended in the Chapter by Gu et al.

2.3.3 Terahertz-Pulse Emission from Semiconductor Quantum Structures

It was demonstrated that the existence of excitonic charge oscillations emits THz radiation, typically in two GaAs/Al$_x$Ga$_{1-x}$As quantum-well structures: an asymmetric double-coupled quantum well (DCQW) consisting of 10 repetitions of a 10 nm GaAs narrow well and a 14.5 nm GaAs wide well separated by a 2.5 nm Al$_{0.2}$Ga$_{0.8}$As barrier [51] and a single quantum well (SQW) consisting of 15 periods of 17.5 nm GaAs wells separated by 15 nm of Al$_{0.3}$Ga$_{0.7}$As barriers [52]. Both samples are used at a temperature of around 10 K. Schematic energy-level diagrams of a DCQW system under a bias field are presented in Fig. 7c. The electric field is applied in the direction perpendicular to the layers between the n-doped substrate and a semitransparent Cr

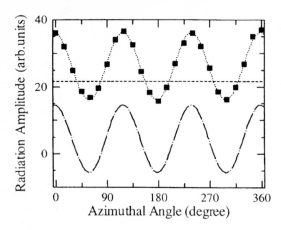

Fig. 8. Angular dependence of the THz radiation amplitude from InAs. *Upper curve* shows THz radiation with a DC offset by the surge current. *Lower curve* shows THz radiation extracted by the surge-current contributions [85]. The amplitude of periodic angular dependence is equal above and below the zero level only for normal incidence [86]

Schottky top contact. In the DCQW, the first ($n = 1$) electron levels in the wide well (WW) and the narrow well (NW) can be energetically aligned with each other by the electric field, while the holes remain localized because the energy levels in the valence band are not aligned at this bias. The aligned levels split into a bonding and an antibonding state ($|2\rangle$ and $|1\rangle$), separated by an energy gap proportional to the tunneling probability of the carriers through the AlGaAs barrier. In SQW these are the heavy- and the light-hole exciton levels. When levels $|1\rangle$ and $|2\rangle$ are excited coherently with a broadband short optical pulse, charge oscillations that are due to quantum beats between these two levels occur with a beat frequency of $\Delta E/h$, where $\Delta E = E_- - E_+$. A time-dependent polarization $P(t)$ is associated with the charge oscillations in the systems, leading to dipole emission with a radiated electric field $E(t) \sim \partial^2 P(t)/\partial t^2$. It should be noted here that the radiated field in DCQW includes the initial transient at middle or relatively high bias fields in addition to the oscillating signal. This transient is attributed to the optical rectification due to the field-induced $\chi^{(2)}$ process. In SQW the THz emission is caused by the charge oscillations following the coherent optical excitation of light- and heavy-hole excitons in an electric field. THz radiation is emitted even when no electric field is present in the sample, suggesting that valence-band mixing leads to a significant FIR transition dipole moment between the light-hole and heavy-hole subbands.

2.3.4 Terahertz-Pulse Emission from Coherent Longitudinal Optical Phonons

The impulsive pumping with femtosecond optical pulses generates coherent phonons in a number of condensed media, such as semiconductors, semimetals and superconductors. Studies of coherent phonons have been performed mostly by measuring the transient reflectivity or transmittance with the use of pump-probe experimental techniques. In these studies, only Raman-active phonon modes were observed and IR-active phonon modes were not observed.

Theoretically, it was first predicted that the IR-active coherent longitudinal optical (LO) phonons excited in polar semiconductors, such as GaAs, emit THz pulses [87]. Experimentally, the THz emission was first observed from the mode excited in Te [54] and then from GaAs [88]. Figure 7d shows a schematic illustration of coherent LO phonons formed in a solid sample. The emitted THz radiation arises from the oscillating polarizations with the radiated electric field $E(t) \sim \partial^2 P(t)/\partial t^2$. The excitation mechanism for coherent LO phonons in a wide-bandgap semiconductor such as GaAs is based on the ultrafast depolarization of the surface depletion field associated with the ultrafast polarization change within the surface-field region caused by the optical injection of carriers. The excitation mechanism for coherent LO phonons in a narrow-gap semiconductor such as Te is based on the ultrafast buildup of a photo-Dember field and the coupling of a polar lattice mode to this field.

Extended discussion of this section will be described in the Chapter by *Gu* et al.

2.3.5 Terahertz-Pulse Emission from a High-T_c Superconducting Bridge

It has been considered that the THz pulses may be emitted from the high-T_c superconducting bridge based on the inverse process of the semiconductor PC switch in which the radiation is generated by optically short-circuiting the switch [58]. The inverse process is to reduce the supercurrent by optically breaking the Cooper pairs of the supercurrent. Figure 7e is a schematic illustration of a high-T_c superconducting bridge analogous to the semiconductor PC switch. Cooling the device below the transition temperature and applying the bias current, the supercurrent flows through the bridge. By the irradiation of the bridge with femtosecond optical pulse, some Cooper pairs are broken and changed to quasiparticles instantaneously that undergo scattering, lose drift velocity and recombine into the Cooper pairs in a very short time in between repetitively arriving optical pulses. The supercurrent therefore, decreases quickly and recovers very rapidly, within about 1 ps. This transient of the supercurrent emits THz pulses of which field E is proportional to $\partial J(t)/\partial t$. The THz pulses are also emitted from the same bridge under an external magnetic field, without bias current [59]. The origin in

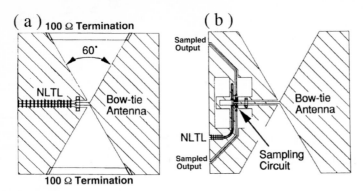

Fig. 9. Simplified integrated circuit layouts of **(a)** picosecond transmitter and **(b)** picosecond receiver [61]

this case is also attributed to a transient of the supercurrent induced in the high-T_c bridge. The Fourier components of the emitted THz pulse extend up to ≈ 3 THz.

More precise descriptions of this section will be shown in the Chapter by *Kida* et al.

2.3.6 All-Electrical Terahertz-Pulse Emission and Detection

An all-electronic system consists of NLTLs (Fig. 7f) and NLTL-gated sampling circuits for generation and detection of EM transients. The NLTL is a transmission line periodically loaded with reverse-biased Schottky varactor diodes serving as voltage-variable capacitors that cause a wave traveling along the line to experience a voltage-dependent propagation velocity, resulting in a shock wave whose main features resemble a step function. This step function is used both as an ultrafast signal source and as a strobe generator for a diode sampling bridge. The formation of shock waves with a sinusoidal input is similar to that occurring with step-function excitation. The negative-going transitions of a sinusoidal input are compressed into shock waves, and the output is a sawtooth waveform. The NLTL output is coupled to a broadband bow-tie antenna having a frequency-independent far-field radiation pattern and antenna impedance, which is used as a transmitter, while the receiver is a bow-tie antenna interfaced to an NLTL-gated sampling circuit (Fig. 9) [61]. The transmitter NLTL is driven between 7 GHz and 14 GHz, while the sampling circuit is driven at a frequency 100 Hz below the transmitter frequency.

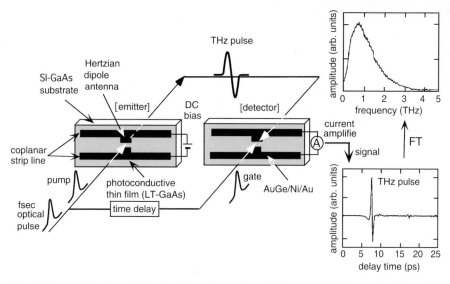

Fig. 10. Standard optical setup for coherent emission and detection of THz pulse with PC antennas. The emitter pumping optical pulses are mechanically chopped. Partially divided pumping beam, after adding time delay, synchronously gate the detector. Emitted THz pulse is recorded as a function of the delay time. Fourier components of the THz pulse are also shown

3 Principles of Terahertz-Pulse Detection

3.1 Terahertz-Pulse Detection with Photoconductive Antennas

The detection of a THz pulse is explained together with the emission of a THz pulse. Figure 10 is a standard optical configuration for coherent emission and detection.

For simplicity of explanation only key components are illustrated clearly and the secondary components such as a chopper, mirrors etc., are excluded. A Ti:sapphire laser generates short optical pulses with the width ranging from $\approx 10\,\mathrm{fs}$ to $\approx 100\,\mathrm{fs}$ at a repetition rate of around 70 MHz to 80 MHz, depending on the laser system used. The femtosecond optical pulse, mechanically chopped at a few kHz are focused on the dipole gap typically with $5\,\mu\mathrm{m}$ separation. The pumping pulses have photon energy above the direct bandgap of the GaAs. The THz pulse is emitted in a dipole-like pattern with the dipole axis oriented along the direction of the applied field at the same repetition rate as the pumping femtosecond optical pulse. The generated THz pulse that propagates through the substrate is collimated using a high-resistivity hyperhemispherical (taking the substrate thickness and its refractive index into consideration) Si lens attached to the back side of the GaAs source chip.

In this arrangement, an off-axis paraboloidal mirror is used to further collimate or focus the THz beam. A symmetric arrangement of paraboloidal

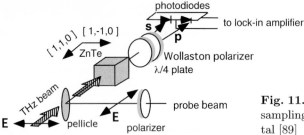

Fig. 11. Details of free-space EO sampling by means of ZnTe crystal [89]

mirror and Si lens is used to focus the beam on the PC antenna that is synchronously gated with a time-delayed femtosecond optical pulses divided partially from the pumping pulses by a beam splitter. The PC-antenna detector is similar to the emitter having a 5 μm gap between 10 μm long dipole. When the carrier lifetime of a PC antenna is much shorter than the THz pulse, it acts as a sampling gate that samples the waveform of the THz pulse. This scheme of detection is called PC sampling. The photocurrent measured at the detector, proportional to the electric field of the focused THz radiation, is amplified with a low-noise current amplifier and is fed to a lock-in amplifier. The THz-pulse waveform is obtained by measuring the average photocurrent versus time delay between the THz pulses and the gating optical pulses. A Fourier analysis of the temporal profile of the received THz pulse reveals the amplitude and phase spectrum. This system allows THz-TDS to be performed with exceptional sensing.

3.2 Terahertz-Pulse Detection with Electro-Optic Crystals

The coherent detection of a THz-pulse beam with EO crystals is based on the linear EO effect (Pockels effect) [89]. An illustration to show the principle is given in Fig. 11.

The incident THz-pulse beam modifies the refractive index ellipsoid (or birefringence) of the EO crystal giving rise to a phase retardation of the linearly polarized optical probe beam. By monitoring the phase retardation, the field strength of the THz pulse is detected. A pellicle beam splitter combines the THz beam and the probe beam so that both may copropagate. The polarization of both the THz and optical probe beams are aligned parallel to the $[1,-1,0]$ direction of a (110) oriented ZnTe sensor crystal. Following the sensor crystal, a quarter-wave plate is used to afford a $\pi/4$ optical bias to the probe beam, which allows the system to be operated in the linear range. A Wollaston polarizer is used to convert the THz-radiation-field-induced phase retardation of the probe beam into an intensity modulation between the two mutually orthogonal linearly polarized beams. A pair of Si p-i-n photodiodes connected in a balanced mode is used to detect the optical intensity modulation. The difference signal of p-i-n photodiodes is fed to a lock-in amplifier referenced at a frequency chopping the pumping

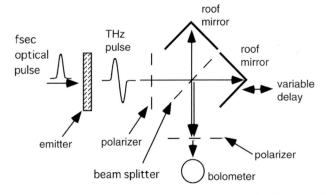

Fig. 12. Optical setup of a Martin–Puplett (M–P) type Fourier transform interferometer that uses free-standing wire-grids as a beam splitter and polarizers. Such a configuration can be used as a Michelson interferometer when the beam splitter is replaced by a dielectric thin film (Mylar film) [91, 92]

optical beam to generate the THz radiation. Because of the instantaneous response of the Pockels effect, the EO crystal well acts as a sampling detector. Hence, this method is called EO sampling. The detection with EO crystals is becoming popular due to its broad-bandwidth capability and ease of implementation. In EO detection there is a clear trade-off between the sensitivity and frequency response that is determined by the choice of crystal and its thickness. A thicker crystal produces a greater interaction length, but on the other hand it reduces the detection bandwidth due to group-velocity mismatch. In addition, the EO technique is very sensitive to laser noise and to low-frequency mechanical and acoustical disturbances. In order to avoid the influence of laser noise the pumping laser is modulated at ≈ 1 MHz using an acousto-opitc modulator [90].

The pumping optical beam to generate THz radiation is partly divided to use for a probe beam. The output signal from the lock-in amplifier versus time-delay scanning of the probe beam reveals the THz-pulse waveform.

3.3 Terahertz-Pulse Detection with Interferometers

In parallel with the PC antenna or the EO detection technique, some methods based on interferometry have been used to detect the THz pulse. They are Fourier transform interferometry, two-source THz optoelectronic interferometry and autocorrelation-type interferometry. A combination of a single THz radiation source and a Fourier transform interferometer, typically in Michelson and Martin–Puplett (MP) mode [91, 92], together with a liquid-helium-cooled (liq. He-cooled) bolometer is shown in Fig. 12.

In this case, the THz radiation source is considered more like the ordinary thermal source than the single-cycle electric pulse emitter, and an interferogram of average THz radiation intensity is recorded. The detected spectral

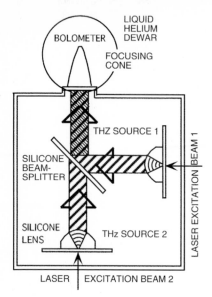

Fig. 13. Two-source THz optoelectronic interferometer [93]

range is affected by an efficiency of the beam splitter, even if the bolometer is sensitive over the whole IR region. The MP mode covers the spectral range below $\approx 200\,\mathrm{cm}^{-1}$ and the Michelson mode with a thin Mylar beam splitter (for example $4\,\mu\mathrm{m}$ thick) covers between $\approx 200\,\mathrm{cm}^{-1}$ and $\approx 700\,\mathrm{cm}^{-1}$. By this method, only THz power information over the wide spectral range is obtained, but the phase information is lost.

The setup of a two-source THz optoelectronic interferometer is shown in Fig. 13 [93].

Two identical THz radiation sources fixed mechanically form a THz two-beam interferometer. An autocorrelation of two THz beams is obtained by scanning the relative time delay between the two excitaiton optical pulses. Two THz beams, combined with a beam splitter, are received with a liquid He-cooled bolometer. As in the Fourier transform interferometer, the information of THz radiation intensity is obtained and phase information is lost.

The autocorrelation type interferometer (Fig. 14) is based on the same idea as the two-source THz optoelectronic interferometer, though a single THz source is used. The femtosecond optical pulses are delivered into a Michelson interferometer.

One arm of the Michelson interferometer is mounted on an alignment adjustable fixed stage, while the other arm is on a translation stage, so that the relative time delay and optical phases of the two pulses can be adjusted. The pair of pulses from the Michelson interferometer strikes the THz emitter (or a sample) of which radiation is detected with a liq. He-cooled bolometer. An autocorrelation of the two emitted THz pulses is recorded by changing the separation of two pumping pulses scanning the translation stage. As in

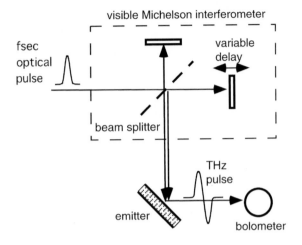

Fig. 14. An autocorrelation-type interferometer

the above two cases, the information of THz radiation intensity is obtained and the phase information is lost. It is an advantage of this method that a THz beam splitter with broad spectral coverage in the THz region is not necessary but one efficient around the incident laser wavelengths is necessary.

3.4 Optical Setups for Terahertz-Pulse Emission and Detection

Components for THz-pulse emission and detection have been described. In this section, two representative THz optical setups are shown. The first is the one constructed with the PC antenna that is shown in Fig. 15a. The femtosecond optical pulses from a mode-locked Ti:sapphire laser are focused by an objective lens (not shown in the figure) on the biased gap of the PC antenna, mounted on the flat side of a Si hyperhemispherical lens, to pump and generate photocarriers. This excitation gives rise to a transient current and emits THz pulse. The Si lens reduces the loss caused by the reflection and refraction of radiation at the surface/air interface. The emitted THz pulse is collimated and focused by a pair of off-axis paraboloidal mirrors onto a sample first, then the diverging THz beam from the sample point is collimated and then focused by a pair of off-axis paraboloidal mirrors onto a PC antenna detector mounted on the back of the Si hyperhemispherical lens with the same diameter. The PC detector is gated by femtosecond optical pulses that is separated from the pump beam by a beam splitter, after passing through time-delay optics. The DC component of photocurrent induced by the incident THz field on the detector is measured by an ammeter; practically connected to a low-noise current amplifier. By delaying the timing of the gate pulse to the pump pulse, the waveform for the THz pulse is obtained. The time resolution is limited by the carrier lifetime of the substrate used for the PC detector. To increase the signal-to-noise ratio, the pump beam is

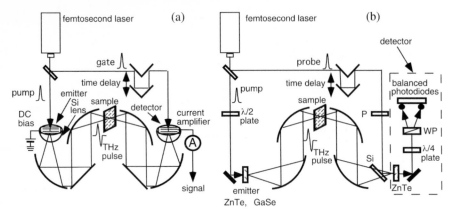

Fig. 15. Representative optical setups for THz-pulse emission and detection, (a) with PC antennas as an emitter and a detector, (b) with NL (EO) crystals

modulated with a mechanical chopper at about 2 kHz, and the output signal from the current amplifier is fed to a lock-in amplifier.

The second is the one that uses the EO crystal as the detector, as can be seen in Fig. 15b. The femtosecond optical pulses from a mode-locked Ti:sapphire laser are focused by an objective lens (not shown in the figure) on the biased gap of the PC antenna or on the NL crystal. This pumping gives rise to a transient current (in the case of the PC antenna) or optical rectification (in the case of a NL crystal that is shown in Fig. 7a) and emits the THz pulse. The emitted THz pulse, after passing through the same optics as in the case of PC antenna, is focused on an EO crystal used for detection after transmitting a pericle beam splitter (or a Si plate). The probe pulses that are separated from the pump pulse beam, pass through time-delay optics, linearly polarized, reflected by a Si plate and then probe the change of the birefringence of the EO crystal caused by the incident THz pulses. The probe pulses that pass through the EO crystal is detected by a pair of photodiodes after passing through a $\lambda/4$ plate and a polarization beam splitter, and the signal from the photodiodes is (after being current amplified) fed to a lock-in amplifier. The waveform of the THz pulse is recorded by delaying the timing of the probe pulse to the pump pulse. The THz-TDS systems have opened up novel aspects in spectroscopy and imaging.

4 Some Experimental Results

Principles and methods of THz-pulse emission and detection have been reviewed. In this section some experimental results [23] with the use of PC antennas and the large aperture photoconductors [94] are described to show practical features. We fabricated some PC antennas with simple form that

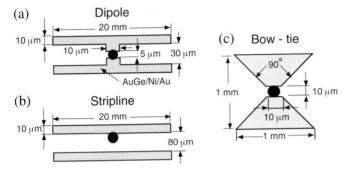

Fig. 16. Schematic view of the PC dipole antenna (**a**), the PC stripline (**b**), and the bow-tie PC antenna (**c**). *Closed circles* show illuminating laser spots

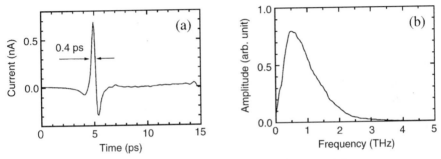

Fig. 17. (**a**) THz pulse emitted and detected with dipole antennas on LT-GaAs. The *vertical axis* indicates the DC signal current detected with the PC detector, (**b**) Fourier-transformed amplitude spectrum of (**a**) [23]

are shown in Fig. 16 and we made emission-detection experiments using the THz-TDS system such as the one shown in Fig. 15a.

The setup is evacuated to avoid water-vapor absorption. The temporal behavior of the emitted THz field looks like that predicted by the calculation based on the classical Drude model and taking the optical pulse width into account (Fig. 5). Figure 17 shows an experimental result with a 30 μm dipole on LT-GaAs as the emitter and a 10 μm dipole on LT-GaAs as the detector.

Both the emitter and the detector were illuminated with the optical pulse having a pulse width of 80 fs at ≈ 780 nm and a repetition rate of 82 MHz. This pulse width and the antenna structure limit the Fourier-transformed amplitude to about 4 THz.

The output amplitude (or power) is influenced by the applied bias voltage, pump laser power, antenna pattern and mobility of the photoconductor.

Figure 18a shows the peak-amplitude dependence of THz radiation on the applied bias voltage and Fig. 18b is the dependence on pump-laser power for a dipole antenna. The amplitude increases linearly with bias voltage but

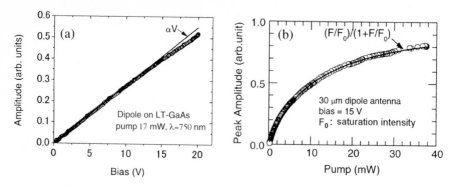

Fig. 18. (a) Bias-voltage dependence of THz radiation amplitude for a dipole on LT-GaAs, (b) Pump-power dependence of THz radiation amplitude for a dipole on LT-GaAs. Saturation behavior is seen

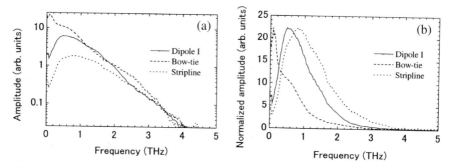

Fig. 19. (a) Emitted amplitude of THz radiation from a dipole, bow-tie, and stripline PC antennas, (b) Normalized amplitude of THz radiation of (a) to compare their spectral coverages

deviates at higher voltage. The pump-power dependence of the amplitude changes in such a way as expressed by $F/(F_0 + F)$ showing a saturation behavior, where F_0 is a saturation intensity. The deviations from the linear dependence in Fig. 18a and the saturation in Fig. 18b are attributed to the screening due to the increased carriers at the gap.

The emission measurements of other PC antennas were also made using the same $10\,\mu m$ dipole as the detector.

The measured Fourier-transformed amplitude is shown in Fig. 19a. Regarding the antenna, the large antenna emits more power than the smaller one. In order to compare the characteristics of the antennas, the normalized amplitudes of each antenna are comparatively displayed in Fig. 19b. It is shown that the stripline extends to higher frequencies than others.

Furthermore, we experimentally found that the emission power is enhanced when the antenna is fabricated on SI-GaAs rather than LT-GaAs.

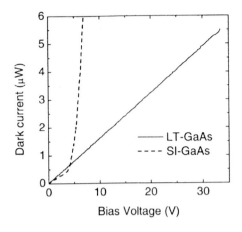

Fig. 20. I–V curves measured for $5\,\mu\text{m}$ gaps of LT-GaAs and SI-GaAs [23]

We assume that this is attributed to the higher mobility of the carriers on SI-GaAs than that in LT-GaAs. But on the other hand, the power increases as the bias voltage becomes high. The I–V curves for $5\,\mu\text{m}$ gap LT-GaAs and SI-GaAs are shown in Fig. 20.

The dark current drastically increases at higher voltages for SI-GaAs. The breakdown voltage for a $5\,\mu\text{m}$ gap on SI-GaAs was $50\,\text{V}$ to $70\,\text{V}$. On the contrary, the dark current increases linearly with the bias voltage and a sample with a $5\,\mu\text{m}$ gap on LT-GaAs has broken at voltages of $\approx 160\,\text{V}$ with increasing voltage.

We roughly estimated that the power of the dipole on LT-GaAs ranges from $0.05\,\mu\text{W}$ to $0.5\,\mu\text{W}$ depending on the dipole size, that of the stripline is sub-μW and those of a bow-tie on LT-GaAs and SI-GaAs are a few μW and the order of ten μW operated at the same bias ($30\,\text{V}$, $100\,\text{V}$ for a stripline) and pump power ($15\,\text{mW}$) conditions.

The large-aperture PC antenna is attractive because it generates a high THz field and it radiates a large THz beam. The emission and detection experiments were done using an SI-GaAs emitter having a $3\,\text{cm}$ gap and using an EO detection scheme. The emitter was pumped with an amplified femtosecond laser, whose pulse width, center wavelength and repetition rate are $150\,\text{fs}$, $775\,\text{nm}$ and $1\,\text{kHz}$, respectively. The pump laser beam was expanded and the emitter was illuminated either with a $9\,\text{cm}^2$ beam or with a $4\,\text{cm}^2$ beam. The temporal behaviors of the emitted THz field are shown in Fig. 21a for these two illuminations and the Fourier-transformed amplitudes are shown in Fig. 21b for $4\,\text{cm}^2$ illuminations. The spectral distribution is similar to that of a bow-tie antenna that emits higher power in the low-frequency regime. Figure 22a shows the dependence of THz radiation amplitude on the applied voltage and Fig. 22b shows the dependence on pump-laser power. Similar to the antenna, the amplitude shows a linear dependence on the applied voltage and it shows a saturation behavior on the illumination laser power expressed by $F/(F + F_0)$.

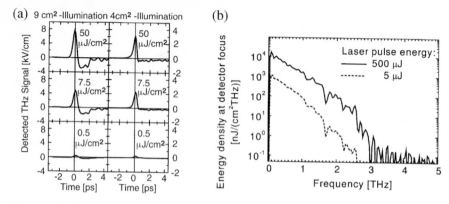

Fig. 21. (a) Temporal behaviors of the emitted THz signals from a large-aperture PC antenna with $9\,cm^2$ and $4\,cm^2$ illuminations, (b) Fourier-transformed spectra of a large-aperture PC antenna for $4\,cm^2$ illumination [94]

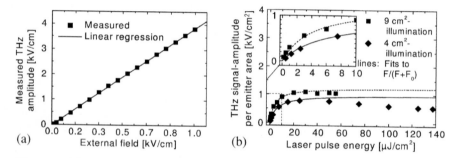

Fig. 22. (a) Applied-external-voltage dependence of the emitted THz radiation amplitude for a $3\,cm$ gap large-aperture PC antenna with $4\,cm^2$ illumination, (b) Pump-power dependence of THz radiation amplitude for a $3\,cm$ gap large-aperture PC antenna with $9\,cm^2$ and $4\,cm^2$ illuminations [94]

5 Conclusions

Since the pioneering work of optoelectronic switching by *Auston* and *Lee* in the mid 1970s a number of optoelectronic THz techniques have been developed. Before the beginning of the 1990s plenty of basic ideas had been established inspite of the inconvenience of using the dye laser. The advent of the Ti:sapphire laser has expanded the population who engage in the field of THz optoelectronics.

The innovations performed after the mid-1990s, i.e., the ultra-broadband capability of the technique extending to mid-to-near IR regions both with EO-sampling and PC-antenna techniques has made the techniques more attractive and useful not only for the basic research but for the practical applications.

Applications are expanding, by means of spectroscopy and imaging in such diverse areas as studies on genetic materials or semiconductor nano-structures, identification of tumours, detection and identification of drugs in mail, detection of explosives or nonmetallic knives in a package, and so on.

More applications are now being researched in wide areas by making use of a variety of features of THz radiation. Among these are the facts that THz radiation is non-ionizing, can penetrate many visually opaque materials, suffers less Rayleigh scattering than NIR and less diffraction than the shortest microwaves, and is highly sensitive to the presence of water and water vapor.

Acknowledgements

This work has been achieved under the leadership by Prof. Dr. N. Fugono (Former Director General at Communications Research Laboratory, *present name* National Institute of Information and Communications Technology: NiCT) who wished to push forward fundamental researches in Japan. The authors are deeply grateful to him for providing us a sophisticated advanced research institute. The authors are particularly indebted to Professor Emeritus of Osaka University S. Nakashima for many useful suggestions. They also wish to express their thanks to Dr. A. Tokuda at Mitsubishi Electric Corp. and Dr. H. Katahama at Sumitomo Mitsubishi Silicon Corp. who prepared us optoelectronic specimens and devices at the initial stage of this research. The authors wish to express their heart-felt thanks to Dr. T. Loeffler and Prof. Roskos at Johann Wolfgang Goethe-University in Frankfurt for allowing us to use the figures included in the Ph. D. Thesis by T. Loeffler prior to publication.

References

[1] D. Auston: Appl. Phys. Lett. **26**, 101 (1975)
[2] C. Lee: Appl. Phys. Lett. **30**, 84 (1977)
[3] D. Auston, C. Lee: *Picosecond Optoelectronic Devices* (Academic, New York 1984) Chap. 4, 5
[4] D. Auston, A. Johnson, P. Smith, J. Bean: Appl. Phys. Lett. **37**, 371 (1980)
[5] P. Smith, D. Auston, A. Johnson, W. Augustyniak: Appl. Phys. Lett. **38**, 47 (1981)
[6] G. Mourou, C. Stancampiano, A. Antonetti, A. Orszag: Appl. Phys. Lett. **39**, 295 (1981)
[7] R. Heidemann, Th. Pfeiffer, D. Jäger: Electron. Lett. **19**, 316 (1983)
[8] D. Auston, P. Smith: Appl. Phys. Lett. **43**, 631 (1983)
[9] D. Auston, K. Cheung, P. Smith: Appl. Phys. Lett. **45**, 284 (1984)
[10] A. DeFonzo, M. Jarwala, C. Lutz: Appl. Phys. Lett. **50**, 1155 (1987)
[11] A. DeFonzo, C. Lutz: Appl. Phys. Lett. **51**, 212 (1987)
[12] P. Smith, D. Auston, M. Nuss: IEEE J. Quantum Electron. **24**, 255 (1988)

[13] M. van Exter, Ch. Fattinger, D. Grischkowsky: Appl. Phys. Lett. **55**, 337 (1989)

[14] M. Ketchen, D. Grischkowsky, T. Chen, C.-C. Chi, I. Duling, III, N. Halas, J.-M. Halbout, J. Kash, G. Li: Appl. Phys. Lett. **48**, 751 (1986)

[15] M. van Exter, D. Grischkowsky: IEEE Trans. Microwave Theory Tech. **38**, 1684 (1990)

[16] D. Spence, P. Kean, W. Sibbett: Opt. Lett. **16**, 42 (1991)

[17] F. Smith, A. Calawa, C.-L. Chen, M. Manfra, L. Mahoney: IEEE Electron Device Lett. **9**, 77 (1988)

[18] S. Gupta, M. Frankel, J. Valdmanis, J. Whitaker, G. Mourou, F. Smith, A. Calawa: Appl. Phys. Lett. **59**, 3276 (1991)

[19] A. Othonos: J. Appl. Phys. **83**, 1789 (1998)

[20] J. Pedersen, V. Lyssenko, J. Hvam, P. Jepsen, S. Keiding, C. Sørensen, P. Lindelof: Appl. Phys. Lett. **62**, 1265 (1993)

[21] P. Jepsen, R. Jacobsen, S. Keiding: J. Opt. Soc. Am. B **13**, 2424 (1996)

[22] C. Ludwig, J. Kuhl: Appl. Phys. Lett. **69**, 1194 (1996)

[23] M. Tani, S. Matsuura, K. Sakai, S. Nakashima: Appl. Opt. **36**, 7853 (1997)

[24] Z. Piao, M. Tani, K. Sakai: Jpn. J. Appl. Phys. **39**, Pt.1 96 (2000)

[25] P. Jepsen, S. Keiding: Opt. Lett. **20**, 807 (1995)

[26] J. Rudd, J. Johnson, D. Mittleman: Opt. Lett. **25**, 1556 (2000)

[27] Y. Cai, I. Brener, J. Lopata, J. Wynn, L. Pfeiffer, J. Federici: Appl. Phys. Lett. **71**, 2076 (1997)

[28] L. Duvillaret, F.-F. Garet, J.-F. Roux, J.-L. Coutaz: IEEE J. Sel. Top. Quantum Electron. **7**, 615 (2001)

[29] J. Darrow, B. Hu, X.-C. Zhang, D. Auston: Opt. Lett. **15**, 323 (1990)

[30] B. Hu, J. Darrow, X.-C. Zhang, D. Auston, P. Smith: Appl. Phys. Lett. **56**, 886 (1990)

[31] L. Xu, X.-C. Zhang, D. Auston: Appl. Phys. Lett. **59**, 3357 (1991)

[32] J. Darrow, X.-C. Zhang, D. Auston, J. Morse: IEEE J. Quantum Electron. **28**, 1607 (1992)

[33] P. Benicewicz, A. Taylor: Opt. Lett. **18**, 1332 (1993)

[34] G. Rodriguez, S. Caceres, A. Taylor: Opt. Lett. **19**, 1994 (1994)

[35] G. Rodriguez, A. Taylor: Opt. Lett. **21**, 1046 (1996)

[36] D. You, R. Jones, P. Bucksbaum, D. Dykaar: Opt. Lett. **18**, 290 (1993)

[37] T. Yajima, N. Takeuchi: Jpn. J. Appl. Phys. **9**, 1361 (1970)

[38] T. Yajima, N. Takeuchi: Jpn. J. Appl. Phys. **10**, 907 (1971)

[39] K. Yang, P. Richards, Y. Shen: Appl. Phys. Lett. **19**, 320 (1971)

[40] D. Auston: Appl. Phys. Lett. **43**, 713 (1983)

[41] D. Auston, K. Cheung, J. Valdmanis, D. Kleinman: Phys. Rev. Lett. **53**, 1555 (1984)

[42] D. Auston, K. Cheung: J. Opt. Soc. Am. B **2**, 606 (1985)

[43] A. Bonvalet, M. Joffre, J. Martin, A. Migus: Appl. Phys. Lett. **67**, 2907 (1995)

[44] D. Grischkowsky, I. Duling, III, J. Chen, C.-C. Chi: Phys. Rev. Lett. **59**, 1663 (1987)

[45] X.-C. Zhang, B. Hu, J. Darrow, D. Auston: Appl. Phys. Lett. **56**, 1011 (1990)

[46] B. Hu, X.-C. Zhang, D. Auston: Appl. Phys. Lett. **57**, 2629 (1990)

[47] S. Chuang, S. Schmitt-Rink, B. Green, P. Saeta, A. Levi: Phys. Rev. Lett. **68**, 102 (1992)

[48] B. Greene, P. Saeta, D. Dykaar, S. Schmitt-Rink, S. Chuang: IEEE J. Quantum Electron. **28**, 2302 (1992)
[49] X.-C. Zhang, D. Auston: J. Appl. Phys. **71**, 326 (1992)
[50] K. Leo, J. Shah, E. Göbel, T. Damen, S. Schmitt-Rink, W. Schäfer: Phys. Rev. Lett. **66**, 201 (1991)
[51] H. Roskos, M. Nuss, J. Shah, K. Leo, D. Miller, A. Fox, S. Schmitt-Rink, K. Köhler: Phys. Rev. Lett. **68**, 2216 (1992)
[52] P. Planken, M. Nuss, I. Brener, K. Goossen, M. Luo, S. Chuang, L. Pfeiffer: Phys. Rev. Lett. **69**, 3800 (1992)
[53] C. Waschke, H. Roskos, R. Schwedler, K. Leo, H. Kurz, K. Köhler: Phys. Rev. Lett. **70**, 3319 (1993)
[54] T. Dekorsy, H. Auer, C. Waschke, H. Bakker, H. Roskos, H. Kurz, V. Wagner, P. Grosse: Phys. Rev. Lett. **74**, 738 (1995)
[55] R. Kersting, K. Unterrainer, G. Strasser, H. Kauffmann, E. Gornik: Phys Rev. Lett. **79**, 3038 (1997)
[56] M. Tani, R. Fukasawa, H. Abe, S. Matsuura, K. Sakai: J. Appl. Phys. **83**, 2473 (1998)
[57] P. Gu, M. Tani, K. Sakai, T.-R. Yang: Appl. Phys. Lett. **77**, 1798 (2000)
[58] M. Hangyo, S. Tomozawa, Y. Murakami, M. Tonouchi, M. Tani, Z. Wang, K. Sakai, S. Nakashima: Appl. Phys. Lett. **69**, 2122 (1996)
[59] M. Tonouchi, M. Tani, Z. Wang, K. Sakai, N. Wada, M. Hangyo: Jpn. J. Appl. Phys. **36**, Pt.2 L93 (1997)
[60] M. Rodwell, D. Bloom, B. Auld: Electron. Lett. **23**, 109 (1987)
[61] Y. Konishi, M. Kamegawa, M. Case, R. Yu, M. W. Rodwell, R. York, D. Rutledge: Appl. Phys. Lett. **61**, 2829 (1992)
[62] D. van der Weide: J. Opt. Soc. Am. B **11**, 2553 (1994)
[63] Q. Wu, X.-C. Zhang: Appl. Phys. Lett. **67**, 3523 (1995)
[64] Q. Wu, X.-C. Zhang: Appl. Phys. Lett. **71**, 1285 (1997)
[65] A. Brodschelm, F. Tauser, R. Huber, J. Sohn, A. Leitenstorfer: *Ultrafast Phenomena XII* (Springer, Berlin, Heidelberg 2001) pp. 215–217
[66] S. Kono, M. Tani, P. Gu, K. Sakai: Appl. Phys. Lett. **77**, 4104 (2000)
[67] S. Kono, M. Tani, K. Sakai: Appl. Phys. Lett. **79**, 898 (2001)
[68] S. Kono, M. Tani, K. Sakai: IEE Proc. Optoelectron. **149**, 105 (2002)
[69] S. Keiding: *THz Pulses*, Dissertation, University of Aarhus, Denmark (1998)
[70] M. Nuss, J. Orenstein: *Millimeter and Submillimeter Wave Spectroscopy of Solids*, Top. Appl. Phys. **74** (Springer, Berlin, Heidelberg 1998) Chap. 2
[71] A. Bonvalet, M. Joffre: *Femtosecond Laser Pulses, Principles and Experiments* (Springer, Berlin, Heidelberg 1998) Chap. 10
[72] J. Shah: *Ultrafast Spectroscopy of Semiconductors and Semiconductor Nanostructures*, 2nd ed. (Springer, Berlin, Heidelberg 1999) pp. 120–131
[73] T. Dekorsy, G. Cho, H. Kurz: *Light Scattering in Solids VIII, Fullerenes, Semiconductor Surfaces, Coherent Phonons*, Top. Appl. Phys. **76** (Springer, Berlin, Heidelberg 2000) Chap. 4
[74] D. Mittleman: *Sensing with Terahertz Radiation* (Springer, Berlin, Heidelberg 2003)
[75] D. Woolard, W. Loerop, M. Shur (Eds.): *Terahertz Sensing Technology, Vol.1: Electronic Devices and Advanced Systems Technology*, vol. 1, Topics Electron. Syst. **30** (World Scientific, New Jersey 2003)

[76] F. Doany, D. Grischkowsky, C.-C. Chi: Appl. Phys. Lett. **50**, 460 (1987)

[77] M. Tani, K. Sakai, H. Abe, S. Nakashima, H. Harima, M. Hangyo, Y. Tokuda, K. Kanamoto, Y. Abe, N. Tsukada: Jpn. J. Appl. Phys. **33**, Pt.1 4807 (1994)

[78] H. Abe, M. Tani, K. Sakai, S. Nakashima: in T. Parker, S. Smith (Eds.): (Conf. Dig. 23rd Int. Conf. Infrared and Millimeter Waves, Colchester, UK 1998) p. 137

[79] D. Rutledge, D. Neikirk, D. Kasilingam: *Infrared and Millimeter Waves*, vol. 10 (Academic, Orlando 1983) Chap. 1

[80] B. Hu, X.-C. Zhang, D. Auston, P. Smith: Appl. Phys. Lett. **56**, 506 (1990)

[81] L. Xu, X.-C. Zhang, D. Auston: Appl. Phys. Lett. **61**, 1784 (1992)

[82] A. Rice, Y. Jin, X. Ma, X.-C. Zhang, D. Bliss, J. Larkin, M. Alexander: Appl. Phys. Lett. **64**, 1324 (1994)

[83] T. Carrig, G. Rodriguez, T. Clement, A. Taylor, K. Stewart: Appl. Phys. Lett. **66**, 10 (1995)

[84] A. Nahata, D. Auston, C. Wu, J. Yardley: Appl. Phys. Lett. **67**, 1358 (1995)

[85] S. Kono, P. Gu, M. Tani, K. Sakai: Appl. Phys. B **71**, 901 (2000)

[86] S. Kono, M. Tani, K. Sakai: in (CLEO, San Francisco, USA 2000) p. 558

[87] A. Kuznetsov, C. Stanton: Phys. Rev. Lett. **73**, 3243 (1994)

[88] A. Leitenstorfer, S. Hunsche, J. Shah, M. Nuss, W. Knox: Phys. Rev. Lett. **82**, 5140 (1999)

[89] Z. Lu, P. Campbell, X.-C. Zhang: Appl. Phys. Lett. **71**, 593 (1997)

[90] Y. Cai, I. Brener, J. Lopata, J. Wynn, L. Pfeiffer, J. Stark, Q. Wu, X.-C. Zhang, J. F. Federici: Appl. Phys. Lett. **73**, 444 (1998)

[91] D. Martin, E. Puplett: Infrared Phys. **10**, 105 (1970)

[92] D. Martin: *Infrared and Millimeter Waves*, vol. 6 (Academic, New York 1982) Chap. 2

[93] S. Ralph, D. Grischkowsky: Appl. Phys. Lett. **60**, 1070 (1992)

[94] T. Loeffler: Dissertation, JWG-University of Frankfurt, Germany (2003)

Generation and Detection of Broadband Pulsed Terahertz Radiation

Shunsuke Kono[1,2], Masahiko Tani[1,3], and Kiyomi Sakai[1]

[1] National Institute of Information and Communications Technology
588-2 Iwaoka, Nihis-ku, Kobe 651-2492, Japan
[2] *Present address:* Fundamental and Environmental Research Laboratories
NEC Corporation
34 Miyukigaoka, Tsukuba 305-8501, Japan
s-kouno@cq.jp.nec.com
[3] Institute of Laser Engineering
Osaka University
2-1 Yamadaoka, Suita, Osaka 565-0871, Japan

Abstract. The generation and detection of ultrashort electromagnetic radiation is reviewed. The detection of ultrashort electromagnetic pulses with the electro-optic sampling techniques is introduced by insisting on its broad detectable bandwidth. The principles and experimental techniques for the optical rectification effect are described for the broadband generation of pulsed terahertz radiation. In addition to these techniques using nonlinear optical effects, the recent findings on the photoconductive antennas enabling detection and generation of ultra-broadband pulsed terahertz radiation are described.

1 Ultra-Broadband Terahertz Radiations with Nonlinear Crystals

One of the research directions in terahertz (THz) optoelectronics is to expand the bandwidth of the coherent generation and detection of the THz radiation. As already mentioned in Sect. 2.3.1 and Sect. 3.2 in the Chapter by *Sakai* and *Tani*, the second-order nonlinear effects like optical rectification and electro-optic (EO) effect have attracted considerable attention for broadband generation and detection of the THz pulses due to its instantaneous response derived from the nonresonant nature of the probe and pump beam. In Sect. 1.1, the developments on the broadband EO sampling technique are introduced including the basic principles of the EO sampling technique. The optical rectification effect for ultra-broadband generation of THz pulse will be discussed in Sect. 1.2.

1.1 Ultra-Broadband Detection of Terahertz Pulses with Electro-Optic Crystals: Principles

The EO sampling is a phase-sensitive detection of electromagnetic (EM) radiation based on a birefringence in an EO crystal induced by the incident

K. Sakai (Ed.): Terahertz Optoelectronics, Topics Appl. Phys. **97**, 31–62 (2005)
© Springer-Verlag Berlin Heidelberg 2005

EM radiation (Fig. 11 in the Chapter by *Sakai* and *Tani*). This birefringence in the crystal is probed as a phase shift of an optical probe beam. In the time-resolved detection of the THz pulse, the transient birefringence is probed with the optical pulses at different time delays.

Contrary to the photoconductive (PC) antenna, the probe beam is non-resonant to the electronic excitation in the EO crystal. Therefore, the birefringence induced by the incident THz pulses can be instantaneously sensed by the probe beam. Due to this nonresonant feature, the EO sampling technique has been expected for the phase-sensitive detection of radiation in the frequency range higher than 5 THz. Actually, the EO sampling technique is commonly used for the phase-sensitive detection of radiation due to its easy experimental preparation and broad detectable bandwidth.

The phase retardation induced in the EO crystal is given by the following equation [1]

$$\Delta \Gamma = \frac{2\pi}{\lambda} d n_{opt}^3 r_{41} E_{THz} ,\tag{1}$$

where d is the EO crystal thickness, n_{opt} is the group refractive index of the EO crystal at the wavelength of the near-infrared (NIR) probe beam, and r_{41} is the EO coefficient. In an actual estimation of the phase retardation, it is necessary to consider the Fresnel transmission coefficient on the EO crystal surface. However, we include this coefficient in the frequency dependence due to the phase-matching condition explained later. The EO effect is common to many kinds of crystals without inversion symmetry. Among such crystals, ZnTe or GaP is commonly used for the EO sampling of the THz pulse because these crystals have large EO coefficients and they are transparent at the wavelength of the incident THz radiation and at that of a mode-locked Ti:sapphire laser as a gating probe beam.

1.1.1 Frequency Response of Electro-Optic Sampling

To study the frequency response of the EO sampling technique, we have to consider the group velocity dispersion of the probe beam and refractive index dispersion at the frequency of the incident THz radiation in the EO crystal. The optical pulse of the probe beam and incident THz beam propagate on the same axis inside the EO crystal, however, due to the difference in the respective dispersions, the mismatch of the group velocity accumulates after the propagation of the two waves through the crystal whose thickness is d. The time difference due to the group velocity mismatch is described by the following equation [2]:

$$\delta (\omega) = \frac{n_g (\lambda_0) - n (\omega)}{c} d ,\tag{2}$$

where $n_g(\lambda_0)$ is the group refractive index at the wavelength of the probe beam, $n(\omega)$ the refractive index at the frequency of the incident THz radiation.

The phase difference in the probe beam is proportional to the crystal thickness as described by $\Delta\Gamma$ in (1). At the same time, due to the different propagation time of the probe pulse and incident THz radiation, the phase difference is also proportional to the time average of the electric field during the group-velocity mismatch time, $\delta(\omega)$. The factor of this time average is expressed by

$$G\left(\omega\right) = \frac{T\left(\omega\right)}{\delta\left(\omega\right)} \int_0^{\delta(\omega)} \exp\left(\mathrm{i}2\pi\omega t\right)\mathrm{d}t = T\left(\omega\right)\frac{\exp\left[\mathrm{i}2\pi\omega\delta\left(\omega\right)\right] - 1}{\mathrm{i}2\pi\omega\delta\left(\omega\right)}. \quad (3)$$

Here, $T(\omega) = 2/[n_{\mathrm{THz}}(\omega) + 1]$, is the Fresnel transmission coefficient with the refractive index of the EO crystal $n_{\mathrm{THz}}(\omega)$ in the range of the target THz frequency. The total frequency response of the EO sampling technique is given by the product of (1) and (3). Although the frequency dependence of (1) is not explicitly described, in a precise estimation of the frequency response, it is necessary to consider the frequency dependences of the respective components in the equation. As already mentioned in Sect. 3.2 of the Chapter by *Sakai* and *Tani*, there is a tradeoff between the crystal thickness and modulation in the EO crystal. The phase modulation is proportional to the crystal thickness, while $G(\omega)$ is significantly dependent on the crystal thickness.

Despite this tradeoff between the crystal thickness and sensitivity, the EO sampling technique was a unique solution for the phase-sensitive detection of THz radiation in the frequency range above 10 THz. Intensive researches have been made on the development of the broadband EO detection.

1.1.2 Some Experimental Examples on Broadband Electro-Optic Detection

The first experimental demonstration of the EO sampling technique beyond the detectable bandwidth of the PC antenna was done by *Wu* and *Zhang* [2]. Using (110) surface GaP crystals, they resolved the THz radiation up to 7 THz. The pulse width of the probe beam was 50 fs.

The detectable bandwidth of the EO sampling technique was much broadened by the same group. *Wu* and *Zhang* demonstrated the EO detection of the THz radiation up to 37 THz. Their light source for pump and probe experiments was a mode-locked Ti:sapphire laser delivering 12 fs light pulses. Figure 1 shows the waveform of the THz pulse from GaAs detected with the 30 μm thick ZnTe crystal [3]. The shortest period was estimated to be 31 fs. The Fourier-transformed spectrum of this detected waveform for a longer scan is shown in Fig. 2. The frequency distribution extends to 37 THz.

Later, *Leitenstorfer* et al. [4] further developed the broadband EO sampling technique for the investigation on the carrier transport in semiconductors. The EO sampling technique enabled direct detection of the phase and amplitude of the infrared (IR) radiation emitted by the accelerating carriers in semiconductors. Exciting the carriers in the semiconductors with 12 fs

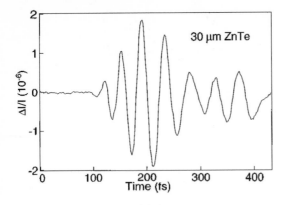

Fig. 1. Waveform of the THz pulse from GaAs surface detected with 30 μm thick ZnTe crystal [3]

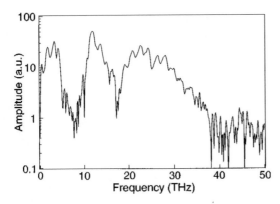

Fig. 2. Fourier-transformed spectra of the THz signal waveform EO-detected with 30 μm thick ZnTe crystal [3]

optical pulses, the IR radiation by the accelerating carriers and interaction of the carriers with the lattice vibration was detected in the time scale of 10 fs. The waveform of the EO signal is shown in Fig. 3. The radiation due to the acceleration of the carriers was clearly detected. Furthermore, under high electric fields, the interaction of the carriers with the lattice vibration (LO phonon) was clearly manifested as the oscillations in the emitted radiation. For this measurement, they prepared thin EO crystals of ZnTe and GaP, 8 μm thick and 13 μm thick, respectively. They investigated in detail the frequency response of the EO sampling technique. The frequency dependence of the refractive index, $n(\omega)$, in the EO crystal and that of the EO coefficient $r_{41}(\omega)$ were carefully evaluated. Figure 4 shows their estimation of $G(\omega)$ and $r_{41}(\omega)$. At the shot-noise limit of the differential detection of the probe beam, estimated to be $\Delta I/I \approx 10^{-8}\,\mathrm{Hz}^{-1/2}$, the spectral bandwidth was estimated to be close to 70 THz.

In addition to the broadband EO sampling technique, broadband coherent radiation sources have also been developed. The combination of the EO sampling technique and broadband optical rectification provides a fully co-

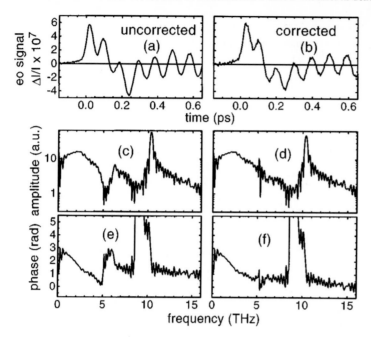

Fig. 3. EO-detected signal waveforms and Fourier-transformed spectra of the THz radiation generated from the accelerating carriers in the biased semiconductors [4]

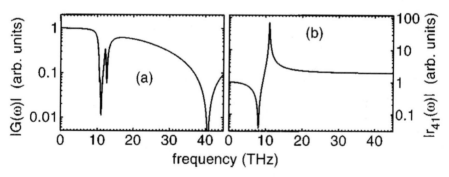

Fig. 4. Frequency dependences of $G(\omega)$ (**a**) and EO coefficient $r_{41}(\omega)$ (**b**) of the 13 μm thick GaP crystal [4]

herent generation and detection system of the far- to mid-IR ranges. In such a situation, the same group also demonstrated the EO detection up to 100 THz with the radiation source using the phase-matched optical rectification technique shown in Fig. 5 [5]. In the next section, the recent developments on the broadband THz radiation sources will be introduced.

L = 20 μm, θ= 60°

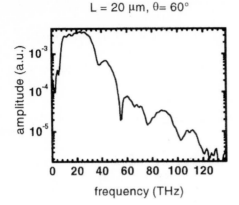

Fig. 5. Amplitude spectrum of the MIR radiation from 20 μm thick GaSe crystal with an incidence angle of 60°. Detector was 10.3 μm thick ZnTe [5]

1.2 Ultra-Broadband Emission Based on Optical Rectification

As introduced in the Chapter by *Sakai* and *Tani*, generating far-infrared (FIR) radiation by the optical rectification dates back to the 1970s as well as the FIR radiation using PC switches. As already pointed out in Shen's famous textbook on nonlinear optics [6], the difference-frequency generation is technically very important to generate the far- to mid-IR radiation, although it does not theoretically differ much from the sum-frequency generation. The developments on the ultrashort pulse lasers have also played a very important role in the generation of the FIR radiation using optical nonlinearity as in the case of THz optoelectronics related to PC switches. In this section, the recent advancement on the far- to mid-IR radiation with the optical rectification effect will be discussed.

1.2.1 Principles of Optical Rectification

In the plane-wave approximation, the electric field generated by the difference-frequency mixing is proportional to the second-order time derivative of the second-order nonlinear polarization given by (5) in the Chapter by *Sakai* and *Tani*. The ultrashort optical pulses contain various spectral components and the beatings between the different frequency components contained in a single ultrashort pulse will generate a broadband, i.e., far- to mid-IR radiation. In order to discuss further the optical rectification, (4) in the same chapter is rewritten by the following equation giving the nonlinear polarization created by a single optical pulse with the complex amplitude of the electric field, E [7]

$$P(\omega) = \varepsilon_0 \chi^{(2)} \int_{-\infty}^{+\infty} E(\omega')E^*(\omega' - \omega)\, d\omega'. \tag{4}$$

In the above equation, the dispersion of the nonlinear coefficient is neglected. The Fourier transformation of this equation gives P in the time domain

$$P(t) = \varepsilon_0 \chi^{(2)} E(t) E^*(t) \propto \chi^{(2)} I(t). \tag{5}$$

This equation in the time domain shows that the pulse width of the generated radiation depends on that of the pump beam.

In the field of THz optoelectronics, the word "optical rectification" is used for the generation of far- to mid-IR radiation using the second-order optical nonlinearity. In a strict sense, the optical rectification effect produces an electric polarization with low-frequency components down to DC in a nonlinear medium. In a plane-wave approximation, the second-order nonlinear term giving the optical rectification effect is proportional to the intensity of the incident light. With an ultrashort optical pulse, its spectral width is broad, so that the various spectral components will beat with one another and generate pulses containing the frequency components from far- to mid-IR radiation. Thus the phenomenon is nothing but the difference-frequency generation. In other words, one of the terms of the second-order nonlinear polarization that contributes to the difference-frequency generation is approximated with a square of the envelope electric field of the incident optical pulses because the frequency components of the carrier wave is much higher than those of the envelope of the optical pulses. Therefore the generation of the far- to mid-IR radiation can be well approximated with the optical rectification effect. In this section, we use the word, "optical rectification", to specify the THz pulse generated by the second-order nonlinear effect.

1.2.2 Experimental Results of Broadband Optical Rectification

Bonvalet et al. [7] first reported the generation of the broadband mid-IR (MIR) radiation by pumping a GaAs surface with 15 fs optical pulses. The 15 fs optical pulses were delivered from the mode-locked Ti:sapphire laser whose center wavelength was at 800 nm. Since GaAs absorbs the pump optical beam mostly at the surface, only the penetrating depth of the wafer surface contributed to the optical rectification effect. The refractive index of GaAs in the MIR range is different from that at the NIR pump pulse, so that the phases of the three interacting waves could not be matched. Even at the expense of the generation efficiency due to this phase mismatching, they observed the MIR radiation beyond the bandwidth of the HgCdTe detector, as shown in Fig. 6.

After this generation of the MIR radiation with ultrashort optical pulses, *Wu* and *Zhang* [3] demonstrated the generation of broadband pulsed THz radiation and detection using EO sampling technique. As cited in Sect. 1.1.2, they adopted a GaAs wafer for the generation of the broadband THz radiation and observed the spectral components up to 37 THz.

Later, *Han* and *Zhang* investigated the optical rectification effect of different materials in order to have broadband THz radiation [8]. They investigated materials with different thickness as following: GaAs, InP, ZnTe, CdTe, GaP, BBO, and LiTaO$_3$. The pulse width of the pump and probe beams was 12 fs.

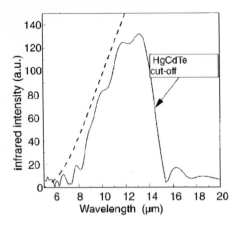

Fig. 6. Intensity spectrum of the MIR radiation from GaAs surface pumped with 15 fs optical pulse. The *dashed line* corresponds to the theoretically expected emission [7]

They proposed ZnTe crystal as one of the most suitable crystals for the generation and detection of broadband radiation with a mode-locked Ti:sapphire laser.

In these experimental demonstrations, the phase-matching condition was taken into account mainly to broaden the emission bandwidth by the optical rectification effect. Reducing the crystal thickness broadens the emission bandwidth although the emission efficiency is reduced.

1.2.3 Phase-Matched Optical Rectification for Broadband Terahertz Pulses

Apart from the optoelectronic THz-pulse techniques just described, generation of femtosecond pulses in MIR region has been of interest for the study of ultrafast dynamics of elementary excitations in molecules and solids [9]. These researches are mainly based on parametric frequency conversion of amplified femtosecond pulses in the visible or NIR regions. Among these experimental researches elongating the converted wavelength into the MIR range, the generation of MIR pulses by the phase-matched optical rectification does not require amplified femtosecond laser pulses [10]. This experimental scheme coincides with the THz time-domain spectroscopy (TDS) in EO sampling shown in Fig. 15b of the Chapter by *Sakai* and *Tani*.

The optical rectification effect with a GaSe crystal provides a tunable MIR radiation because of the birefringence of the crystal. *Kaindl* et al. [10, 11] reported the generation of the MIR femtosecond pulses with the phase-matched optical rectification effect in GaSe as shown in Fig. 7. The wavelength of the generated pulses can be tuned from 7 μm to 20 μm by tilting the angle of the GaSe crystal. In this report, the autocorrelation of the generated radiation was measured by an interferometer with a HgCdTe detector whose sensitivity was up to 20 μm.

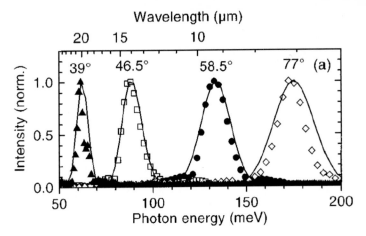

Fig. 7. Normalized intensity spectra of femtosecond MIR pulses generated by phase-matched optical rectification effect in GaSe for different phase-matching angles [11]

It is quite natural to combine these MIR radiation sources with the EO sampling technique to characterize the generated pulses since the detectable bandwidth of the EO sampling technique was theoretically estimated up to 70 THz [4]. *Huber* et al. reported the generation of the femtosecond MIR pulses and their phase-resolved detection as shown in Fig. 8. The generated radiation was at 41 THz, corresponding to 7 μm at the center wavelength and the pulse duration was less than 50 fs [12]. In the same group, *Brodschelm* et al. [5] also reported the MIR radiation with a GaSe crystal tunable from 6 μm to 40 μm with the EO detection system resolving the frequency components up to 100 THz.

The phase-matched optical rectification was further developed beyond the wavelength tunability of the generated radiation. *Eickemeyer* et al. [13] demonstrated controlling the shape of the MIR pulse radiation by modulating the phases of the pumping femtosecond pulses. They detected the controlled shape of the generated radiation with the EO sampling technique. Their demonstration opens up a possibility of coherent-control experiments in the MIR range.

The generation of broadband THz radiation by a nonphase-matched optical rectification effect will be promising for static spectroscopy in the MIR range. With the EO sampling technique, a fully coherent generation and detection system can be a novel candidate for MIR spectroscopy where FTIR is commonly used. The importance of the phase-matched optical rectification does not need to be repeated. The studies on elementary excitations in solids and molecules will be developed in the real-time domain with the combination of the coherent detection and generation of the IR pulse radiations.

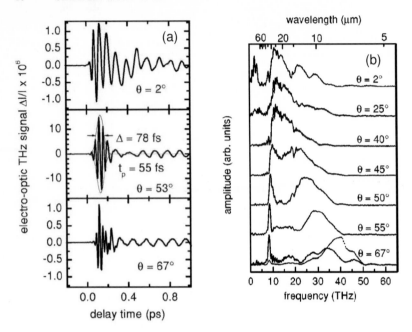

Fig. 8. (a) Waveforms of THz pulses generated by phase-matched optical rectification effect in 90 μm thick GaSe. The pulse width of the pump beam was 10 fs. The waveforms were detected by the EO sampling technique with a 10.3 μm thick ZnTe crystal, (b) Normalized amplitude spectra of EO-detected THz signal waveforms from GaSe crystal with different phase-matching angles [12]

2 Ultra-Broadband Terahertz Pulses with Photoconductive Antennas

The optical nonlinear effects treated in the previous sections are now commonly used in the generation and detection of THz pulses. The nonlinear optical techniques are especially advantageous in their broadband properties and well-investigated frequency properties. The authors' group recently investigated the broadband properties of the PC antenna. The recent development on the broadband detection of the PC antenna is precisely discussed in Sect. 2.1. In Sect. 2.2, the latest report on the broadband emission of the THz radiation with PC antenna is introduced.

2.1 Ultra-broadband Detection of Terahertz Pulses with Photoconductive Antennas

Until very recently, the EO sampling technique has been the only method for the phase-sensitive detection of the THz radiation whose frequency is higher than 10 THz. Although the PC antenna was a key device to initiate the recent development of THz optoelectronics, the PC sampling method has become somewhat old-fashioned. In such a situation, the authors found a novel aspect of the PC antenna: ultra-broadband detection. The detectable bandwidth of the PC antenna can be extremely expanded with an optical gate pulse of 15 fs. After a brief introduction of the principles of the PC detection, this section is devoted to the recent developments on the ultra-broadband detection with PC antenna.

2.1.1 Detection Principles of Photoconductive Antennas

As already mentioned in Sect. 3.1 of the Chapter by *Sakai* and *Tani*, the detection principle of a PC antenna is the inverse process of the THz-pulse generation. In the process of THz-pulse emission, photocarriers created by the ultrashort optical pulses are accelerated by the bias electric field between the antenna electrodes. This acceleration of carriers generates THz-pulse emission. Inversely, in the detection of THz radiation with PC antennas, the current meter is connected instead of the bias voltage. Photocreated carriers are accelerated by the electric field of the incident THz radiation and this acceleration of the photocarriers is detected as a current. This current is proportional to the product of the incident THz field and existing photo-created carriers. The temporal change of the incident pulse can be traced by changing the arrival time of the optical gate pulse. The photocurrent by the incident radiation at a time delay t is described by the following equation [14]

$$ J(t) = e\mu \int_{-\infty}^{\infty} E(t')N(t'-t)\,\mathrm{d}t' \,, \tag{6} $$

where $E(t')$, $N(t')$, e and μ are the incident electric field of the THz radiation, number of photocreated carriers, elementary electric charge, and electron mobility, respectively.

In Sect. 2.1 of the Chapter by *Sakai* and *Tani*, the generation of a short EM pulse is given by (1) as the time derivative of the photocurrent. Equation (6) in this chapter corresponds to the time integration of (1) in the Chapter by *Sakai* and *Tani*. Equation (2) in the same chapter shows that the photocurrent density is given by the convolution of the optical pulse intensity and carrier density. In the description hereafter, the number of photocreated carriers, $N(t)$, includes this convolution for simplicity.

When the PC antenna is considered as a sampling detector, the temporal increase and decrease of $N(t)$ should be as short as possible: $J(t)$ would exactly reflect $E(t)$ if $N(t)$ were a δ-function. In reality, $N(t)$ will be restricted by several factors, such as gating pulse width, carrier lifetime of PC material, and momentum relaxation of photocarriers. This is one of the reasons to explore the materials with short carrier lifetime as already described in the Chapter by *Sakai* and *Tani*. With the motivation to shorten the carrier lifetime, low-temperature-grown GaAs (LT-GaAs) has been well developed.

Actually, the reported bandwidth of the PC antennas has been so far up to 6 THz by *Ralph* et al. [15] with a radiation damaged silicon-on-sapphire (RD-SOS) PC antenna, and 7 THz by *Gu* et al. [16] with an LT-GaAs PC antenna.

2.1.2 Experimental Findings of the Broadband Terahertz Detection with a Photoconductive Antenna

At the same time, however, it was pointed out by *Ralph* and *Grischkowsky* that the detectable bandwidth could be determined by the rise in carrier number influenced by the optical gate pulse, not by the carrier lifetime [17]. The equation giving the photocurrent of the PC antenna, $J(t)$, is transformed into the following equation according to the convolution theorem of the Fourier transformation:

$$J(\omega) \propto N(\omega) \cdot E_{\mathrm{THz}}(\omega), \tag{7}$$

where, $J(\omega)$, $N(\omega)$, and $E_{\mathrm{THz}}(\omega)$ are the Fourier transforms of $J(t)$, $N(t)$, and $E_{\mathrm{THz}}(t)$, respectively. According to this equation, the frequency distribution of $N(t)$ should be exactly taken into account to consider the PC-antenna response. In particular, when the rise time of $N(t)$, which is mainly determined by the pulse width of the gating beam, is much shorter than the carrier lifetime, $N(t)$ can be approximated by a step function. In this case, the PC antenna will work as an integrating detector of the ultrashort EM pulses. Based on this principle, there were reports on the detection of the subpicosecond EM pulses by the PC antennas fabricated on semiconductors with long carrier lifetime [18, 19].

Figure 9 shows the subpicosecond EM pulses detected with the PC antenna fabricated on semi-insulating GaAs (SI-GaAs) reported by *Sun* et al. [18]. The carrier lifetime was subnanosecond. The radiation source was an unbiased GaAs wafer and the current from the PC antenna shows the change in the subpicosecond time scale. They treated the change in the photocarrier number as a step function and concluded that the time derivative of the photocurrent will give the incident THz-radiation waveform.

Tani et al. [19] further expanded these measurements by using PC antennas fabricated on SI-GaAs and SI-InP. The waveforms shown in Fig. 10 are the waveforms of the photocurrent detected by the PC antennas fabricated on the SI-GaAs, SI-InP, and LT-GaAs, respectively. The carrier lifetime of SI-GaAs was 36 ps, that of SI-InP 74 ps, while that of LT-GaAs was 0.4 ps. As shown in the dashed trace of Fig. 10, the time derivative of the waveform detected with the SI-InP PC antenna is very similar to that detected with the LT-GaAs PC antenna. Furthermore, the Fourier-transformed spectra of these waveforms shown in Fig. 11 are almost the same bandwidth. According to these experiments, when the carrier lifetime is long, the PC antenna was concluded to work as an integrating detector. In these experiments, the

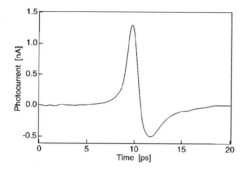

Fig. 9. Photocurrent signal waveform of THz pulse from GaAs wafer detected with a long lifetime GaAs PC antenna [18]

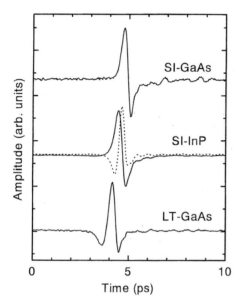

Fig. 10. Signal waveforms of THz pulse detected with a PC antenna fabricated on SI-GaAs (**a**), SI-InP (**b**), and LT-GaAs (**c**). *Dotted curve* represents a time differentiation of the waveform detected with a SI-InP PC antenna [19]

temporal resolution of the integrating detector is determined by the rise time of the carrier number corresponding to the pulse width of the gating probe beam.

In the integrating detection of the PC antenna, the temporal change in the electric field of the incident THz radiation shorter than the gating pulse width will be averaged. These experiments show that the frequency response of the PC antenna is independent of the carrier lifetime. To increase the time resolution of the PC antenna with the integrating property, the rise time of the carrier number should be reduced. The detectable bandwidth of the PC antenna was supposed to be extended by shortening the pulse width of the gating beam as much as possible.

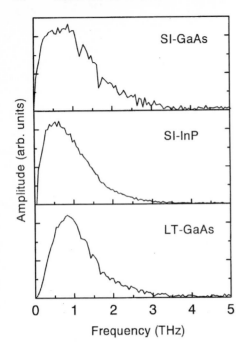

Fig. 11. Fourier-transformed amplitude spectra of THz pulses detected
with a PC antenna fabricated on
SI-GaAs (**a**), SI-InP (**b**), and LT-
GaAs (**c**) [19]

2.1.3 Demonstration of Broadband Photoconductive Detection

The first experimental demonstration of broadband detection with PC antennas was done by *Kono* et al. Figure 12a shows the waveform of the THz
radiation from a SI-InP surface detected with an LT-GaAs PC antenna gated
with 15 fs optical pulses [20]. The measurements were made using the standard THz-TDS system shown in Fig. 15a in the Chapter by *Sakai* and *Tani*,
by replacing the PC emitter with an unbiased wafer of SI-InP. Besides the
large peak around the 0.5 ps time delay, which was attributed to the surface
current due to the optical pumping, very fast oscillations were overlapped on
the entire waveform. The shortest period was estimated to be 45 fs. These fast
oscillations are attributed to the optical rectification effect on the InP surface as observed in the GaAs [3]. The PC antenna used in this measurement
was a 30 μm long dipole antenna with a 5 μm gap at the center, fabricated
on the 1.5 μm thick LT-GaAs layer grown at 250 °C on the GaAs substrate.
The carrier lifetime was estimated to be 1.4 ps by a transient reflectance
measurement.

Figure 13 shows the Fourier-transformed spectrum of the detected THz-
signal waveform in Fig. 12b measured for a longer trace. The frequency
components up to 20 THz were detected. In this spectrum, some spectral
structures were identified. The large absorption band from 7 THz to 9 THz
is the phonon resonance in the 0.4 mm thick GaAs substrate of the PC antenna. Some sharp lines due to water-vapor absorption were identified with

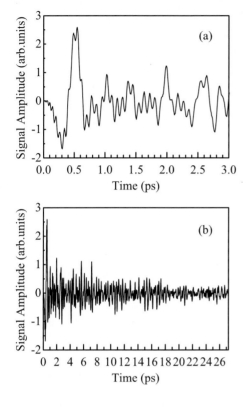

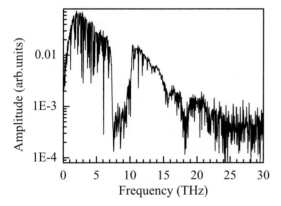

Fig. 12. Beginning (a) and entire (b) signal waveforms of THz pulse detected with an LT-GaAs PC antenna gated with 15 fs optical pulses. The THz-radiation source was an unbiased SI-InP wafer [20]

Fig. 13. Fourier-transformed spectrum of the detected THz-signal waveform shown in Fig. 12b [20]

the values in the literature within the spectral resolution of 36.6 GHz. The absorption centered at 15.5 THz is due to the phonon oscillation in the Si hemispherical lens.

However, the optical rectification effect in the wide-bandgap semiconductor surfaces is not as efficient as that in the EO crystals. Thus the upper limit

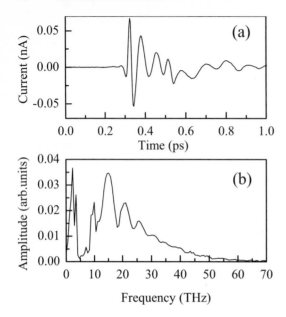

Fig. 14. (a) PC-antenna-detected signal waveform of THz pulse generated from a 100 µm thick ZnTe crystal, (b) Fourier-transformed spectrum of the signal waveform shown in (a) [21]

of the detectable bandwidth was explored assuming the detectable bandwidth to be broader than 20 THz.

The PC-antenna detection in the high-frequency region was investigated by changing the radiation source from the semiconductor wafer to an EO crystal. Figure 14a shows the PC-antenna-detected THz waveform radiated from the 100 µm thick ZnTe crystal. The PC antenna was gated with the 15 fs optical pulses [21]. The pulse width of the pumping beam incident on the ZnTe crystal was the same. The shortest period of the oscillation cycle was estimated to be 15 fs.

The Fourier-transformed spectrum of this waveform is shown in Fig. 14b. The frequency distribution exceeds 60 THz. This is the highest frequency detected with the PC antenna so far reported [21]. The detectable bandwidth of the PC antenna was found to be almost ten times broader than before. The absorption band around 5 THz is due to the transverse-optical phonon absorption in the ZnTe emitter. The peaks at 14.3 THz, 19.9 THz, and 25.2 THz are due to the phase-matching condition of the optical rectification effect in the ZnTe emitter. In this measurement, the PC antenna was reversed from the usual antenna geometry where the detected radiation was incident on the side of the electrodes. The phonon absorption in the GaAs antenna substrate could be reduced by reversing the antenna. This configuration also gives another experimental advantage that is explained later.

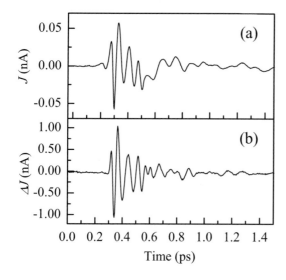

Fig. 15. Signal waveforms of THz pulses from ZnTe crystal excited with 15 fs optical pulses. (**a**) detected with an LT-GaAs PC antenna gated with 15 fs optical pulses, (**b**) detected with EO sampling technique. The EO crystal was 12 μm thick [22]

2.1.4 Spectral Response of Photoconductive Antennas in Broadband Detection

These measurements of the broadband THz radiation with the PC antenna show that the detectable bandwidth is comparable to that of the EO sampling technique using a thin EO crystal explained in Sect. 1.1. The frequency response of the EO sampling technique was already investigated and well explained by the phase-matching condition between the gating probe pulse and incident THz radiation. For the spectroscopic application of the PC antenna, its frequency response should preferably be simple. In the broadband detection regime, the PC antenna qualitatively works as an integrating detector as explained by the above description. For more quantitative investigation of the frequency response of the PC antenna in this broadband detection regime, the spectral response of the PC antenna was compared with that of the EO sampling technique [22].

Figure 15 shows the waveforms of the THz radiation emitted from the 100 μm thick ZnTe crystal excited with 15 fs optical pulses: (a) detected with the PC antenna, (b) detected with the 12 μm thick ZnTe crystal. The oscillation cycles in the respective waveforms are similar to each other, however, the fast structures shown in Fig. 15a were rather rounded compared with those in Fig. 15b. In both of the measurements, the pulse width of the gating or probing beam was also 15 fs. The position of the PC antenna and EO crystal was carefully adjusted to the focal point of the THz radiation for the precise comparison, as depicted in the schematic representation shown in Fig. 16b. The reversing of the PC antenna meant that the PC antenna could be easily replaced with the ZnTe crystal without any change in the optical path, as shown in Fig. 16a,c.

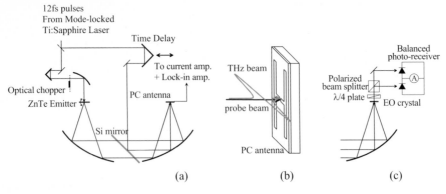

Fig. 16. (a) Schematic representation of time-resolved PC sampling experimental setup, (b) Antenna geometry. The THz pulse and gating beam were incident on the side of the electrodes. (c) Schematic representation of EO sampling

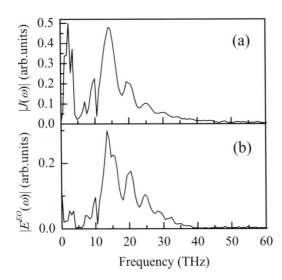

Fig. 17. Fourier-transformed spectra of (a) PC-detected signal waveform shown in Fig. 15a and EO-detected signal waveform shown in Fig. 15b

Figure 17 shows the Fourier-transformed spectra of the respective waveforms shown in Fig. 15. The frequency components below 5 THz in the PC-detected Fourier-transformed spectrum shown in Fig. 17a, is more enhanced than those in the EO-detected Fourier-transformed spectrum shown in Fig. 17b. On the other hand, the decay in the frequency components higher than 20 THz in the PC-detected spectrum is faster than the EO-detected spectrum.

For the quantitative comparison of these spectra, the frequency distribution of the carrier number, $N(\omega)$, is necessary. In order to determine $N(\omega)$, the following assumptions were made: (1) the rise time of $N(t)$ is given by

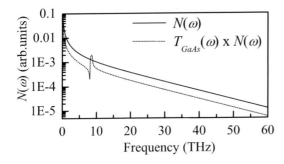

the pulse width of the gating beam Δt, (2) the decay of $N(t)$ is given by the carrier lifetime τ_c, (3) the momentum relaxation of the photocarriers is ne-glected for simplicity. The rise of $N(t)$ is proportional to the time integration of the optical pulse intensity $I(t)$:

$$I(t) = I_0 \operatorname{sech}^2 (1.76t/\Delta t). \tag{8}$$

The above assumptions are taken into account to describe $N(t)$ as follows:

$$N(t) = N_0 \exp(-t/\tau_c) [\tanh(1.76t/\Delta t) + 1]/2. \tag{9}$$

Here, N_0 is the maximum number of photocarriers.

The Fourier-transformed spectra of this equation, $N(\omega)$, is shown in Fig. 18 with the gating pulse width of 15 fs. This frequency response shows that the frequency components higher than 10 THz rapidly decrease. To de-termine the actual antenna response, the frequency dependence of the Fresnel coefficient $T(\omega)$ of the GaAs was also taken into account (shown by a dotted trace in Fig. 18).

In order to compare the spectra shown in Fig. 17, the PC-detected spec-trum is converted into an EO-detected spectrum by taking $N(\omega)$ and $G(\omega)$ into account. Here, $G(\omega)$ is the frequency response of the EO sampling tech-nique described in Sect. 1.1.1. In this conversion, $G(\omega)$ is calculated with a 12 μm thick ZnTe crystal.

Figure 19 shows the result of this conversion. The converted PC-detected spectrum shown by the solid trace well reproduces the EO-detected spectrum shown by the dashed trace. The dashed-dotted trace shows $G(\omega)$ used in this conversion. The frequency components higher than 40 THz in the converted spectrum are artifacts in this conversion. The signals on the noise floor are enhanced by the division with $N(\omega)$ that rapidly decreases with the increase in the frequency. This comparison shows that the frequency response of the PC antenna is mainly determined by $N(\omega)$. According to the assumptions of $N(\omega)$, $N(\omega)$ in the high-frequency region is determined by the pulse width of the gating beam because the gating pulse width is a hundred times shorter than the carrier lifetime. The frequency response of the PC antenna in the high-frequency regime is mainly governed by the rise in carrier number, the

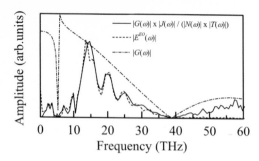

Fig. 19. *Solid curve* represents the PC-detected spectrum converted into an EO-detected spectrum by taking $N(\omega)$, $T_{\mathrm{GaAs}}(\omega)$, and $G(\omega)$ into account. *Dashed curve* represents the EO-detected spectrum shown in Fig. 17b. *Dashed-dotted curve* represents the EO response function, $G(\omega)$, with 12 μm thick ZnTe crystal [22]

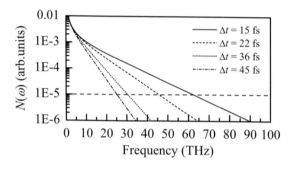

Fig. 20. PC response function $N(\omega)$ for different gating pulse widths

gating pulse width. Thus the detectable bandwidth of the PC antenna is found to be determined by the gating pulse width, independent of the carrier lifetime.

2.1.5 Photoconductive Antenna Response with Different Gating Pulse Widths

Figure 20 shows $N(\omega)$ with the different pulse widths of the gating probe beam. The decay of the frequency response above 5 THz critically depends on the pulse width of the gating beam as expected. On the other hand, the frequency response below 5 THz is almost independent on the gating pulse width. The PC antenna works as a sampling detector in sufficiently low-frequency regimes. The horizontal dashed line indicates the noise level of the PC detection empirically determined. Figure 20 predicts that even with the gating pulse width of 45 fs, the detectable bandwidth of the PC antenna should be extended up to 30 THz with an intense enough radiation source.

This prediction was confirmed by detecting the THz radiation from a 100 μm thick ZnTe crystal pumped by 15 fs optical pulses with the PC antenna gated with the probe beam of different pulse widths. The PC-detected

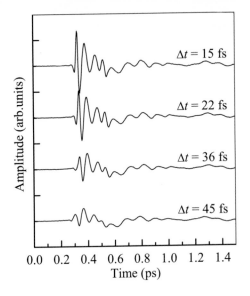

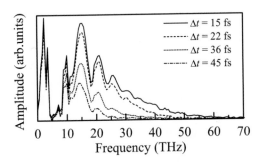

Fig. 21. Waveforms of THz pulses detected with an LT-GaAs PC antenna gated with the beam having different pulse widths. The THz-pulse source was pumped with 15 fs optical pulses for all the waveforms

Fig. 22. Fourier-transformed spectra of the respective waveforms shown in Fig. 21

waveforms gated with the beams having various pulse widths are shown in Fig. 21. As the pulse width of the probe beam was broadened, the sharp peaks in the waveforms were rounded. This behavior is consistent with Fig. 20, showing that the high-frequency components rapidly decrease with the broadening of the gating pulse width.

Figure 22 shows the Fourier-transformed spectrum of the respective waveforms shown in Fig. 21. The frequency components below 5 THz are almost the same for all the spectra, while those above 5 THz decrease with the broadening of the gating pulse widths. As the THz radiation source was the same for all the waveforms, the decay of the high-frequency components of the spectra is consistent with $N(\omega)$ of the different gating pulse widths, shown in Fig. 20. As expected from the theoretical estimation, the frequency components up to 30 THz were detected with the PC antenna gated with the pulse width of 45 fs.

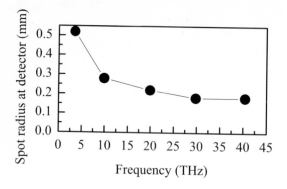

Fig. 23. Frequency dependence of the THz-beam spot radius on the detectors

2.1.6 Effects of Antenna Structure

In the early days of the PC antenna development, the frequency response of PC antennas was well investigated by *Jepsen* et al. [23]. In their treatments, they considered the response of the photocreated carriers based on the Drude–Lorentz model and geometrical factors derived from the antenna structure. The analysis based on the Drude–Lorentz model well explains the actual emission spectrum of the Hertzian dipole antenna that shows a broad spectral distribution with a summit in the low-frequency region.

Furthermore, the frequency response of the PC antenna was analyzed in terms of the geometrical response function (GRF) [23]. The GRF is a function that describes the overlap between the focal spot of the THz beam and the dipole antenna. As the focal size depends on the wavelength of the incident radiation, GRF has a frequency dependence. Especially in the case of the broadband detection, the incident THz pulses are described as a superposition of the different frequency components. Thus GRF is necessarily considered. The focal size of the incident THz pulses at different frequency components was estimated by measuring the THz beam radii between the two off-axis paraboloidal mirrors. The estimated beam radii were converted into the focal sizes at the detectors assuming that the THz beam was a Gaussian beam. Figure 23 shows the estimated focal size of the THz beam on the detectors as a function of the frequency. This estimation shows that the spot size of the THz beam does not significantly depend on the frequency in the frequency range above 10 THz.

In the above comparison of the PC and EO sampling, the relative position between the THz beam and gating or probing beam were the same for both of the detection techniques. The area on the EO crystal sensitive to the THz radiation was determined by the focal size of the probe beam. The focal size of the gating probe beam was estimated to be about 30 μm. Thus the active area on the EO crystal was similar to the size of the antenna dipole. Therefore, even with the considerable change of the spot size in the frequency range lower than 10 THz, the effect of GRF is not significant for the present comparison of the spectral response.

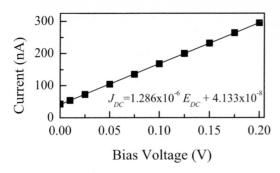

Fig. 24. Photocurrent of the LT-GaAs PC antenna as a function of the DC bias voltage under a certain gating-beam illumination condition [21]

2.1.7 Signal-to-Noise Characteristics in Broadband Photoconductive Detection

The signal-to-noise ratios (SNR) of the signal waveforms measured by the respective detection techniques were similar to each other, as shown in Fig. 15. In order to evaluate our experimental comparison in terms of SNR, we estimate the peak electric field of the THz radiation detected in our measurements [21].

The photocurrent from a biased PC antenna under a gating beam illumination is proportional to the applied biased voltage E_{DC}

$$J_{\mathrm{DC}} = \alpha E_{\mathrm{DC}}. \tag{10}$$

The relation between the antenna responses under DC bias voltage given by (6) and (10) is given by the following equation

$$\beta = \frac{J_{\mathrm{THz}}(t)}{J_{\mathrm{DC}}} = \frac{\int_{-\infty}^{\infty} E_{\mathrm{THz}}(t')\, N(t'-t)\, dt'}{\int_{-\infty}^{\infty} E_{\mathrm{DC}} N(t')\, dt'}. \tag{11}$$

Here, $E_{\mathrm{THz}}(t)$ assumes a normalized waveform of the incident THz radiation and E_{DC} is the DC electric field equal to unity. With these two factors, α and β, and the reflection loss of the incident THz radiation at the antenna surface, the photocurrent from the PC antenna at a delay time τ_{d} is approximated by the DC electric field by the following equation

$$J_{\mathrm{THz}}(\tau_{\mathrm{d}}) = T\alpha\beta\, E_{\mathrm{DC}}. \tag{12}$$

Here, T is the Fresnel transmission coefficient already introduced. The α is deduced from Fig. 24 showing the current characteristics as a function of the bias voltage under the same gating beam illumination as the waveform measurements. According to the above formulae, the peak electric field of the THz radiation was estimated to be about $20\,\mathrm{V/cm}$ at the peak current of $J_{\mathrm{THz}} = 38\,\mathrm{pA}$ obtained from one of the typical signal waveforms.

The phase retardation induced by this peak electric field in an EO crystal can be estimated with (1). Using this estimated phase retardation, the

SNR of the EO sampling technique is estimated to be 470 with the following assumptions: 10 μm thick ZnTe EO crystal, shot-noise limited measurement.

The typical SNR of the PC-detected signal waveform shown in our measurements was estimated to be about 125. Thus the estimated SNR in the EO sampling technique with the same electric field of the THz radiation is about four times larger than the actual SNR in our measurement with the PC antenna. This quantitative comparison shows that the experimental comparison of both detection methods was quite reasonable, as shown in Fig. 15.

2.1.8 Experimental Techniques for the Application of Photoconductive Antennas for Broadband Detection

To analyze PC-detected spectra in the broadband detection, the calibration of the obtained spectra with the frequency response of the PC antenna is necessary. This calibration is done by dividing the measured spectrum by $N(\omega)$. This division was made in the experimental comparison of the PC and EO sampling. However, the noise components around the noise floor were unexpectedly enhanced as shown as the artifacts higher than 40 THz in Fig. 19. Not only for noise components, but also for weak signals, the division by $N(\omega)$ is not preferable since $N(\omega)$ rapidly decreases with the increase in frequency. In order to avoid the inaccuracy derived from the numerical analysis, another experimental method can be employed: the shaker modulation technique [21]. The shaker modulation technique gives a modulation on the optical path length in one of the arms in the THz-TDS system. The output of the lock-in amplifier with shaker modulation will be proportional to the time derivative of the detected signal. As described in Sect. 2.1.2–Sect. 2.1.5, the PC antenna will work as an integrating detector in the high-frequency region. If $N(t)$, the temporal change in carrier number, is approximated with a step function, the time derivation of the signal waveform with the shaker modulation will give the original waveform of the THz pulses incident on the PC antenna. Figure 25 shows the Fourier-transformed spectra, (a) detected with a chopper modulation multiplied by ω (b) detected with the shaker modulation, of the THz pulse from the same 100 μm thick ZnTe crystal. According to the theorems of the Fourier transformation, the time derivation of the signal waveform is transformed into the frequency spectrum multiplied by the frequency ω. The figure shows that the noise floor in the spectrum obtained by the shaker modulation is almost ten times lower than that obtained by the chopper modulation multiplied by ω. This experimental demonstration shows that the shaker modulation technique is advantageous for the spectroscopic analysis of the PC-detected spectrum in the broadband detection as far as the following conditions are fulfilled: 1. the shaker modulation does not increase the noise specific to the modulation technique itself, 2. the noise levels in the signal waveforms are the same under both modulation techniques.

The experimental advantage of the shaker modulation is explained in another way. The $N(\omega)$ is described by $1/\omega$ when $N(t)$ is a step function. In the

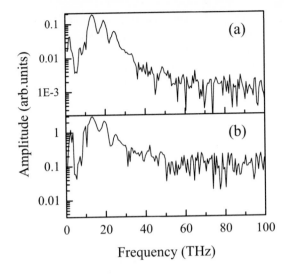

Fig. 25. Fourier-transformed spectra of the THz pulses from a 100 μm thick ZnTe crystal detected with an LT-GaAs PC antenna; (**a**) shows the PC-detected spectrum under shaker modulation; (**b**) shows the PC-detected spectrum multiplied by ω under chopper modulation [21]

calibration of the Fourier-transformed spectrum, chopper-modulated spectra are necessarily divided by $N(\omega)$, while shaker-modulated spectra are divided by $N(\omega)/(1/\omega)$. By shortening the pulse width of the gating beam, $N(\omega)$ in the actual experimental condition approaches $1/\omega$. Thus the inaccuracy derived from the numerical calibration in the frequency response can be reduced with the shaker-modulation technique.

2.1.9 Discussions

Recent development on the EO sampling technique has been quite promising for the phase-sensitive detection of the radiation in the high-frequency region up to 100 THz. Here, we mention some experimental interests on the PC antenna for the broadband detection.

The PC antenna compliments some disadvantages specific to the EO sampling technique. In the EO sampling technique, there are frequency bands insensitive to the incident radiation due to the phonon resonances. On the other hand, the PC antenna can avoid this phonon absorption from 6 THz to 10 THz by the following methods: 1. fabricating the PC antenna on the different PC material, for example, high-resistivity silicon, 2. reversing the PC antenna to receive the incident radiation on the antenna electrodes side.

For the broadband detection of THz radiation, the pulse width of the gating probe beams should be as short as possible for the detection methods, PC and EO sampling. However, in the EO sampling technique, the crystal thickness should be estimated by taking into account of the phase-matching condition dependent on the target frequency range. The frequency response of the PC antenna, simply dependent on the gating pulse width, is advantageous in the frequency analysis of spectra. Especially the combination with the shaker-modulation technique will compensate the frequency response rapidly decreasing in the high-frequency range.

The robustness against the optical noises is another interesting point in the PC detection. In the broadband detection of the radiation, the thickness of the EO crystal should be reduced to $\approx 10\,\mu$m. The reduction of the thickness also reduces the phase change in the probe beam. Therefore, the sensitivity of the photodetectors is quite important to make an efficient measurement. On the other hand, the electric field of the incident THz radiation is directly converted to the photocurrent in the PC antenna. This means that PC antennas are rather robust to the optical noises compared with the EO sampling technique.

2.2 Ultra-Broadband Generation of Terahertz Radiation with Photoconductive Antennas

In Sect. 2.1, the broadband detection with PC antenna was shown. The analysis on the frequency response of the PC antenna shows that the classical Drude–Lorentz model is still valid with the gating pulse width of 15 fs. In other words, the detectable bandwidth of PC antennas is determined by the gating pulse width, not by the RC time constant, nor by carrier lifetime. This, in turn, implies that a PC antenna will work as an emitter generating broadband THz radiation when the pulse width of the pumping beam is short enough. In this section, the recent achievement on the generation of broadband THz radiation with a PC antenna is described.

2.2.1 Generation of Broadband Terahertz Radiation with Photoconductive Antennas

The first demonstration of the generation of the broadband THz radiation higher than 10 THz with a PC antenna was done by *Tani* et al. [24]. The PC antenna used to generate the broadband THz radiation was fabricated on a 1.5 μm thick LT-GaAs layer grown at 250 °C. The LT-GaAs layer was grown on the 0.4 mm thick Si substrate in order to reduce absorption typically seen in GaAs substrate. The antenna structure consists of 10 μm wide coplanar striplines separated by 30 μm. The carrier lifetime of the LT-GaAs was estimated to be 0.6 ps by a transient reflectivity measurement. The flat surface of a Si lens was attached to the substrate side of the PC antenna so that the Si lens and the Si substrate may form a hyperhemispherical lens. The optical pump pulse was focused onto the biased PC gap with an off-axis paraboloidal mirror. The excitation power was 240 mW. The pulse width of the pumping beam was estimated to be 18 fs. The generated radiation was detected with the THz-TDS system. In this system, an LT-GaAs PC antenna was used as a detector. The detectable bandwidth of the PC antenna was shown to be comparable to the EO sampling technique using a 10 μm-order thickness EO crystal as explained in Sect. 2.1.4. In this measurement too, the PC antenna detector was reversed, so that the generated THz radiation and the gating beam were both incident on the side of the antenna electrodes.

Figure 26a shows an observed waveform of the THz radiation generated with the PC antenna with a bias voltage of 12 V, pumped with 18 fs optical pulses. The Fourier-transformed spectrum of this waveform is shown in Fig. 26b. The frequency distribution extends beyond 10 THz. An absorption dip around 8 THz and peak at 8.8 THz were due to the phonon resonance in GaAs layers of the PC emitter and detector. Other sharp absorption peaks are due to water-vapor absorption. The amplitude of the generated THz radiation was found to be linear to the bias voltage. Thus, the generated THz radiation was concluded to be purely from the photoconductive current. The nonlinear optical effect, such as optical rectification effect, was negligible in this measurement.

Quite recently, ultra-broadband generation of THz radiation from LT-GaAs PC antennas excited with 12 fs pulses was reported by *Shen* et al. [25, 26]. They demonstrated two types of generation and detection schemes: one is a combination of PC generation and EO detection [25], the other is that of PC generation and PC detection [26]. The bow-tie-type LT-GaAs PC antenna used by *Shen* et al. consisted of wide-gapped electrodes whose separation was 0.4 mm. The bias voltage applied to the PC antenna was modulated with an amplitude of ±120 V. In the scheme of the PC generation and EO detection, the bandwidth of the generated THz radiation was over 30 THz that is the highest frequency generated with PC antennas. Figure 27 shows the EO-detected waveform and its corresponding Fourier amplitude spectrum. In the other scheme, using PC generation and PC detection, the detected bandwidth was 15 THz. This bandwidth is consistent with the observation by *Tani* et al. The reduction of the bandwidth in this scheme was due to the detection frequency response of a PC antenna explained in Sect. 2.1.4. As they also adopted PC detection, the amplitude of the THz radiation generated by the PC antenna was compared to the THz radiation generated by a 100 μm thick ZnTe crystal reported in [22]. The amplitude of the radiation by the PC antenna was concluded to be two orders of magnitude larger than the amplitude of the radiation generated by the EO crystal. This intense radiation also enabled detection of the broadest bandwidth of 30 THz with the EO sampling technique. Thus, the combination of the PC generation and PC detection is interesting for practical application of the THz spectroscopy system.

2.2.2 Frequency Distribution of Broadband Terahertz Radiation from Photoconductive Antennas

In Sect. 2.1, the frequency response of the PC antenna in the broadband detection was mainly analyzed with the temporal change in carrier number. It was found that the fast rise of the carrier number by short enough optical

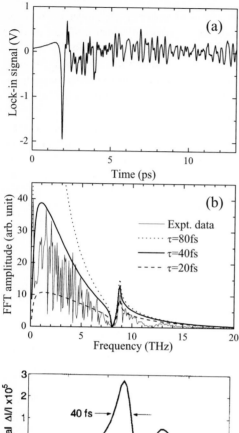

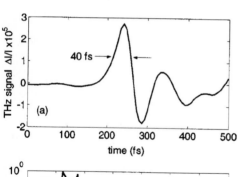

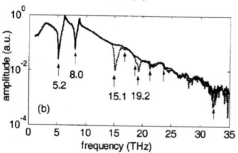

Fig. 26. (a) Signal waveform of THz pulse generated by a PC antenna made on LT-GaAs gated with 18 fs optical pulses. Detector was another LT-GaAs PC antenna gated with 18 fs optical pulses, (b) *Thin solid curve* represents the Fourier-transformed spectrum of the signal waveform shown in (a). Dotted curve ($\tau =$ 80 fs), *thick solid curve* ($\tau =$ 40 fs), and *dashed curve* ($\tau =$ 20 fs) represent numerical estimations of the frequency distribution of the THz pulses generated by the PC antenna with different carrier momentum relaxation times [24]

Fig. 27. (a) THz waveform generated by a PC antenna made on LT-GaAs gated with 12 fs optical pulses. This waveform was detected by the EO sampling technique with 20 μm thick ZnTe crystal, (b) *Solid curve* represents the Fourier-transformed amplitude spectrum corresponding to the waveform shown in (a). *Dotted curve* represents the Fourier-transformed spectrum measured in the presence of a PTFE sample [25]

pulses was critical to broaden the detectable bandwidth. *Tani* et al. further included the antenna factors considered in the model developed by *Jepsen* et al. [23] to characterize the broadband THz radiation spectrum by the PC antenna. The following factors were taken into account: the carrier lifetime τ_c, the carrier momentum relaxation time τ, the optical phonon resonance effect in the LT-GaAs emitter and detector, and the frequency-dependent factor of the geometrical structure of the PC antenna.

As introduced in the Chapter by *Sakai* and *Tani*, the transient current $J(t)$ in a PC emitter with DC bias E_b and pumped with a δ-function-like optical pulse at time zero is given by the following equations

$$J(t) = 0 \qquad\qquad (t \leq 0)$$

$$J(t) = -eN_0 \exp(-t/\tau_c)\, \frac{\tau e E_b}{m}\, [1 - \exp(-t/\tau)] \qquad (t > 0). \qquad (13)$$

An actual pump pulse has a finite pulse width, given by $I(t)$, therefore the real transient current is described by the convolution of $J(t)$ with $I(t)$. Therefore, the transient current created by the pump pulse is given by the convolution of $J(t)$ with $I(t)$

$$J^{\text{real}}(t) = J(t) \otimes I(t). \qquad (14)$$

The time derivative of this transient current is proportional to the radiated electric field

$$E_{\text{THz}}(t) \propto \frac{\partial J^{\text{real}}(t)}{\partial t}. \qquad (15)$$

The frequency distribution of the generated THz radiation is given by the Fourier transform of $J^{\text{real}}(\omega)$. The frequency components of the generated radiation are given by the following equation

$$E_{\text{THz}}(\omega) \propto \omega J^{\text{real}}(\omega) = \omega J(\omega) I(\omega). \qquad (16)$$

With this equation, the intensity profile of the excitation pulse $I(\omega)$ is given by Fourier transformation of the sech^2 function as described in Sect. 2.1.4. In this model treatment, the carrier momentum relaxation time is only an adjustable parameter. The dotted, thick, and dashed curves in Fig. 26b are the estimated frequency distribution with the carrier momentum relaxation time $\tau = 20$ fs, 40 fs, and 80 fs, respectively. The model calculation shows that the momentum relaxation time of $\tau = 40$ fs reproduces the observed spectrum. According to their model calculation, however, they found that the low-frequency components are significantly affected by the carrier relaxation time while the high-frequency components are not really affected. Therefore, the broadband characteristics of the PC antenna are mainly determined by the pulse width of the pump beam. In Sect. 2.1, the frequency response of the PC antenna in broadband detection was analyzed by considering the pulse width of the gating beam, not the carrier momentum relaxation time. The analysis in the former section was consistent with their finding. This experimental conclusion was supported by the recent theoretical analysis done by *Duvillaret* et al. [27].

2.2.3 Comparison of Radiation Bandwidth

In Sect. 1.2, the generation of the broadband THz radiation with a thin nonlinear crystal was introduced. As the PC antenna is also capable of generating broadband THz radiation, it is worthwhile to compare the broadband radiation efficiency of the PC antenna with that of the optical rectification effect. The frequency distribution of the observed spectrum shown in Fig. 26b is much narrower than that generated by the optical rectification effect using a thin nonlinear crystal.

In the optical rectification effect, the generated THz radiation is proportional to the second-order time derivative of the nonlinear (NL) polarization, given by (5) in the Chapter by *Sakai* and *Tani*, which is repeated here in the following form

$$E^{\mathrm{NL}}\left(t\right) \propto \frac{\partial^2 P\left(t\right)}{\partial t^2}. \tag{17}$$

The nonlinear polarization $P(t)$ is proportional to the intensity of the excitation pulse as shown in (5) in Sect. 1.2.1 of this chapter. Therefore, the generated radiation by the optical rectification effect is proportional to the second-order time derivative of the excitation intensity $I(t)$. The Fourier transformation gives the relation between the bandwidth of the generated radiation and the frequency distribution of the optical pump pulse described by the following equation

$$E^{\mathrm{NL}}\left(\omega\right) \propto \omega^2 G\left(\omega\right) P\left(\omega\right) \propto \omega I\left(\omega\right). \tag{18}$$

Here, the phase-matching condition, $G(\omega)$, is included. In the third term of the equation, $G(\omega)$ is approximated to be proportional to the inverse of the frequency, $\propto 1/\omega$ in the high-frequency region. This estimation shows that the bandwidth of the generated THz radiation with the optical rectification effect is qualitatively proportional to the product of the frequency bandwidth of the optical pump pulse and frequency.

On the other hand, the THz radiation generated with the PC antenna is proportional to the first-order time derivative of the transient current

$$E^{\mathrm{PC}}\left(t\right) \propto \frac{\partial J^{\mathrm{real}}\left(t\right)}{\partial t} \propto \frac{\partial N\left(t\right)}{\partial t} \propto \frac{\partial\left(\int I\left(t\right)\mathrm{d}t\right)}{\partial t} \propto I\left(t\right). \tag{19}$$

The transient current is approximately proportional to the time integration of the optical pump pulse because the pulse width of the pumping beam is much shorter than the carrier lifetime in the case of the broadband THz generation with PC antennas. Therefore, as in the treatment done in the broadband PC detection, the generated THz radiation with a PC antenna is proportional to the optical pulse intensity $I(t)$.

Finally, the radiation bandwidths of the optical rectification and PC antenna are compared by the following equations

$$\text{Nonlinear:} \quad E^{\text{NL}}(\omega) \propto \omega I(\omega),$$
$$\text{Photoconductive:} \quad E^{\text{PC}}(\omega) \propto I(\omega). \tag{20}$$

The above relation shows that the optical rectification effect is advantageous in the generation of the broadband THz radiation by the factor of frequency ω.

The recent developments on the PC antenna enable the generation of the broadband THz radiation beyond 10 THz. In spite of such developments, the above analysis shows the advantage of the optical rectification effect in the broadband THz generation. However, as pointed out in the previous section, PC antennas can be fabricated on various PC materials. As seen in the spectra shown in Sect. 1.2.2, the nonlinear crystal emitters like ZnTe, GaP can not generate the radiation from 5 THz to 10 THz due to the phonon absorption of the emitters themselves. The PC emitters can compensate this inactive frequency region due to the phonon absorption by replacing the antenna materials with a material other than GaAs, for example, RD-SOS. Recent reports on the broadband generation of THz radiation using a PC emitter proves that the combination of PC emitter and PC receiver has advantages for the practical application of THz spectroscopy because of its intense radiation and its simple spectral response [25, 26]. These recent reports support the view that the PC emitters can be a promising radiation source and further investigations on the antenna materials are expected.

3 Conclusions

Since the invention of the photoconductive antenna in the mid-1970s, the generation and detection of ultrashort electromagnetic radiation has been extended from the far- to mid-infrared region with the advancement of ultrashort lasers. Nonlinear optical effects have played an important role to develop the coherent generation and detection of broadband THz radiation. In this Chapter, the nonlinear optical techniques broadening the bandwidth of the coherent THz radiation were introduced. In addition to the broadband nonlinear optical effects, new aspects of the PC antenna, broadband PC detection and generation, were discussed in detail. In terms of the bandwidth, the PC antenna is not as advantageous as the nonlinear optical techniques. However, the PC antenna has a variety of fabrication techniques and materials and may provide some complementary functions to the EO sampling and optical rectification techniques. These experimental techniques will be important for further applications of pulsed THz radiation.

References

[1] S. Boyd: *Nonlinear Optics* (Academic, San Diego 1992)
[2] Q. Wu, X.-C. Zhang: Appl. Phys. Lett. **70**, 1784 (1997)
[3] Q. Wu, X.-C. Zhang: Appl. Phys. Lett. **71**, 1285 (1997)
[4] A. Leitenstorfer, S. Hunsche, J. Shah, M. Nuss, W. Knox: Appl. Phys. Lett. **74**, 1516 (1999)
[5] A. Brodschelm, F. Tauser, R. Huber, J. Sohn, A. Leitenstorfer: *Ultrafast Phenomena XII* (Springer, Berlin, Heidelberg 2001) pp. 215–217
[6] Y. Shen: *Principles of Nonlinear Optics* (Wiley, New York 1984)
[7] A. Bonvalet, M. Joffre, J. Martin, A. Migus: Appl. Phys. Lett. **67**, 2907 (1995)
[8] P. Han, X.-C. Zhang: Appl. Phys. Lett. **73**, 3049 (1998)
[9] R. Kaindl, M. Wurm, K. Reimann, P. Hamm, A. Weiner, M. Woerner: J. Opt. Soc. Am. B **17**, 2086 (2000)
[10] R. Kaindl, D. Smith, M. Joschko, M. Hasselbeck, M. Woerner, T. Elsaesser: Opt. Lett. **23**, 861 (1998)
[11] R. Kaindl, F. Eickemeyer, M. Woerner, T. Elsaesser: Appl. Phys. Lett. **75**, 1060 (1999)
[12] R. Huber, A. Brodschelm, F. Tauser, A. Leitenstorfer: Appl. Phys. Lett. **76**, 3191 (2000)
[13] F. Eickemeyer, R. Kaindl, M. Woerenr, T. Elsaesser, A. Weiner: Opt. Lett. **25**, 1472 (2000)
[14] S.-G. Park, M. Melloch, A. Weiner: Appl. Phys. Lett. **73**, 3184 (1998)
[15] S. Ralph, S. Perkowitz, N. Katzenellenbogen, D. Grischkowsky: J. Opt. Soc. Am. B **11**, 2528 (1994)
[16] P. Gu, M. Tani, K. Sakai, T.-R. Yang: Appl. Phys. Lett. **77**, 1798 (2000)
[17] S. Ralph, D. Grischkowsky: Appl. Phys. Lett. **60**, 1070 (1992)
[18] F. Sun, G. Wagoner, X.-C. Zhang: Appl. Phys. Lett. **67**, 1656 (1995)
[19] M. Tani, K. Sakai, H. Mimura: Jpn. J. Appl. Phys. **36**, L1175 (1997)
[20] S. Kono, M. Tani, P. Gu, K. Sakai: Appl. Phys. Lett. **77**, 4104 (2000)
[21] S. Kono, M. Tani, K. Sakai: IEEE Proc. Optoelectron. **149**, 105 (2002)
[22] S. Kono, M. Tani, K. Sakai: Appl. Phys. Lett. **79**, 898 (2001)
[23] P. Jepsen, R. Jacobsen, S. Keiding: J. Opt. Soc. Am. B **13**, 2424 (1996)
[24] M. Tani, M. Nakajima, S. Kono, K. Sakai: in F. Kaertner (Ed.): (2002 IEEE/LEOS Annual Meeting Conference Proceedings, Glasgow, Scotland 2002) pp. 532–533
[25] Y. Shen, P. Upadhya, E. Linfield, H. Beere, A. Davies: Appl. Phys. Lett. **83**, 3117 (2003)
[26] Y. Shen, P. Upadhya, H. Beere, E. Linfield, A. Davies, I. Gregory, C. Baker, W. Tribe, M. Evans: Appl. Phys. Lett. **85**, 164 (2004)
[27] L. Duvillaret, F. Garet, J.-F. Roux, J.-L. Coutaz: IEEE J. Sel. Top. Quantum Electron. **7**, 615 (2001)

Terahertz Radiation
from Semiconductor Surfaces

Ping Gu[1,2] and Masahiko Tani[1,3]

[1] National Institute of Information and Communications Technology
558-2 Iwaoka, Nishi-ku, Kobe, Hyogo 651-2492, Japan
[2] *Present address:* Faculty of Education, Wakayama University
930 Sakaedani, Wakayama-shi, Wakayama 640-8510, Japan
guping@center.wakayama-u.ac.jp
[3] *Present address:* Institute of Laser Engineering, Osaka University
2-6 Yamadaoka, Suita, Osaka 565-0871, Japan
tani@ile.osaka-u.ac.jp

Abstract. In this chapter, terahertz radiation from semiconductor surfaces by excitation of femtosecond laser pulses is discussed with special attention, paid to the influence of lattice vibrations and plasma oscillations in semiconductors. The effect of dielectric dispersion near the optical-phonon resonance, the coupling of the THz electrical transient with longitudinal optical phonons and semiconductor plasmons is considered.

1 Introduction

It was found by *Zhang* et al. [1] in 1990 that ultrashort electromagnetic radiation with terahertz (THz) bandwidth can be generated by illuminating semiconductor surfaces with femtosecond laser pulses. Soon, the investigations of THz radiation from various semiconductors followed with study into the emission efficiency as well as the emission mechanism [2, 3, 4, 5, 6, 7, 8, 9]. Currently, it is commonly understood that the emission of THz radiation from semiconductor surfaces is primarily due to the surge current normal to the surface [1, 10] and/or the second-order nonlinear optical processes in the semiconductors [11, 12, 13]. The radiation intensity is generally low compared to the other THz radiation sources, such as biased photoconductive (PC) antennas, or phase-matched nonlinear optical crystals. However, after it was revealed that the InAs surface is an efficient THz emitter and that the efficiency is further enhanced by applying an external magnetic field [14, 15], THz radiation from an InAs surface attracted much interest because of the potential as a simple and powerful THz radiation source in practical applications, such as spectroscopy and imaging. The special topic on the enhancement of THz radiation from InAs under a magnetic field is treated in the Chapter by *Ohtake* et al.

In this Chapter, we discuss THz radiation from semiconductor surfaces by excitation of femtosecond laser pulses with special attention on the influence of lattice vibrations and plasma oscillations in semiconductors.

K. Sakai (Ed.): Terahertz Optoelectronics, Topics Appl. Phys. **97**, 63–97 (2005)
© Springer-Verlag Berlin Heidelberg 2005

In Sect. 2, we describe the several basic mechanisms of THz radiation from semiconductor surfaces. Three mechanisms are specified and explained: the optical rectification effect, the surface depletion field, and the photo-Dember effect. A detailed study is presented for THz radiation from InAs and InSb, for which the contributions from the optical rectification and the photo-Dember effect are both significant.

In Sect. 3, we describe THz radiation from coherent LO-phonons and LO-phonon–plasmon coupled modes in polar semiconductors. The mechanism of emission of THz radiation from coherent phonons is explained in relation to the excitation mechanism of coherent phonons with specific examples of Te, PbTe, and CdTe. The effect of the dielectric dispersion near the optical-phonon resonance is discussed. A study of THz radiation from doped InSb by the authors is also presented as an example of THz radiation from the LO-phonon–plasmon coupled modes.

2 Terahertz Radiation from Semiconductor Surfaces

In this section we describe the basic mechanisms of emission of THz radiation from semiconductor surfaces. Emission mechanisms of THz radiation from bulk semiconductor surfaces are categorized into two types. One is the nonlinear optical processes without real carrier excitation and the other is the ultrafast surface surge-current with real carriers due to the surface built-in field or photo-Dember effect. The THz emission by the nonlinear optical process is explained as the optical rectification of ultrashort laser pulses, which creates a transient polarization on the semiconductor surface. The THz radiation is emitted in proportion to the second time derivative of the electronic polarization due to the optical rectification (in the far-field approximation). This process is equivalently explained in the frequency domain as the difference-frequency mixing (DFM) between spectrum components of the femtosecond laser pulse in the semiconductors [12]. For the surge-current model, two origins are considered: one is the acceleration of photoexcited carriers by the surface-depletion field, and the other is the photo-Dember effect originating from the difference between the diffusion velocities of the electrons and holes. Because electrons gain much greater velocity due to their higher mobility than that of holes, electrons diffuse more rapidly from the surface inward of the semiconductor, creating an effective surge current normal to the surface.

A typical experimental setup for measurements of THz radiation generated from a semiconductor surface is illustrated in Fig. 1. Near infrared (NIR) femtosecond light pulses from a mode-locked Ti: sapphire laser ($\lambda \approx 800\,\text{nm}$, $h\nu \approx 1.5\,\text{eV}$, $\delta t < 100\,\text{fs}$) excite the semiconductor surface at a finite incident angle, such as 45°, and THz radiation emitted in the direction of optical reflection is collected and focused to a detector by using a pair of paraboloidal mirrors. For the time-domain measurement, an ultrafast sampling detector,

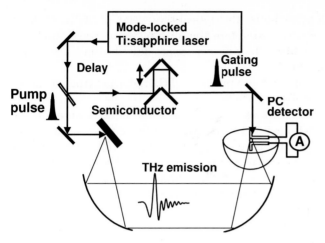

Fig. 1. Experimental setup for the generation of THz radiation from the semiconductor surface and its detection

Table 1. Sample properties

Sample	InSb	InAs	GaAs	InP
Bandgap E_g	0.17 eV	0.36 eV	1.43 eV	1.34 eV
Electron mobility μ_e $(\mathrm{cm}^2/(\mathrm{V}\cdot\mathrm{s}))$	76 000	30 000	8600	4000
Hole mobility μ_h $(\mathrm{cm}^2/(\mathrm{V}\cdot\mathrm{s}))$	3000	240	400	650
Electron mass m_e^*	$0.015 m_e$	$0.027 m_e$	$0.067 m_e$	$0.073 m_e$
Hole mass m_h^*	$0.18 m_e$	$0.33 m_e$	$0.5 m_e$	$0.40 m_e$
Refractive idex $(\lambda = 800\,\mathrm{nm})$	4.47	3.73	3.68	3.47
Refractive idex (at 1 THz)	1.93	3.78	3.61	3.52
Absorption depth $(\lambda = 800\,\mathrm{nm})$	94 nm	142 nm	749 nm	305 nm
Excess energy ΔE $(T = T_e + T_h = \frac{2}{3}\frac{\Delta E}{\kappa})$	1.38 eV (10 500 K)	1.18 eV (9000 K)	0.12 eV (900 K)	0.21 eV (1600 K)
Electron temp. T_e	9800 K	8300 K	790 K	1350 K
Hole temp. T_h	700 K	700 K	110 K	250 K

such as a PC antenna is used, and the optical time delay of the probe laser pulse to the pump is scanned to obtain the THz radiation waveforms (the details of the PC antennas and their properties are described in the Chapter by *Sakai* et al.).

For the explanation of the mechanisms of THz radiation from semiconductor surfaces, we describe the studies carried out by the authors on several representative semiconductors. The sample semiconductors used in the studies are listed in Table 1 with their basic properties and some important experimental parameters: an n-type and p-type of InSb (100) with a carrier concentration of of $n \approx 10^{16}\,\mathrm{cm}^{-3}$, an n-type InAs (111) and p-InAs (100) with a carrier concentration of $n \approx 10^{18}\,\mathrm{cm}^{-3}$, and an n-type and p-type of InP (100) with a residual carrier concentration of $n \approx 10^{18}\,\mathrm{cm}^{-3}$.

2.1 Optical Rectification

In the case of high-intensity excitation, the optical rectification effect becomes dominant [11]. The nonlinear optical process was first suggested by *Chuang* et al. [11, 16, 17] as the emission mechanism of THz radiation from semiconductor surfaces. They suggested an effective $\chi^{(2)}$ process, which is actually a $\chi^{(3)}$ process involving the DC surface-depletion field, in addition to the incident pump laser field. It was later found that the radiation emitted from InP by excitation of high-intensity laser pulses of 2.0 eV light was due to the bulk optical rectification (a $\chi^{(2)}$ process) [12]. Many other semiconductors without the inversion crystal symmetry, such as the zincblende-type semiconductors, have also been reported to emit THz radiation due to the bulk optical rectification effect [17, 18]. The THz radiation field generated by the optical rectification, $\boldsymbol{E}_{\mathrm{THz}}(t)$, is proportional to the second-order nonlinear polarization (in near field), which is described by the following equations:

$$\boldsymbol{P}(t) = \frac{1}{2\pi} \int_{-\infty}^{+\infty} \boldsymbol{P}^{(2)}(\varOmega) \mathrm{e}^{-\mathrm{i}\varOmega t}\, \mathrm{d}\varOmega\,, \tag{1}$$

$$P_i^{(2)}(\varOmega) = \sum_{j,k} \varepsilon_0 \chi_{ijk}^{(2)}(\varOmega = \omega_1 - \omega_2)$$
$$\cdot \int_{-\infty}^{\infty} E_j(\omega_1 = \varOmega + \omega_2) E_k(\omega_2)\, \mathrm{d}\omega_2\,, \tag{2}$$

where $\chi_{ijk}^{(2)}$ is the second-order nonlinear susceptibility tensor for a difference frequency, $\varOmega = \omega_1 - \omega_2$, and $E_j(\omega_1)$ $(E_k(\omega_2))$ is the amplitude spectral component of the pump laser at frequency ω_1 (ω_2) in the $j(k)$ direction. Here, i, j, and k are the dummy indices for x-, y-, and z-directions in the crystallographic axis system. The integral is extended to the negative frequency by using the definition, $E(-\omega) = E^*(\omega)$ (the asterisk means the complex conjugate).

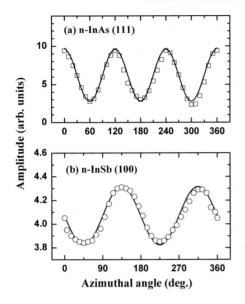

Fig. 2. Azimuthal-angle dependence of THz-emission amplitude from semiconductors with (**a**) n-InAs (111) (*squares*) and (**b**) n-InSb (100) (*circles*) surfaces at a 45° incidence angle of the excitation light on the sample. The *solid lines* are the theoretical curves fitted to the data

In the far field, the observed THz field amplitude, $E_{\mathrm{THz}}^{\mathrm{ob}}$, is proportional to the projection of the second time derivative of the nonlinear polarization to the polarization direction of detection, e (a unit vector normal to the observation direction), at the observation point:

$$E_{\mathrm{THz}}^{\mathrm{ob}}(t) = e \cdot E_{\mathrm{THz}} \propto e \cdot \frac{\partial^2 P(t)}{\partial t^2} \,. \tag{3}$$

The spectral amplitude, $E_{\mathrm{THz}}^{\mathrm{ob}}(\Omega)$, at a frequency Ω is thus given by

$$E_{\mathrm{THz}}^{\mathrm{ob}}(\Omega) \propto \Omega^2 e \cdot P(\Omega) \,. \tag{4}$$

The most unambiguous evidence for the contribution of the $\chi^{(2)}$ process is a strong dependence of the emitted THz radiation intensity on the crystal orientation to the pump-laser polarization. By rotating a sample about its surface normal, the relative contribution of the azimuthal-angle-dependent DFM component to the total THz radiation can be estimated.

Figure 2 shows the azimuthal-angle dependence of the peak amplitude in the time-resolved waveform of THz radiation (a) from n-InAs(111) and (b) from n-InSb(100) surfaces measured in the experimental configuration shown in Fig. 1 when the samples were rotated about their surface normal. A laser with a 40 fs pulse width, a 76 MHz repetition rate, and a central wavelength of 800 nm was focused on the sample semiconductor surfaces at 45° by using a lens with a focal length of about 25 cm. The pump laser was p-polarized, and the average power was about 80 mW. The excitation intensity on the samples were estimated to be about 60 MW/cm^2.

Taking the surface normal as the X-axis and the reflection plane as the XY-plane in the laboratory frame, the nonlinear polarization induced in the semiconductor due to the optical rectification for (111) and (100) surfaces is expressed by (5) and (6), respectively [19]:

$$
\boldsymbol{P} = \begin{pmatrix} P_x(\Omega) \\ P_y(\Omega) \\ P_z(\Omega) \end{pmatrix}
$$

$$
= 2\varepsilon_0 d_{14} \langle E^2(\Omega) \rangle \begin{pmatrix} -\frac{1}{\sqrt{3}}\cos^2\phi + \frac{2}{\sqrt{6}}\sin^2\phi \\ \frac{2}{\sqrt{6}}\cos^2\phi\cos 3\theta - \frac{2}{\sqrt{6}}\cos\phi\sin\phi \\ \frac{2}{\sqrt{6}}\cos^2\phi\sin 3\theta \end{pmatrix}, \tag{5}
$$

$$
\boldsymbol{P} = \begin{pmatrix} P_x(\Omega) \\ P_y(\Omega) \\ P_z(\Omega) \end{pmatrix}
$$

$$
= 2\varepsilon_0 d_{14} \langle E^2(\Omega) \rangle \begin{pmatrix} \cos^2\phi\sin 2\theta \\ \sin 2\phi\sin 2\theta \\ \sin 2\phi\cos 2\theta \end{pmatrix}, \tag{6}
$$

$$
\langle E^2(\Omega) \rangle = \int_{-\infty}^{\infty} E(\omega_1 = \Omega + \omega_2) E(\omega_2)\, \mathrm{d}\omega_2
$$

$$
= \int_{-\infty}^{\infty} E(\omega_1 = \Omega - \omega_2) E^*(\omega_2)\, \mathrm{d}\omega_2 . \tag{7}
$$

Here, ϕ is the angle between the surface normal and the pump-laser beam refracted inside the sample, θ is the azimuthal angle of the sample orientation around the X-axis (the angle between the Y-axis in laboratory frame and the crystallographic $(11-\bar{2})$ direction), $E(\omega_i)$ is the amplitude component of the pump laser for frequency ω_i, and $d_{14} = \chi_{14}^{(2)}/2$ is the nonlinear susceptibility coefficient for the difference frequency, Ω, in the contracted notation, $\langle E(\Omega) \rangle$ is the autocorrelation function of $E(\omega)$.

Using (3) and considering the refraction at the interface between the semiconductor/air interface, the p-polarized THz field amplitude observed in the direction of optical reflection is given by the following equation:

$$
E_{\mathrm{THz}}^{\mathrm{ob}} \propto \boldsymbol{e}\boldsymbol{P} = (-\sin\phi_{\mathrm{THz}}, \cos\phi_{\mathrm{THz}}, 0)(P_X, P_Y, P_Z)^t
$$

$$
= -P_X \sin\phi_{\mathrm{THz}} + P_Y \cos\phi_{\mathrm{THz}} , \tag{8}
$$

where ϕ_{THz} is the refraction angle of THz radiation inside the semiconductor (see Fig. 3). The refraction angles for the optical and THz beams are determined by the generalized Snell's law as

$$
\sin 45^\circ = n_{\mathrm{opt}} \sin\phi = n_{\mathrm{THz}} \sin\phi_{\mathrm{THz}} , \tag{9}
$$

where n_{opt} and n_{THz} is the refractive index for the pump laser and THz radiation in the semiconductor, respectively. For the pump-laser wavelength of

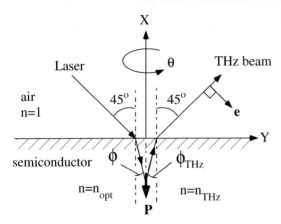

Fig. 3. Optical geometry for generation and detection of THz radiation from a semiconductor surface

800 nm, ϕ is 10.9°, 9.0° and 11.8° for InAs, InSb, and InP, respectively. For THz radiation, ϕ_{THz} is estimated to be 10.9°, 21.5°, and 11.9° for InAs, InSb, and InP, respectively. Using these values, the azimuthal-angle dependence of the radiation amplitude due to the optical rectification can be written as follows:

$$E_{\mathrm{THz}}^{\mathrm{ob}} \propto \quad 1.093 d_{14}(\cos 3\theta - 0.103) \qquad \text{for (111)-InAs}, \tag{10a}$$

$$E_{\mathrm{THz}}^{\mathrm{ob}} \propto \quad 0.1823 d_{14} \sin 2\theta \qquad \text{for (100)-InAs}, \tag{10b}$$

$$E_{\mathrm{THz}}^{\mathrm{ob}} \propto -0.069 d_{14} \sin 2\theta \qquad \text{for (100)-InSb}, \tag{10c}$$

$$E_{\mathrm{THz}}^{\mathrm{ob}} \propto \quad 0.199 d_{14} \sin 2\theta \qquad \text{for (100)-InP}. \tag{10d}$$

The nonlinear contribution is proportional to the azimuthal angular modulation of $\cos 3\theta$ with a small DC offset for (111)-oriented crystals, and $\sin 2\theta$ for (100)-oriented crystals in the experimental configuration shown in Fig. 1. The constants in front of d_{14} in (10a–d) represent the geometrical factors of the nonlinear contribution for the 45° pump-incidence on the samples. Although nonlinear optical rectification coefficients, $d_{14}(\Omega \approx 0)$, for InSb, InAs and InP are not known, the second-harmonic generation coefficients, $d_{14}(2\omega)$, may be helpful in indicating the relative magnitude of $d_{14}(\Omega \approx 0)$ among the samples: $d_{14}(2\omega)$ at 1.06 μm is 520 pm/V ± 47 pm/V, 364 pm/V ± 47 pm/V, and 167 pm/V for InSb, InAs, and InP, respectively [20]. Figure 2a shows the azimuthal-angle dependence measured for the (111)-InAs sample. A pronounced angle dependence is observed (squares), which agrees well with the dependence described by (10a). From a theoretical curve fitting (solid line), we estimated the ratio of the nonlinear contribution to the total radiation amplitude to be 40%.

The azimuthal-angle dependence for the (100)-oriented n-InSb sample is also shown in Fig. 2b by open circles, for which the angle dependence is not so significant because of the small geometrical factor (≈ -0.069) for the (100) surface. The ratio of the nonlinear contribution to the total radiation amplitude is estimated to be 6% from the theoretical curve fitting (solid line). It should be noted that for the (100)-oriented zincblende crystals the small nonlinear contribution arises from the offnormal incidence of the pump-laser beam (45° in this case), and the nonlinear electro-optic effect is always zero for the normal incidence of the pump-laser beam. The azimuthal-angle dependences for the other (100)-oriented samples were as weak as that observed for the n-InSb sample. From the results of the azimuthal-angle dependence, it is concluded that, for the offnormal incidence, the nonlinear contribution to THz radiation is finite but not very significant for the (100)-oriented zincblende-type semiconductor surfaces, while it can be very significant for the (111)-oriented ones.

The contribution to the THz emission from the surge current alone can be investigated by choosing a proper azimuthal angle, where the nonlinear contribution becomes null: For an azimuthal angle of 26° for (111)-InAs and 0° (or 90°) for all (100)-oriented samples, the nonlinear contribution becomes null (The θ-dependent factor vanishes).

2.2 Surge Current

2.2.1 Surface Depletion Field

In semiconductors with a wide bandgap, such as GaAs ($E_g = 1.43\,\mathrm{eV}$) or InP ($E_g = 1.34\,\mathrm{eV}$), the surface bands of a semiconductor lie within its energy bandgap, and thus Fermi-level pinning occurs, leading to band bending and formation of a depletion region, where the surface built-in field exists [21]. After optical excitation, the electrons and holes are accelerated in opposite directions under the surface-depletion field, forming a surge current in the direction normal to the surface. The direction and magnitude of the surface-depletion field depend on the dopant or impurity species and the position of the surface states relative to the bulk Fermi level. In general, the energy band is bent upward in n-type semiconductors (Fig. 4a) and downward in p-type semiconductors (Fig. 4b).

Therefore, the built-in surface field in p-type semiconductors drives the photogenerated carriers, and thus the transient surge current, in the opposite direction to that in n-type semiconductors as shown in Fig. 4a. In the far-field approximation, the emitted THz-radiation-field amplitude, $E_{\mathrm{THz}}(t)$, is proportional to the time derivative of the surge current, $J(t)$:

$$E_{\mathrm{THz}}(t) \propto \frac{\partial J(t)}{\partial t} . \qquad (11)$$

When the depletion-surface field is the dominant mechanism for the surge current, the polarity of the emitted THz radiation waveform is opposite between that of the n-type and that of the p-type semiconductors.

On the other hand, when the photo-Dember effect is the dominant mechanism for the surge current, the polarity of the THz radiation will remain the

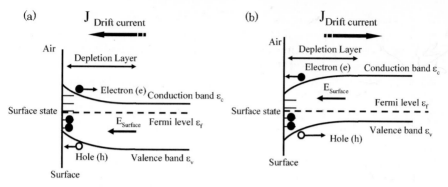

Fig. 4. Band diagram and the schematic flow of drift current in **(a)** n-type and **(b)** p-type semiconductors

same irrespective of the type of semiconductor, as explained in Sect. 2.2.2. Therefore, by comparing the polarity of the THz waveforms emitted from an n-type semiconductor and that of a p-type semiconductor, we can determine which mechanism is dominant in the semiconductor.

The time-domain THz-radiation waveforms of n-InP(100) and p-InP(100) are shown in Fig. 5. The azimuthal angle, θ, of the crystal orientation to the pump-laser polarization was set near 90° for these (100)-oriented samples to suppress the contribution from the nonlinear optical rectification effect. It is clearly shown in Fig. 5 that the THz waveform flips its polarity between n-type and p-type InP. The contribution of the photo-Dember effect is expected to be negligible because the absorption depth of an 800 nm pump laser in InP is relatively long ($\approx 0.3\,\mu m$) and the excess carrier energy is small ($\approx 0.3\,eV$) as indicated in Table 1, neither of which is a preferred condition to form a strong photo-Dember field, as discussed in Sect. 2.2.2. Therefore, the surge current due to the surface-depletion field is considered to be the dominant mechanism for THz radiation from InP. The same discussion applies for other wide-bandgap semiconductors, such as GaAs.

2.2.2 Photo-Dember Effect

InAs and InSb are very interesting semiconductors because of their high electron mobilities: $\approx 30\,000\,cm^2/(V\cdot s)$ for InAs and $\approx 76\,000\,cm^2/(V\cdot s)$ for InSb. Recently, InAs is attracting much attention as a THz emitter since it has been found that the THz emission power from InAs is significantly enhanced under magnetic fields [15]. The surface-depletion voltage of the narrow-bandgap semiconductors is generally not so large because of the small bandgap energy. For the excitation of narrow-bandgap semiconductors with NIR light ($h\nu \approx 1.5\,eV$), the absorption depth is very small ($\approx 100\,nm$) [22], and the excess energy of photoexcited carriers is very large. All these conditions in the narrow-bandgap semiconductors enhance the photo-Dember

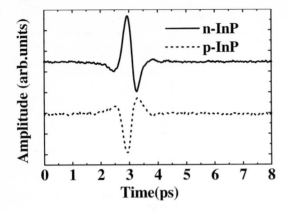

Fig. 5. Time-domain wave-forms of the THz radiation from n- and p-InP. The az-imuthal angle of crystal orien-tation was 90° for the (100)-oriented samples so that the contribution from the optical rectification effect was sup-pressed

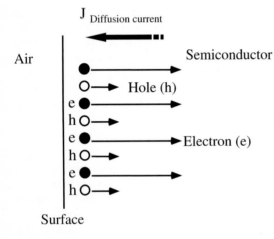

Fig. 6. Schematic flow of diffusion current by photoex-cited carriers near the sur-face of a semiconductor

effect, which is known to generate current or voltage in semiconductors due to the difference of the electron and hole diffusion velocities.

The diffusion current due to the photo-Dember effect after photoexcita-tion near a semiconductor surface is illustrated in Fig. 6. Because the electron mobility is always larger than that of holes, the direction of diffusion current due to the photo-Dember effect is the same with any kind of semiconductor and irrespective of the doping type (n or p). Therefore, the THz radiation emitted from the surface surge current due to the photo-Dember effect will show the same polarity for n-type and p-type semiconductors.

The diffusive currents of the electrons (J_n) and holes (J_p) are, respectively, described by the following equations [23],

$$J_n \sim -eD_e \frac{\partial \Delta n}{\partial x}, \tag{12a}$$

$$J_p \sim eD_h \frac{\partial \Delta p}{\partial x}, \tag{12b}$$

where e is the electron unit charge, Δn and Δp the density of photocreated electrons and holes, D_e and D_h diffusion coefficient of electrons and holes, respectively. The diffusion coefficient D is defined by the Einstein relation, $D = k_B T \mu / e$, where k_B is the Boltzman constant, T is the temperature of the corresponding carrier, and μ is the mobility of electrons or holes. The THz radiation due to the Dember current $J_{\mathrm{dif}} = J_n + J_p$ is thus proportional to the difference in the mobility and temperature for the electrons and holes, and the gradient of the carrier density.

The time-domain waveforms for the (100)-oriented n-InSb and p-InSb are shown in Fig. 7a, and the (111)-oriented n-InAs and the (100)-oriented p-InAs in Fig. 7b. The azimuthal angle, θ, of the crystal orientation was set near to $26°$ for the n-InAs (111) sample, and $0°$ for the other (100)-oriented samples so that the contribution from the optical rectification effect was suppressed. Because the THz waveforms of the n-type and p-type of both samples have the same polarity, and their polarity is the same as that observed for the n-InP shown in Fig. 5, whose direction of surface-depletion field is expected to be the same as that of the photo-Dember field, the dominant mechanism for the surge current on InSb and InAs surfaces is considered to be an ultrafast build-up and relaxation of the diffusion current due to the photo-Dember effect.

We now extend our discussion by using the equation of the steady-state photo-Dember voltage (V_D) [24]

$$V_D = \frac{k_B(T_e b - T_h)}{e} \frac{1}{b+1} \ln\left(1 + \frac{(b+1)\Delta n}{n_0 b + p_0}\right). \tag{13}$$

Here, $b = \mu_e / \mu_p$ is the mobility ratio of the electrons (μ_e) and holes (μ_p), n_0 and p_0 are the initial density of the electrons and holes, T_e and T_h are the temperature of photoexcited electrons and holes, respectively. This equation tells us that the photo-Dember effect is enhanced by larger electron mobility $(\mu_e \propto b)$, and higher electron excess energy $(\propto T_e)$. The narrow-bandgap semiconductors have the preferred conditions necessary to create a large photo-Dember field, that is, the very large electron mobilities and large excess carrier energies. Moreover, the photo-Dember field $(V_D/d$, d: absorption depth) in narrow-bandgap semiconductors is further enhanced by the small absorption depth. On the other hand, the surface-depletion field is expected to be small because of the small bandgap energy, in contrast to the widegap semiconductors. For 800 nm light, the absorption depth for InSb and InAs is estimated to be 94 nm and 142 nm, respectively, while that of InP is 304 nm. For 800 nm light excitation $(h\nu = 1.55\,\mathrm{eV})$, The excess energy, ΔE, in InSb and InAs is $1.38\,\mathrm{eV}$ and $1.18\,\mathrm{eV}$, respectively, which are much bigger values than that of InP $(\Delta E = 0.21\,\mathrm{eV})$, because of their narrow bandgaps (see Table 1). These conditions indicate the existence of a large photo-Dember field for InSb and InAs under 800 nm optical excitation. Thus, the main source of THz radiation for InAs and InSb is considered to be the

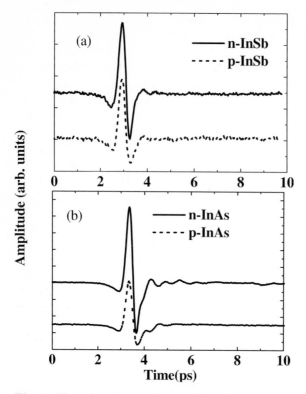

Fig. 7. Time-domain waveforms of the THz radiation (**a**) from n- and p-InSb, and (**b**) n- and p-InAs. The azimuthal angle of crystal orientation was 26° for n-InAs (111), and 90° for the other (100)-oriented samples to suppress the contribution from the optical rectification effect

photo-Dember field, rather than the screening of the surface depletion field. The radiation amplitude of the n-type InAs and n-type InSb is a little bigger than their counterpart p-type samples. This suggests that there might be an enhancement of THz radiation due to the surface-depletion field because the direction of the surface depletion field and the photo-Dember field is the same in n-type semiconductors.

From (13), we can deduce several important properties for the emission of THz radiation by the photo-Dember effect.

1. Low residual electron and hole concentrations (n_0 and p_0) enhance the photo-Dember effect.
2. The photo-Dember voltage, V_D, and thus the emitted THz-field amplitude, is expected to be proportional to the pump intensity, I, ($V_D \propto \Delta n \propto I$) in a low-intensity regime and proportional to $\ln(I)$ [$\propto \ln(\Delta n)$] in a high-intensity regime.

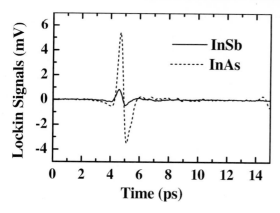

Fig. 8. Time-domain waveforms of InAs (*dotted curve*) and InSb (*solid curve*) at 800 nm excitation

3. The photo-Dember voltage does not depend strongly on the electron–hole mobility ratio, b, when it is large enough (> 10). This is the case for InSb ($b = 95$) and InAs ($b = 125$).

Because of the small absorption depth of InSb (92 nm), the photo-Dember field ($E_D \propto V_D/d$) should be almost twice that of InAs (142 nm) for 800 nm light excitation. In addition, the electron mobility, μ_e, of InSb is about twice that of InAs. Because the emitted THz-radiation-field amplitude is proportional to the acceleration field and electron mobility ($E_{THz} \propto \Delta J \propto e\mu_e E_D$), it is expected that the THz emission-efficiency of InSb is about four times better in amplitude and therefore 16 times better in power than that of InAs. However, the experimental results do not support this prediction.

Figure 8 shows the THz-radiation waveforms observed for InAs (dotted curve) and InSb (solid curve) at 800 nm excitation. The THz power observed from InSb was merely one-hundredth of that observed from InAs (without a magnetic field). The unexpectedly low THz emission of InSb might be explained by the reduction of the transient mobility due to intervalley scatterings of electrons to the L-valley, where the electron mobility is expected to be extremely low. Because the energy of the L-valley from the top of the valence-band edge (the Γ-point) is 1.03 eV in InSb, the electrons are easily scattered into the L-valley by the 800 nm light excitation ($h\nu = 1.55$ eV). On the other hand, as the energy of the L-valley is 1.53 eV (the energy barrier is higher than this value) in InAs, the electrons are scarcely scattered to the L-valley by the 800 nm light excitation.

Since the carrier dynamics such as the diffusions and intervalley scatterings strongly depends on the excitation wavelength, the wavelength dependence of THz radiation will provide important information on the emission mechanism of THz radiation from semiconductor surfaces. Figure 9 shows the waveforms of THz radiation from n-InAs (111) at 780 nm excitation with a power of 3.3 mW (solid curve) and at 1.55 μm excitation with a power of 9.0 mW (dotted curve). By normalizing the amplitude signal with the respective pump and probe-beam power, the THz-emission efficiency at 1.55 μm

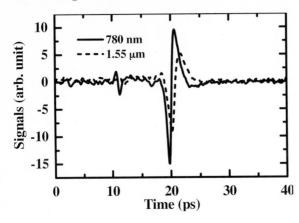

Fig. 9. Time-domain waveforms of n-InAs (111) excited by 780 nm and 1.55 μm light

was estimated to be almost one order of magnitude lower than that at 780 nm. The large absorption depth ($d = 590$ nm) and smaller excess energy ($\Delta E = 0.44$ eV) at a 1.55 μm excitation would give about one-sixth of the photo-Dember field expected at a 780 nm excitation ($d = 140$ nm, $\Delta E = 1.23$ eV) if the simple theory of the photo-Dember effect is valid (note that the photon number, thus the photoexcited carrier number, for 1.55 μm light was doubled with the same power of 780 nm light). Thus, the photo-Dember model agrees reasonably well with the experimental results at different wavelengths for InAs. In contrast to the case of InAs, *Howells* et al. [25] reported the emission efficiency of THz radiation from InSb increases by about six times in amplitude (normalized by the photon numbers) at 1.9 μm excitation compared to that at 800 nm. This observation is not consistent with the photo-Dember model, which predicts a considerable decrease of the THz emission efficiency at 1.9 μm compared to that at 800 nm because of the large absorption depth for 1.9 μm light. *Takahashi* et al. [26] also reported the THz radiation power from InSb increases by about several tens of times at 1560 nm excitation compared to that at 780 nm. However, these results can be explained if we assume a significant reduction of the transient mobility due to the L-valley scattering of electrons, as *Howells* et al. also suggested in their paper.

In fact, *Nuss* et al. [27] reported a significant drop in mobility due to intervalley scattering in GaAs. Although further experimental studies are needed to come to a definite conclusion. We believe the intervalley scattering plays an important role in the THz-emission mechanism in semiconductors.

3 Terahertz Radiation from Coherent Phonons and Plasmons

In the previous section, we have described three basic mechanisms for THz radiation emitted from semiconductor surfaces: optical rectification, ultrafast screening of the surface-depletion field, and ultrafast build-up of the photo-Dember field, induced by femtosecond laser irradiation on semiconductor surfaces.

In this section, we describe emission of THz radiation from oscillating polarizations associated with coherent lattice vibrations and/or plasma oscillations induced by femtosecond laser irradiation on semiconductor surfaces. By excitation with a femtosecond laser pulse, an electrical transient is created by the ultrafast screening of the surface-depletion field and/or ultrafast building up of the photo-Dember field as discussed in Sect. 2.2.2. This electrical shockwave triggers coherent lattice vibrations, which are called coherent phonons. Such coherent lattice vibrations induce changes of macroscopic optical constants, such as the polarizability or the linear dielectric susceptibility (the dielectric constant). Therefore, the coherent motion of lattice vibrations can be observed as the transient change of transmission or reflection at the sample surface of a second laser pulse (the probe pulse) that is synchronized with the first pump laser pulse with a variable time delay.

When the coherent phonons are infrared active, the lattice vibrations induce oscillations of macroscopic polarization. As illustrated in Fig. 10, transverse optical phonons (TO phonons) do not produce macroscopic charge density or macroscopic polarization because the integration of the polarization induced by the coherent phonons over a finite thickness, which corresponds to the absorption depth of the excitation laser, is averaged out to null (Fig. 10a). On the other hand, the longitudinal optical phonons (LO phonons) directed normal to the surface can create a macroscopic polarization, which is observed as an oscillating charge on the surface (Fig. 10b). The macroscopic polarization induced by coherent LO phonons thus emits electromagnetic radiation (THz radiation) at the LO-phonon frequency.

3.1 Coupling of Transverse Electromagnetic Waves with Longitudinal Coherent Phonons and/or Plasmons

3.1.1 Coupling to LO Phonons

Some readers might be uncomfortable with the above explanation of THz radiation from the longitudinal polarization due to LO phonons since LO phonons, in general, do not interact with electromagnetic waves, which are essentially transverse waves. This conceptual conflict is reconciled by interpreting the coherent LO phonon described above as the upper branch of the phonon–polariton mode at low wave numbers, $q \simeq 0$; When the wave number of optical phonons, q, is exactly zero, we can not distinguish the transverse mode from the longitudinal mode (and vice versa), and thus the two modes are degenerated, having the same frequency, ω_{LO}, as the frequency of unperturbed LO phonons [28]. In the excitation conditions with a finite incident angle, the excited longitudinal polarization gets a small but finite wavevector component in the transverse direction due to the phase delay determined by

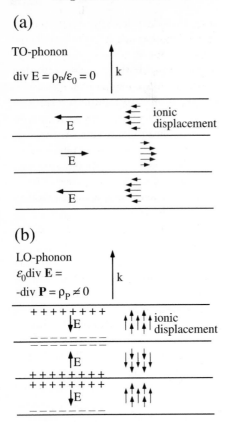

Fig. 10. (a) Charge density of TO phonon is effectively zero. (b) Charge density of LO phonon is not zero, and macroscopic polarization is formed perpendicular to the semiconductor surface

the timing of the pump-laser incidence on the semiconductor surface. Therefore, the coherent LO phonon described above is not a pure LO phonon but rather considered as a hybrid mode of LO and TO phonons, whose transverse components couple to the electromagnetic waves and are able to emit electromagnetic waves. When the wave vector of the coupled polarization in the transverse direction is small but nonzero, its frequency is close to, but slightly higher than, the frequency of the unperturbed LO phonon (ω_{LO}). Such an upshift of the frequency of THz radiation from an optical coherent phonon was actually observed for Te by *Dekorsy* et al. [22].

3.1.2 Coupling to Plasma Oscillations

When a semiconductor with a finite carrier density is excited impulsively with femtosecond lasers, coherent plasmon oscillations are initiated in a similar manner as the coherent optical phonons. The plasma starts to oscillate coherently after the excitation with a femtosecond laser pulse, due to surface-field screening or build-up of a photo-Dember field in a direction normal to the surface. Plasmon oscillations are longitudinal and thus can not couple to the transverse oscillations of an electromagnetic field or can not emit electromagnetic waves in a homogeneous or symmetrical excitation geometry. With a finite incident angle of excitation of a semiconductor surface, however, the plasma oscillations can emit electromagnetic waves in the same way as discussed for the THz radiation from coherent LO phonons. The frequency of unperturbed plasma oscillations (plasma frequency) is dependent on the carrier density, N, and is given by $\omega_p = \sqrt{e^2 N / \varepsilon \varepsilon_0 m^*}$ ($-e$: electron charge, ε: dielectric permeability of semiconductor, ε_0: dielectric constant of vacuum, and m^*: effective electron mass). When the plasma frequency is close to the LO-phonon frequency, the two vibrations are coupled and give rise to coupled-mode vibrations, that is, the LO-phonon–plasmon modes. These coupled modes are also observed as coherent emission of radiation from semiconductor surfaces after femtosecond laser irradiation.

THz radiation is observed not only from bulk semiconductor surfaces but also from semiconductor surfaces with quantum-well structures. Such THz radiation originates from charge oscillations due to two coupled quantized states, or Bloch oscillations in superlattice structures. Readers who are interested in THz radiation from semiconductor quantum-well structures are recommended to refer to a recent review on coherent THz radiation in semiconductors by *Gornik* and *Kersting* [29], which also gives a good review of THz radiation from bulk semiconductors.

3.1.3 Coherent-Phonon Excitation Mechanisms

For the later discussion, it is worth mentioning the excitation mechanisms for coherent phonons observed in the pump-and-probe measurements. In a photoreflectance pump-probe measurement, the change of the reflectance or the change of polarization of probe-laser pulses reflected from the sample surfaces is detected with variable time delay against the excitation of the sample by the pump pulse [30]. Since the coherent phonons induce the change of the dielectric constant or linear susceptibility, coherent phonons are observed as the oscillatory behavior of the intensity or the polarization of the reflected probe laser light, whose oscillation frequency and the damping rate are directly related to the frequency and relaxation rate of coherent lattice vibrations, respectively. In such pump-probe measurements, all the observed phonon modes were Raman-active, and no Raman-inactive[4] phonon modes have been observed so far. This fact suggests that the Raman activity is essential for the excitation and detection of the coherent phonons observed in the optical pump-probe measurements. Therefore, the impulsive-stimulated Raman scattering is widely accepted as the excitation mechanism for the coherent phonons [31, 32, 33]. However, the Raman-scattering process does

---

[4] Note that an optical phonon mode can be both infrared-active and Raman-active when the crystal does not have the inversion symmetry, depending on the mode symmetry.

not emit electromagnetic radiation at THz frequencies. Therefore, the excitation mechanism for coherent phonons that emit THz radiation needs to be sought in other processes, such as the ultrafast surface-field screening and the ultrafast build-up of the photo-Dember field, as discussed above.

In the first part of this section we describe THz radiation from coherent phonons in semiconductors with several experimental results obtained by the authors. In the latter part of this section, we describe THz radiation from coherent plasma oscillations and coupled oscillations of coherent phonon and plasmon excited in semiconductors.

3.2 THz Radiation from Coherent Phonons

3.2.1 THz Radiation from Coherent Phonons in Telluride Compound Semiconductors

The first observation of the THz emission from coherent phonons was reported by *Dekorsy* et al. for an LO phonon of an only-infrared(IR)-active mode (A_2-mode) in Te [10, 22]. Similar results were also reported by *Tani* et al. [34] for PbTe and CdTe, whose optical phonon modes are only-IR-active and IR-and-Raman-active, respectively. In the following, we review the experimental results obtained by *Tani* et al. [34] for the three semiconductors, that is, Te, PbTe and CdTe. These telluride semiconductors have low-frequency optical phonons due to the heavy mass of tellurium atom and are appropriate for observation of THz emission from coherent phonons using detectors with a limited detection bandwidth, such as a PC antenna ($< 6\,$THz).

3.2.2 Experimental Setup

They used a mode-locked Ti: sapphire laser, whose wavelength, pulse width, and repetition rate were $\approx 800\,$nm, $\approx 70\,$fs and $82\,$MHz, respectively. Laser pulses were irradiated onto the sample surface at an incident angle of $45°$ and was loosely focused to a diameter about $1\,$mm. The photoexcited carrier densities for Te, PbTe, and CdTe were estimated as $5.2 \times 10^{17}\,$cm^{-3}, $3.3 \times 10^{17}\,$cm^{-3}, and $1.2 \times 10^{16}\,$cm^{-3}, respectively, by using the absorption depth of the pump laser ($12\,$nm, $18\,$nm, and $0.5\,\mu$m for nondoped Te, PbTe, and CdTe, respectively). Because these photocarrier densities are not so high that the phonon–plasmon coupling effect becomes significant, the THz emission they observed for undoped samples should thus be related to almost pure LO-phonon modes. The radiation emitted in the direction of the optical reflection was collected and focused onto a PC sampling detector by a pair of parabolic mirrors.

3.2.3 Sample Properties

The properties of samples used by *Tani* et al. are listed in Table 2. All samples were monocrystalline. The sample of Te had a surface perpendicular to the c-axis. The samples of PbTe and CdTe had surfaces perpendicular to the $\langle 100 \rangle$ direction. Te has several optical-phonon modes because of the low symmetry of its crystal structure (hexagonal): a Raman-active A_1 mode (3.6 THz), two Raman-and-IR-active E modes ($E'_{\text{TO/LO}}$: 2.76/3.09 THz, $E''_{\text{TO/LO}}$: 4.22/4.26 THz), and one only-IR-active A_2 mode (A_2,TO/LO: 2.6/2.82 THz). PbTe has the NaCl-like crystal structure and has a TO- and an LO-phonon mode at 0.96 THz and 3.42 THz, respectively. CdTe has the zincblende crystal structure and has TO- and an LO-phonon modes at 4.20 THz and 5.08 THz, respectively. [35]

Table 2. Sample properties

Sample	Crystal structure	Optical-phonon mode	Raman activity	IR activity
Te (c-axis $\perp$ surface)	hexagonal	$A_1 = 3.6$ THz	yes	no
		A_2: TO/LO $= 2.6/2.82$ THz	no	yes
		E': TO/LO $= 2.76/3.09$ THz	yes	yes
		E'': TO/LO $= 4.22/4.26$ THz	yes	yes
PbTe (100)	rock-salt	TO/LO $= 0.96/3.42$ THz	no	yes
CdTe (100)	zincblende	TO/LO $= 4.20/5.08$ THz	yes	yes

3.2.4 Spectra of Te, PbTe, and CdTe

Figures 11a–c show the time-domain signal waveforms detected by the PC sampling detector for THz radiation from Te, PbTe, and CdTe surfaces, respectively. For all the samples the radiation was p-polarized, suggesting that it was emitted by the electric polarization perpendicular to the sample surface. Figures 12a–c show the Fourier-transformed amplitude spectrum of each waveform in Fig. 11.

Te   In Fig. 11a, the first cycle of the oscillations corresponds to the radiation due to the transient electric polarization caused by the ultrafast build-up of the photo-Dember field at Te surface. The subsequent oscillations persisting for about 3 ps correspond to THz radiation from the coherent LO-phonon of the A_2-mode in Te [10, 22]. The Fourier-transformed amplitude spectrum shows a spectral peak at 2.83 THz $\pm$ 0.05 THz with a FWHM width of 0.4 THz (Fig. 12a). No coherent radiation with a frequency corresponding to the TO-phonon mode is observed. Instead, there is a spectral dip at the TO-phonon frequency (2.6 THz). Radiation from the other IR-active modes ($E'_{\text{TO/LO}}$ and $E''_{\text{TO/LO}}$), whose polarization axes are parallel to the

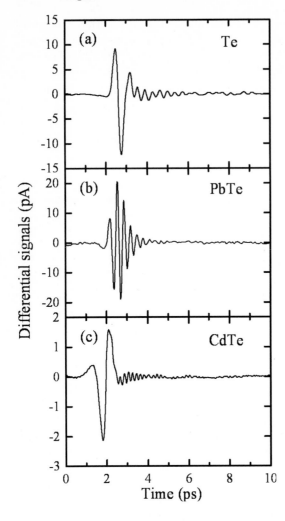

Fig. 11. Differential signals detected by the PC dipole antenna for the emission from (a) Te, (b) PbTe, and (c) CdTe surfaces

surface (perpendicular to the *c*-axis), were not observed since these modes do not couple to the field normal to the surface induced by the photo-Dember effect.

PbTe For PbTe, strongly damped oscillations (Fig. 11b) have been observed without significant initial radiation burst, which is observed for the other two samples. The Fourier-transformed spectrum (Fig. 12b) shows a broad spectral peak located at 3.2 THz ± 0.1 THz and having a 1.3 THz bandwidth (FWHM). This coherent radiation is attributed to the emission from the LO-phonons excited in PbTe (3.4 THz). The slight shift to the lower frequency may be due to a rapid decrease of the detector responsivity at higher frequencies. As is the case of the only-IR-active A_2 mode in Te, any coherent phonon oscillation was

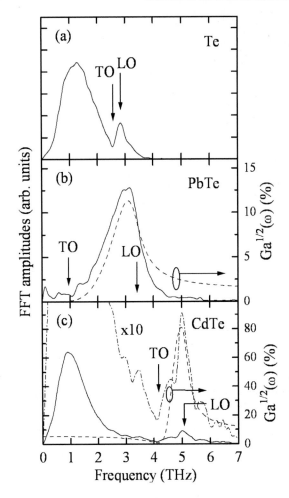

Fig. 12. Fourier-transformed amplitude spectra for the signal waveforms in Figs. 11a–c. The TO- and LO-phonon frequencies of each sample are indicated by *vertical arrows*. The calculated emission efficiencies $G_a^{1/2}$ for radiation from PbTe and CdTe into the air are indicated by *dashed lines*. The *dashed-dotted line* in Fig. 12c is the spectrum for CdTe magnified by a factor of 10 in order to show the structure at higher frequencies

not observed in the transient reflectivity measurement for the undoped bulk PbTe[5].

The driving force for the excitation of the coherent phonons in PbTe, as in the case of Te, can be attributed to the ultrafast build-up of the photo-Dember field [22]. Since the optical phonons in PbTe are Raman-inactive, the Raman process is excluded from the excitation mechanism for the coherent phonons. The displacive excitation mechanism [36] is also excluded since this excitation scheme is applicable only to totally symmetric modes such as A_1 mode. Since the surface built-in

---

[5] Oscillations in the transient reflectivity signal corresponding to the LO-phonon frequency were observed for a heavily doped n-PbTe ($n > 10^{19}$ cm^{-3}), for which the parity selection rule is expected to be broken by the high surface field or stress introduced by the high-density carriers, giving rise to the Raman activity.

field for PbTe is expected to be weak because of the small bandgap energy (E_g = 0.29 eV) and the large dielectric constant (ε_s = 388), the instantaneous screening of the surface field by the photoexcited carriers [37] cannot be a strong driving force for excitation of coherent phonons in PbTe. On the other hand, from the large inhomogeneity in photocarrier density due to the small absorption length and the large diffusivity of the highly energetic photoexcited carriers (with excess energy of ΔE = 1.26 eV), a strong and fast build-up of the photo-Dember field normal to the surface is expected, as in the case of Te [22].

CdTe For CdTe, fast oscillations after the strong radiation burst are observed (Fig. 11c). The Fourier-transformed amplitude spectrum (Fig. 12c) shows a spectral peak at 5.1 THz $\pm$ 0.05 THz with a FWHM width of 0.51 THz. This radiation component is attributed to the emission by coherent LO-phonons in CdTe. There is a spectral dip at the TO-phonon frequency (4.2 THz) as well. Because the detector responsivity is low around 5 THz the amplitude of the radiation from the coherent LO-phonon is estimated to be as large as the amplitude of the first radiation burst. As expected from the Raman activity of the optical phonon modes in CdTe, oscillations of both the coherent TO- and LO-phonons were observed in the transient photoreflectance measurement (not shown). As in other polar semiconductors with wide bandgaps (> 0.4 eV), such as GaAs or InP, the excitation of coherent phonons in CdTe can be attributed to the ultrafast screening of the surface built-in electric field by the photocarriers [37].

3.2.5 Spectral Dip at TO-Phonon Frequency

In addition to the pronounced coherent emission at the LO-phonon frequency, a striking feature persistent in the observed spectra of THz radiation shown above is the almost complete lack of spectral components at the TO-phonon frequencies. These spectral dips at the TO-photon frequencies are not well explained by the absorption at TO-phonon frequencies alone.

Although coherent emission from pure TO-phonon oscillations is not probable, coherent THz emission near the TO-phonon frequency is expected from the lower-frequency mode of the LO-phonon–plasmon (L_- mode) at carrier densities higher than 10^{18} cm^{-3} [38]. (The details of the LO-phonon–plasmon coupled modes are discussed in Sect. 3.3). Such an LO-phonon–plasmon coupled mode has been actually observed for n-GaAs in the optical transient reflectivity measurements [39, 40], which showed LO-phonon–plasmon coupling-mode frequencies determined by the total carrier density including the contribution from both the photoexcited carriers and the residual carriers by doping. The total lack of the emission spectrum near the TO-phonon frequency or L_- mode thus contradicts the observation of these modes in the optical transient reflectivity measurements. The reason may be attributed to the emission property of the coherent phonons from the sample surfaces.

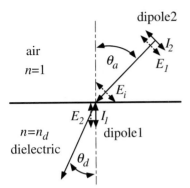

dipole2

air
$n=1$

θ_a

I_2

E_1

E_i

E_2 I_1

$n=n_d$
dielectric

dipole1

θ_d

Fig. 13. Geometry for calculations of the radiation pattern in air from a dipole on a dielectric substrate using the reciprocity theorem. E_i is the incident electric field from element I_2, θ_a is the incident angle and θ_d is the refracted angle

The absorption of the THz radiation through the thin emitting volume is too small to explain the spectral dip observed at TO-phonon frequencies: the absorption of radiation at TO-phonon frequencies by propagation through a distance equal to the optical absorption depth is estimated to be $< 1\%$ for Te and 1% to 2% for PbTe. For CdTe, the absorption of the radiation at the TO-phonon frequency is estimated to be about 10% for a propagation length of 150 nm, which is half the thickness of the depletion layer calculated using the Schottky-barrier model. Thus, we need to find other mechanisms for the spectral dip at TO-phonon frequencies.

3.2.6 Radiation Pattern and Emission Efficiency

Let us consider the radiation pattern of THz radiation from a semiconductor surface excited by a femtosecond laser pulse in order to clarify the mechanism of the spectral dip observed at the TO-phonon frequency.

It is well known that an emitting antenna on a thick dielectric substrate emits most of its radiation to the dielectric side, rather than to the air side. The ratio of the total emitted power into the dielectric, $\overline{P_d}$, to that into air, $\overline{P_a}$, is approximated by $(1/2)\varepsilon_d^s$, where ε_d is the dielectric constant of the substrate and s is a constant depending on the type of antenna and its orientation. For example, s equals 3/2 for an infinitesimal dipole antenna parallel to the dielectric surface [41]. Since, in the frequency region near optical phonons, the value of the dielectric constant changes significantly with frequency, the fraction of emitted radiation to free space, $\overline{P_a}/(\overline{P_a}+\overline{P_d})$, also changes significantly with frequency.

In the present case, the radiation source is the transient electric polarization at the semiconductor surface and is considered to be distributed infinitesimal dipoles (Hertzian dipoles) with their polarization perpendicular to the surface and the relative phase among them determined by the incidence of the optical wavefront of the pump-laser pulses. In this case, most of the radiation is emitted in the direction determined by the generalized Snell's law given by (9), which we rewrite for an incident angle of the pump laser,

θ_i,

$$\sin\theta_i = \sin\theta_a = \Re\left[\tilde{n}_d(\omega)\right]\sin\theta_d = n_d(\omega)\sin\theta_d \,, \tag{14}$$

where $\theta_a(=\theta_i)$ is the direction (polar angle) of the radiation emitted into free space, θ_d is the direction of the radiation emitted into the dielectric, $\tilde{n}_d(\omega) = n_d(\omega) - i\kappa_d(\omega) = \sqrt{\varepsilon_d(\omega)}$ is the complex refractive index of the dielectric (semiconductor) at the radiation frequency ω (Fig. 13).

All other emissions except the emission in directions to θ_a and θ_d become null because of the destructive interference between the field components from the distributed dipoles. The effect of diffraction is not expected to be significant when the excitation spot at the surface is large enough compared to the wavelength of the relevant THz radiation. For the present case, the pump-laser spot on the semiconductor surface was about 1 mm, which is large enough for the wavelength of the relevant radiation ($< 0.3\,\mathrm{mm}$).

The far-field radiation pattern of an elementary antenna on a dielectric surface can be calculated easily by using the *Lorentz reciprocity theorem* [41], by which we can avoid having to solve the complex boundary-value problem directly. To obtain the radiation pattern of a dipole (dipole 1) on a dielectric surface that is oriented perpendicular to the surface, we assume a test dipole (dipole 2) in air, as depicted in Fig. 13, in which a current element I_1 in the dipole 1 produces a field component E_1 parallel to a current element I_2 in the test dipole 2 and the current element I_2 in turn produces a field component E_2 parallel to I_1. According to the Lorentz reciprocity theorem, if $I_1 = I_2$, then $E_1 = E_2$. Because we know the radiation field of dipole 2, we can easily calculate E_2, which is the transmitted field component parallel to I_1 of the incident field on the interface, E_i. We can then write

$$E_1 = E_2 = t^{(p)}\sin\theta_d E_i \,, \tag{15}$$

where $t^{(p)}$ is the transmission coefficient for the p-polarized wave, θ_d is the refracted angle given by (14), and E_i is the far field of the dipole 2 in air. With an incident angle θ_a the Fresnel transmission coefficient $t^{(p)}$ is given by

$$t^{(p)}(\theta_a) = \frac{2\cos\theta_a}{\tilde{n}_d\cos\theta_a + \cos\theta_d} \,. \tag{16}$$

Using these equations, we find that the radiation power $P_a(\theta_a)$, emitted by dipole 1 to the air at an angle θ_a and per unit solid angle, as

$$P_a(\theta_a) = \frac{c\varepsilon_0 E_1^2}{2} = |t^{(p)}(\theta_a)|^2\sin^2\theta_d\frac{c\varepsilon_0 E_i^2}{2}$$

$$= P_0\frac{3}{8\pi}\frac{\sin^2\theta_d\cos^2\theta_a}{|\tilde{n}_d\cos\theta_a + \cos\theta_d|^2} \equiv P_0\frac{3}{8\pi}g_a(\theta_a)\,, \tag{17}$$

where c is the velocity of light and $P_0 = 4\pi c\varepsilon_0 E_i^2/3$ is the total power emitted by a dipole when it is placed in free space. Here, $g_a(\theta_a)$ represents the radiation pattern to the air-side half-space for an infinitesimal dipole placed just beneath the dielectric surface and oriented to the surface normal. Similar consideration for the dielectric side gives

$$P_d(\theta_d) = P_0 \frac{3}{8\pi} \frac{n_d^3 |\tilde{n}_d|^2 \sin^2 \theta_d \cos^2 \theta_d}{|\tilde{n}_d \cos \theta_a + \cos \theta_d|^2} \equiv P_0 \frac{3}{8\pi} g_d(\theta_d). \qquad (18)$$

Here, $g_d(\theta_d)$ represents the radiation pattern to the dielectric-side half-space for an infinitesimal dipole placed just beneath the dielectric surface and oriented to the surface normal.

The above equations for $P_a(\theta_a)$ and $P_d(\theta_d)$ are valid for an element dipole at the air/dielectric interface. The radiation pattern of the distributed dipoles excited with a pump laser at an incident angle, θ_i, is given by using *the principle of pattern multiplication* in the conventional antenna theory, which states that the electric field or intensity pattern of an array consisting of similar elements is the product of the pattern of a single-element antenna (the element pattern) and the pattern of an array of isotropic point sources with the same locations, relative amplitudes and phases as the original array (the array factor) [42].

In the present case, the element pattern is the radiation pattern, $g_a(\theta)$ and $g_d(\theta)$, given by (17) and (18). The array factor, $f(\theta, \phi)$, in this case is equal to the optical beam pattern reflected or refracted at the semiconductor/air interface when we neglect the diffraction effect for THz radiation, and is approximated by a delta function centered at spherical angles $(\theta = \theta_a, \phi = 0)$ and $(\theta = \theta_d, \phi = 0)$, where the polar angle θ is defined as the angle from the surface normal and the azimuthal angle ϕ is defined as the angle from the reflection plane. In the present experimental geometry, $\theta_a = \theta_i = 45°$ and $\theta_d(\omega) = \arcsin[\sin 45°/n_d(\omega)]$.

Thus, the ratio of the radiation power emitted into free space to the total emitted radiation power is approximately given by

$$\begin{aligned} G_a(\omega) &= \frac{\int_{air} f_a(\theta, \phi) g_a(\theta) \, d\Omega}{\int_{air} f_a(\theta, \phi) g_a(\theta) \, d\Omega + \int_{diel} f_d(\theta, \phi) g_d(\theta) \, d\Omega} \\ &= \frac{g_a(\theta_a)}{g_a(\theta_a) + g_d(\theta_d)} = \frac{\cos^2 \theta_a}{\cos^2 \theta_a + n_d^3 |\tilde{n}_d|^2 \cos^2 \theta_d} \\ &= \frac{1}{1 + n_d^3 |\tilde{n}_d|^2 \cos^2 \theta_d / \cos^2 \theta_a}, \end{aligned} \qquad (19)$$

where the angular integrations, $\int_{air} d\Omega$ and $\int_{diel} d\Omega$, are taken for the air-side half-space and dielectric-side half-space, respectively, and the relation between θ_a and θ_d is given by (14). To obtain the second equation, we assumed the array factors, $f_a(\theta, \phi)$ and $f_d(\theta, \phi)$, are delta functions centered at the direction to the optical reflection and refraction, $\delta(\theta_a - \theta)\delta(\phi)$ and $\delta(\theta_d - \theta)\delta(\phi)$, respectively. To a rough approximation, (19) reduces to

$$G_a(\omega) \sim \tilde{n}_d^{-5} = \varepsilon_d^{-5/2}. \qquad (20)$$

This relation corresponds to $G_a \propto \varepsilon_d^{-s} = \varepsilon_d^{-3/2}$ given for a dipole antenna on a dielectric substrate oriented parallel to the dielectric surface. The reason why the power coefficient, s, is larger for the vertical dipole is attributed to the fact that the vertical dipole emits more power in the direction parallel to the surface and thus is more subject to the total reflection at the air/dielectric interface. Equation (20) suggests that the THz-emission efficiency from semiconductor surfaces is strongly influenced by the dispersion property of the semiconductor.

For the frequency-dependent refractive index , $\tilde{n}_d(\omega)$, we assume

$$\tilde{n}_d^2(\omega) = \varepsilon_d(\omega) = \varepsilon_\infty + \frac{(\varepsilon_s - \varepsilon_\infty)\omega_{TO}^2}{\omega_{TO}^2 - \omega^2 - i\omega\Gamma}, \tag{21}$$

where ε_∞ and ε_s denote the high-frequency and static dielectric constants, ω_{TO} is the TO-phonon frequency, and Γ is the phonon damping parameter. Although the contribution from free carriers (the Drude term) is neglected in (21), the inclusion of carrier effects only smears out the TO-phonon resonant dispersion of $\varepsilon_d(\omega)$ and hence $G_a(\omega)$ as well.

We carried out the calculations of $G_a(\omega)$ for PbTe, with $\varepsilon_\infty = 32.8$, $\varepsilon_s = 388$, and $\Gamma = 26\,\mathrm{cm}^{-1}$ [43] and for CdTe with $\varepsilon_\infty = 7.1$, $\varepsilon_s = 10.2$, and $\Gamma = 8.7\,\mathrm{cm}^{-1}$ [44]. The square root of the calculated $G_a(\omega)$ is shown by the dashed lines in Fig. 12. The curves show a peak around the LO-phonon frequency, where the dielectric constant, $|\varepsilon_d| = n_d^2 + \kappa_d^2$, takes a minimum, and a dip or a reduction of $G_a^{1/2}(\omega)$ around the TO-phonon frequency, where the dielectric constant takes a maximum. Thus, the enhanced amplitude of radiation from the coherent LO-phonons is explained by the increased outcoupling efficiency of the radiation from the crystal to free space due to the small dielectric constant around the LO-phonon frequency. The spectral dip at the TO-phonon frequency is explained by the strong reduction of the outcoupling efficiency $[G_a(\omega_{TO}) \ll 1]$ due to the large dielectric constant around the TO-phonon frequency. The lack of an initial radiation burst from PbTe is also explained by a significantly low outcoupling efficiency over a range of frequency below the TO-phonon frequency ($< 1\,\mathrm{THz}$), resulting in suppression of the emission of radiation by the photo-Dember field. Although the calculation was not carried out for Te because of the complication due to the presence of many optical phonon modes, the same considerations also apply to the spectral dip at the TO-phonon frequency of the A_2-mode of Te, where the dielectric constant takes a large value ($|\varepsilon_d| > 100$) [45].

3.3 Coupling of LO Phonon and Plasmon

Now we discuss THz radiation from plasma oscillations coupled to coherent phonon oscillations. Excitation by the femtosecond laser pulses drives not only the coherent lattice vibrations but also the collective motion of carriers, that is, the plasmon oscillations in semiconductors. In the absence of coupling,

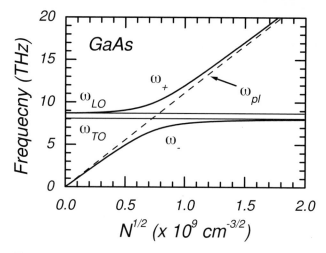

Fig. 14. Frequency of the LO-phonon–plasmon mode in GaAs plotted against the carrier density: $\omega_{\mathrm{LO}} = 8.76\,\mathrm{THz}$, $\omega_{\mathrm{TO}} = 8.06\,\mathrm{THz}$, $\varepsilon_\infty = 10.6$

the plasmon is expected to oscillate with the plasma frequency as mentioned earlier:

$$\omega_{\mathrm{p}} = \sqrt{4\pi e^2 N/\varepsilon\varepsilon_0 m^*}\,, \tag{22}$$

where, N is the total density of electrons including both the intrinsic electrons and photoexcited electrons, and m^* is the reduced mass of electron.

In polar materials, LO-phonons and -plasmons couple and give rise to the LO-phonon–plasmon modes, whose frequencies are dependent on the carrier density N. In Fig. 14, the oscillation frequencies of the coupled modes without damping are shown as a function of the square-root of plasma density, $N^{1/2}$, for GaAs. The frequency of TO-phonons and LO-phonons in GaAs is $\omega_{\mathrm{TO}} = 8.06\,\mathrm{THz}$ and $\omega_{\mathrm{LO}} = 8.76\,\mathrm{THz}$, respectively. When the carrier density is low and the frequency of the plasmon oscillations is far below the LO-phonon frequency, the interaction between plasmons and LO-phonons is weak. In this case, the upper branch of the LO-phonon–plasmon coupled modes (L_+ mode with frequency $\omega = \omega_+$) has phonon-like characteristics with its frequency close to the pure LO-phonon ($\omega_+ \sim \omega_{\mathrm{LO}}$), and the lower branch of the coupled modes (L_- mode with frequency $\omega = \omega_-$) has plasmon-like characteristics with its frequency close to the pure plasmon ($\omega_- \sim \omega_{\mathrm{pl}}$). As the carrier density increases, as does the plasmon frequency, L_-, L_+ modes anticross near the LO-phonon frequency due to the LO-phonon–plasmon interaction. At higher carrier density the L_+ mode frequency increases linearly with $\sqrt{N}$, while the L_- mode frequency approaches that of the TO-phonon ($\omega_- \sim \omega_{\mathrm{TO}}$). In this case, the L_+ mode has plasmon characteristics with its frequency close to the plasma frequency ($\omega_+ \sim \omega_{\mathrm{pl}}$), while the L_- mode is considered as the longitudinal phonon mode, whose frequency is softened due to the screening

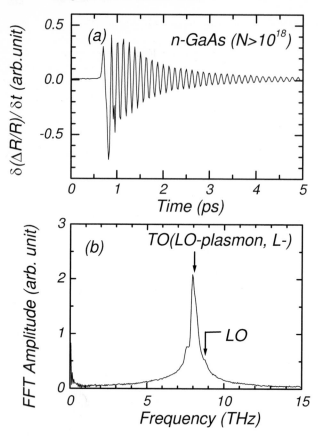

Fig. 15. (a) Differential transient reflectivity for n-GaAs ($N > 10^{18}\,\mathrm{cm}^{-3}$) and (b) its Fourier-transformed spectrum

effect of the carriers. Note that although the frequency of L_- mode, ω_-, is near the TO-phonon frequency, it is not a transverse mode but a longitudinal mode of lattice vibrations.

The frequencies of the LO-phonon–plasmon coupled modes are generally carrier-density dependent. However, in the transient reflectance measurements we usually observe density-independent oscillations at TO-phonon and LO-phonon frequencies: at low carrier density the oscillations at the LO-phonon frequency are dominant, while at high carrier density the oscillations at the TO-phonon frequency are dominant.

Kuznetsov and *Stanton* [38] investigated the LO-phonon–plasmon interaction in detail and made a theoretical calculation on GaAs. They included the effect of inhomogeneous density distribution in the calculation and found that the oscillations near the TO-phonon and the LO-phonon frequencies are preserved in the averaging process over different carrier densities due to the high density of the modes at these frequencies, agreeing with the experimental observation. Figure 15a shows the differential photoreflectance of a heavily doped n-GaAs ($N > 10^{18}$ cm^{-3}). In the Fourier-transformed spectrum (Fig. 15b) we observe a dominant spectral peak at the TO-phonon frequency, which is interpreted as the L_- mode of the LO-phonon–plasmon coupled mode. The same discussion as for GaAs is valid for other polar semiconductors. Therefore, we expect to observe the THz emission originating from the L_- mode with the TO-phonon frequency at high carrier densities ($N > 10^{18}$ cm^{-3}). Such a coherent emission near the TO-phonon frequency is suppressed due to the high refractive index and is thus difficult to observe. However, for moderately doped polar semiconductors the L_- or the L_+ mode frequencies are offresonant from the TO-phonon frequency and thus are observed as coherent THz emission with femtosecond laser excitation.

3.3.1 InSb Phonon–Plasmon THz Emission

Gu et al. [46] carried out THz-emission spectroscopy for doped InSb films and observed coherent THz radiation corresponding to L_- and/or the L_+ phonon–plasmon modes. In the experiment they used a 1.4 μm thick InSb film with an intrinsic carrier concentration of $n = (6.70 \times 10^{16})$cm^{-3} fabricated on GaAs (100) substrates by MOCVD. Raman-scattering measurements showed that the TO- and LO-phonon modes of the film were at frequencies of 5.39 THz and 5.72 THz, respectively. The electron plasma frequency at zero wave number ($q = 0$) is calculated to be 4.45 THz, which is slightly lower than the LO-phonon frequency of InSb. From the intrinsic carrier concentration the L_- and L_+ modes are calculated to be 3.94 THz and 6.10 THz at $q = 0$. To detect high-frequency THz radiation, ultrashort laser pulses less than 30 fs and tight focusing of the gate laser beam on a 5 μm LTG-GaAs PC gap of the detector antenna were used [19].

Figure 16 shows the THz waveforms detected by a PC antenna in the (a) reflection (THz radiation in the optical-reflection direction is detected) and (b) transmission geometry (THz radiation transmitted through the substrate is detected) for the same InSb film at an optically excited carrier density $n_{ex} \approx (5 \times 10^{16})$ cm^{-3}. The inset in Fig. 16a is the waveform magnified by a factor of 3 and shows the oscillatory structure after a 3.5 ps delay time. As shown in the inset, a beating is observed, indicating the presence of two modes with different frequencies. The corresponding Fourier-transformed amplitude spectra of the THz radiation are shown in Figs. 17a and b. The first large and broad spectral peak at around 1 THz corresponds to the first cycle of the THz pulse originating from the surge current induced by the photo-Dember effect. To remove the contribution from the THz radiation due to the surge current, which forms the slowly varying background in the spectrum, we carried out a Fourier transformation with a time window later than the third maximum (≈ 3.5 ps) in the waveform in Figs. 16a and b. The spectral peak profiles by the windowed Fourier transformations are also shown

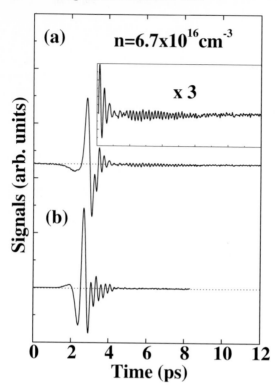

Fig. 16. Time-domain waveforms of the THz radiation detected in (**a**) reflection and (**b**) transmission geometry for the same InSb film. The *inset* is the waveform of (**a**) magnified by a factor of 3 at a delay time of more than 3.5 ps

in Fig. 17 with a magnified scale. By fitting a Lorentzian function to these spectral peak profiles, the second spectral peak in Fig. 17a and b is found at 3.94 THz $\pm$ 0.05 THz, and the third peak in Fig. 17a at 6.10 THz $\pm$ 0.05 THz, respectively.

No significant coherent radiation was observed at the bare TO- and LO-phonon frequencies (indicated by vertical arrows in Fig. 17a). The measured frequencies of 3.94 THz and 6.10 THz agreed well with those of the lower (L_-) and upper (L_+) branches of the phonon–plasmon coupled modes expected from the theoretical calculation with the doped carrier density of (6.70×10^{16})cm^{-3}. These results indicate that the coherent radiation we observed originates from L_- and L_+ modes. The absence of L_+ mode radiation in the transmission geometry is due to the strong absorption in the 0.7 mm thick GaAs substrates: the internal absorption is estimated to be over 90% at 6.10 THz and less than 40% at 3.94 THz in the GaAs substrate.

In Fig. 17a, the magnitude of the THz-radiation spectrum at the L_- mode frequency (3.94 THz) is almost half that at 1 THz where the surge-current contribution is dominant. The sensitivity of a PC-antenna detector decreases rapidly with increase of frequency: The sensitivity of the PC-antenna detector at 4 THz and 6 THz is, respectively, estimated to be one-fifth and one-fifteenth

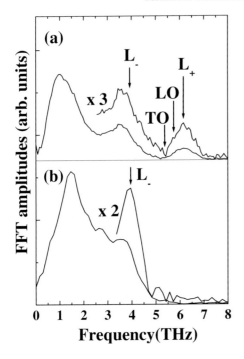

Fig. 17. (**a**) and (**b**) are the Fourier-transformed spectra of the data in Figs. 16a and b. The *broken lines* in (**a**) and (**b**) are the frequency spectra of L_- and L_+ modes magnified by a factor of 2. The frequencies of the TO- and LO-phonons, and the L_+ and L_- modes are denoted by *vertical arrows*

of that at 1 THz. Therefore, the spectral intensities due to the LO-phonon–plasmon coupling modes are actually very intense.

Figure 18 shows the THz-radiation spectra obtained with pump laser powers of 20 mW (photocarrier density $n_{ex} \approx (2.5 \times 10^{16})\,cm^{-3}$), 40 mW ($n_{ex} \approx (5.0 \times 10^{16})\,cm^{-3}$), and 60 mW ($n_{ex} \approx (7.5 \times 10^{16})\,cm^{-3}$) in the (a) reflection geometry and (b) transmission geometry, for a limited window to remove the contribution of the THz radiation component due to the surface surge current. Figures 18a and b show that the peak frequencies of the coherent THz radiation corresponding to the L_- and L_+ modes do not change with the excitation power, while the spectral amplitudes of the these modes increase with the pump power. This is rather surprising when we suppose that the L_- and L_+ modes frequencies are determined by the total carrier density (the sum of the intrinsic and photoexcited carriers). This result clearly indicates that the oscillation frequencies of the phonon–plasmon coupling modes depend only on the intrinsic carrier concentration and not on the density of photogenerated carriers. Similar results were reported by *Kersting* et al. [47] for the coherent THz radiation from plasma oscillations in GaAs, the frequency of which was independent of the photogenerated carrier density. In addition, the decay time of oscillations for the L_- mode does not change with excitation power. This means that the dephasing of the L_- modes is not affected by the optically excited carrier dynamics such as

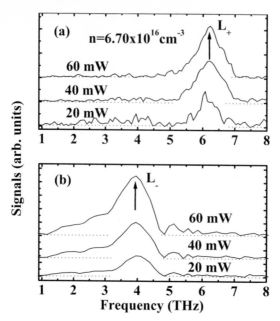

Fig. 18. Spectra measured (**a**) in the reflection geometry and (**b**) transmission geometry at optical pump powers of 60 mW, 40 mW and 20 mW. To remove the contribution of the THz-radiation component due to the surface surge current, the Fourier transformations of waveforms were carried out with limited time windows. The frequencies of the L_+ mode and L_- mode calculated for the intrinsic carrier density are indicated by the *vertical arrows*

carrier–carrier scattering. A similar result was obtained for the L_+ mode, too.

On the other hand, a time-domain optical pump-probe measurement [48] and CW-Raman scattering measurement [49] have shown that the photogenerated carriers affect the frequency and damping time of the coupling modes. The origin of this discrepancy is not clear at present, and further investigations are needed.

4 Conclusions

The THz radiation from semiconductor surfaces is attractive not only as a simple THz-radiation source but also as a THz-radiation source for high frequencies (> 3 THz), where other radiation sources such as PC antennas are not efficient.

Not only as a simple THz-radiation source, THz radiation from semiconductor surfaces can be used as the probe for the investigation of carrier-transport dynamics near to semiconductor surfaces [50, 51, 52]: Since the field amplitude is proportional to the carrier acceleration, information on the ultrafast carrier transport such as the peak velocity in the overshoot regime and the influence of the lattice vibrations are obtained from the time-resolved measurement of the emitted radiation (with a proper correction for the detector response). For example, *Leitenstorfer* et al. [50] measured the time-resolved THz field amplitude from InP and GaAs surface with a time resolution of 20 fs. They measured electron peak velocities of (6×10^7) cm/s and (8×10^7) cm/s for GaAs and InP, respectively, under an electric field of 90 kV/cm. These overshoot velocities are much higher than the drift velocities in the steady-state regime, and the one for InP was larger than that of GaAs, reflecting the larger energy separation between the central valley and the side minima in the InP conduction band compared to that in GaAs.

Optical phonon dynamics are also investigated by measuring the THz radiation from the polar semiconductor surface because the carrier transport is influenced by the lattice vibration or the electron plasma coupled with LO-phonons. THz radiation from coherent phonons or coherent plasmons are interesting phenomena, which are still not well understood. For example, the emission frequencies for L_- and L_+ modes are dependent only on the intrinsic carrier density, and are not dependent on the excitation carrier density. The radiation power from coherent phonons and plasmon are quite efficient, considering the low sensitivity at high frequencies of the PC antennas used as the detector, which contradict the theoretical prediction by *Kuznetsov* and *Stanton* [38].

References

[1]   X.-C. Zhang, B. Hu, J. Darrow, D. Auston: Appl. Phys. Lett. **56**, 1011 (1990)
[2]   X.-C. Zhang, J. Darrow, B. Hu, D. Auston, M. Schmidt, P. Tham, E. Yang: Appl. Phys. Lett. **56**, 2228 (1990)
[3]   X.-C. Zhang, B. Hu, S. Xin, D. Auston: Appl. Phys. Lett. **57**, 753 (1990)
[4]   X.-C. Zhang, D. Auston: J. Appl. Phys. **71**, 326 (1992)
[5]   X.-C. Zhang, Y. Jin, K. Yang, L. Schowalter: Phys. Rev. Lett. **69**, 2303 (1992)
[6]   M. Li, F. Sun, G. Wagoner, M. Alexander, X.-C. Zhang: Appl. Phys. Lett. **67**, 25 (1995)
[7]   J. Perdersen, I. Balslev, J. Hvam, S. Keiding: Appl. Phys. Lett. **61**, 1372 (1992)
[8]   B. Hu, X.-C. Zhang, D. Auston: Appl. Phys. Lett. **57**, 2629 (1990)
[9]   Y. Jin, X. Ma, G. Wagoner, M. Alexander, X.-C. Zhang: Appl. Phys. Lett. **65**, 682 (1994)
[10] T. Dekorsy, H. Auer, C. Waschke, H. Bakker, H. Roskos, H. Kurz, V. Wanger, P. Grosse: Phys. Rev. Lett. **74**, 738 (1995)
[11] S. Chuang, S. Schmitt-Rink, B. Greene, P. Saeta, A. Levi: Phys. Rev. Lett. **68**, 102 (1995)
[12] B. Greene, P. Saeta, D. Dykaar, S. Schmitt-Rink, S. Chuang: IEEE J. Quantum Electron. **28**, 2302 (1992)
[13] S. Howells, L. Schlie: Appl. Phys. Lett. **67**, 3688 (1995)

[14] X.-C. Zhang, Y. Jin, T. Hewitt, T. Sangsiri, L. Kingsley, M. Welner: Appl. Phys. Lett. **63**, 2003 (1993)

[15] N. Sarukura, H.Ohtake, S. Izamida, Z. Liu: J. Appl. Phys. **84**, 1 (1998) according to a recalibration of the bolometer sensitivity by *Sarukura* et al., the total radiation power from an InAs surface under a magnetic field of 1 T is corrected to be about 50 μW with a pump power of 1 W. 1398 (2000) I.; P.N. Saeta, D.R. Dykaar, S. Schmitt-Rink, and S.L. Chuang, IEEE J. Quantum Electron. **28**, 2302 (1992)

[16] A. Bonvalet, M. Joffre, J. Martin, A. Migus: Appl. Phys. Lett. **67**, 2907 (1995)

[17] A. Rice, Y. Jin, X. Ma, X.-C. Zhang, D. Bliss, J. Larkin, M. Alexander: Appl. Phys. Lett. **64**, 1324 (1994)

[18] X.-C. Zhang, Y. Jin, K. W. X. Ma, A. Rice, D. Bliss, J. Larkin, M. Alexander: Appl. Phys. Lett. **64**, 622 (1994)

[19] P. Gu, M. Tani, S. Kono, K. Sakai: J. Appl. Phys. **91**, 5533 (2002)

[20] S. Singh: *Handbook of Lasers* (CRC, Cleveland 1971)

[21] For review papers, see, for example, X.-C. Zhang and D.H. Auston: J. Appl. Phys. **71**, 326 (1992), S.C. Howells and L.A. Schlie: Appl. Phys. Lett. **67**, 3688 (1995), T. Kondo, M. Sakamoto, M. Tonouchi, and M. Hangyo: Jpn. J. Appl. Phys. **38**, L1035 (1999), M. Hangyo, M. Migita, and K. Nakayama: J. Appl. Phys. **90**, 3409 (2001), M.B. Johnston, D.M. Whittaker, A. Corchia, A.G. Davies, and E.H. Linfield: J. Appl. Phys. **91**, 2104 (2002), M.B. Johnston, D.M. Whittaker, A. Corchia, A.G. Davies, and E.H. Linfield: Phys. Rev. B **65**, 165301 (2002)

[22] T. Dekorsy, H. Auer, H. Bakker, H. Roskos, H. Kurz: Phys. Rev. B **53**, 4005 (1996)

[23] S. Kono, P. Gu, M. Tani, K. Sakai: Appl. Phys. B **71**, 901 (2000)

[24] W. Mönch: *Semiconductor Surface and Interface* (Springer, Berlin, Heidelberg 1993) p. 68

[25] S. Howells, S. Herrera, L. Schlie: Appl. Phys. Lett. **65**, 2946 (1994)

[26] H. Takahashi, Y. Suzuki, M. Sakai, A. Ono, N. Sarukura, T. Sugiura, T. Hirosumi, M. Yoshida: Appl. Phys. Lett. **82**, 2005 (2003)

[27] M. Nuss, D. Auston, F. Capasso: Phys. Rev. Lett. **58**, 2355 (1987)

[28] P. Y. Yu, M. Cardona: *Fundamentals of Semiconductors*, 3rd ed. (Springer, Berlin, Heidelberg 2001) pp. 110–113

[29] Coherent THz emission in semiconductors, in K. Tsen (Ed.): *Ultrafast Physical Properties in Semiconductors*, Semicond. Semimet. **67** (Academic, New York 2001)

[30] For the pump-probe experiment of coherent phonons, see for example, W. Kütt: Adv. Solid State Phys. **32**, 113 (1992), W. Kütt, W. Albrecht, and H. Kurz: IEEE J. Quantum Electron. **28**, 2434 (1992), H.J. Zeiger, J. Vidal, T.K. Cheng, E.P. Ippen, G. Dresselhaus, and M.S. Dresselhaus: Phys. Rev. B **45**, 768 (1992)

[31] Y.-X. Yan, J. E.B. Gamble, K. Nelson: J. Chem. Phys. **83**, 5391 (1985)

[32] S. Ruhman, A. Joly, K. Nelson: J. Chem. Phys. **86**, 6563 (1987)

[33] S. Ruhman, A. Joly, K. Nelson: IEEE J. Quantum Electron. **24**, 460 (1988)

[34] M. Tani, R. Fukasawa, H. Abe, S. Matsuura, K. Sakai: J. Appl. Phys. **83**, 2473 (1998)

[35] P. Grosse, W. Richter: *Numerical Data and Functional Relationships in Science and Technology*, Landoldt-Börnstein, New Series **17** (Springer, Berlin, Heidelberg 1983)

[36] H. Zeiger, J. Vidal, T. Cheng, E. Ippen, G. Dresselhaus, M. Dresselhaus: Phys. Rev. B **45**, 768 (1992)

[37] T. Pfeifer, T. Dekorsy, W. Kütt, H. Kurz: Appl. Phys. A **55**, 482 (1992)

[38] A. Kuznetsov, C. Stanton: Phys. Rev. B **51**, 7555 (1995)

[39] G. Cho, T. Dekorsy, H. Bakker, R. Hovel, H. Kurz: Phys. Rev. Lett. **77**, 4062 (1996)

[40] M. Hase, K. Mizoguchi, H. Harima, F. Miyamaru, S. Nakashima, R. Fukasawa, M. Tani, K. Sakai: J. Lumin. **76/77**, 68 (1998)

[41] D. Rutledge, D. Neikirk, D. Kasilingam: Infrared Millimeter Waves **10**, 1 (1983)

[42] W. Stutzman, G. Thiele: *Antenna Theory and Design* (Wiley, New York 1981) Chap. 3

[43] S. Perkowitz: Phys. Rev. B **12**, 3210 (1975)

[44] S. Perkowitz, R. Thorland: Phys. Rev. B **9**, 545 (1974)

[45] P. Grosse, W. Richter: Phys. Stat. Sol. **41**, 239 (1970)

[46] P. Gu, M. Tani, K. Sakai, T.-R. Yang: Appl. Phys. Lett. **77**, 1798 (2000)

[47] R. Kersting, K. Unterrainer, G. Strasser, H. Kauffmann, E. Gornik: Phys. Rev. Lett. **79**, 3038 (1997)

[48] For review paper, see for example, G.C. Cho, T. Dekorsy, H.J. Bakker, R. Hävel, and H. Kurz: Phys. Rev. Lett. **77**, 4062 (1996), T.T. Dekorsy, H. Auer, C. Waschke, H.J. Bakker, H.G. Roskos, and H. Kurz: Phys. Rev. Lett. **74**, 738 (1995), M. Vobebörger, H.G. Roskos, F. Wolter, C. Waschke, and H. Kurz: J. Opt. Soc. Am. B **13**, 1045 (1996)

[49] C. Collins, P. Yu: Solid State Commun. **51**, 123 (1984)

[50] A. Leitenstorfer, S. Hunsche, J. Shah, M. Nuss, W. Knox: Phys. Rev. B **61**, 16642 (2000)

[51] J.-H. Son, T. Norris, J. Whitaker: J. Opt. Soc. Am. B **11**, 2519 (1994)

[52] Y. Shimada, K. Hirakawa, S.-W. Lee: Appl. Phys. Lett. **81**, 1642 (2002) ; N. Tsukada: Jpn. J. Appl. Phys. **33**, 4807 (1994), Appl. Opt. **36**, 7853–7859 (1997); J. Appl. Phys. **80**, 4214–4216 (1996); Opt. Lett. **15**, 323–325 (1990, 1995), Opt. Lett. **62**, 1265 (1993)

Enhanced Generation of Terahertz Radiation from Semiconductor Surfaces with External Magnetic Field

Hideyuki Ohtake[1,2], Shingo Ono[1], and Nobuhiko Sarukura[1]

[1] Institute for Molecular Science
   38 Nishigonaka, Myodaiji, Okazaki, Aichi 444-8585, Japan
[2] *Present address:* AISIN SEIKI CO., LTD.
   17-1 Kojiritsuki, Hitotsugi, Kariya, Aichi 448-0003, Japan
   ohtake@imra.co.jp

Abstract. In this Chapter, a scheme to enhance generation of terahertz radiation from semiconductor surfaces irradiated by femtosecond laser with an external magnetic field is described. The emitted terahertz radiation power from InAs is enhanced more than 2 orders by applying a magnetic field, and achieves approximately 100 μW.

1 Introduction

Following the generation of terahertz (THz) pulse radiation from semiconductor surfaces [1, 2], *Zhang* et al. studied emission of THz radiation from those in the presence of external magnetic field [3]. In some experimental results, it is noteworthy that the emission of THz radiation is enhanced by applying the external field produced with the use of relative handy magnets.

Susequent investigations have made this effect clear in succession, including its variation as a function of the strength and geometry of magnetic field [4, 5, 6, 7], of the semiconductor materials [8, 9, 10, 11], and of the temperature [12, 13, 14]. It has been found that InAs is the most efficient emitter in the presence of magnetic field. The radiation power from InAs is achieved approximately 100 μW at the magnetic field strenght of 3 T. By using this enhancement scheme, a compact emitter system is demonstrated.

2 Terahertz Radiation from Semiconductor Surfaces in a Magnetic Field

2.1 Emission Mechanism of Terahertz Radiation

Generation of THz radiation from a semiconductor in a magnetic field is basically explained based on the motion of an electron subjected to the surface electric field and the Lorentz force as

K. Sakai (Ed.): Terahertz Optoelectronics, Topics Appl. Phys. **97**, 99–116 (2005)
© Springer-Verlag Berlin Heidelberg 2005

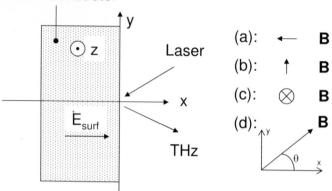

Fig. 1. Orientation of the semiconductor, surface field E_{surf}, and external magnetic field B. (a): $B \parallel x$, (b): $B \parallel y$, (c): $B \parallel z$, (d): a magnetic field is in the xy-plane and the angle between B and x is θ

$$a = \frac{dv}{dt} = -\frac{e}{m^*}(E_{\text{surf}} + v \times B),\tag{1}$$

where a, e, m^*, v, E_{surf}, and B are acceleration of electron, elementary electric charge, effective mass, velocity of electron, an electric field on the surface and a magnetic field.

In our model, four different directions of magnetic field are considered, as depicted in Fig. 1.

- Case (a): $E_{\text{surf}} = (-E, 0, 0)$, $B = (-B_0, 0, 0)$.
  In the case of (a), (1) is written as follows,

$$\begin{pmatrix} a_x \\ a_y \\ a_z \end{pmatrix} = -\frac{e}{m^*}\left[\begin{pmatrix} -E \\ 0 \\ 0 \end{pmatrix} + \begin{pmatrix} 0 \\ -B_0 v_z \\ B_0 v_y \end{pmatrix}\right] = \frac{e}{m^*}\begin{pmatrix} E \\ B_0 v_z \\ -B_0 v_y \end{pmatrix},\tag{2}$$

where $a_x = \mathrm{d}^2x/\mathrm{d}t^2$, $a_y = \mathrm{d}^2y/\mathrm{d}t^2$, $a_z = \mathrm{d}^2z/\mathrm{d}t^2$, E is an electric field that is written as $E(x, t)$, and B_0 is a static magnetic field. As well as the above case, other cases for (b), (c), and (d) are described as follows.

- Case (b): $E_{\text{surf}} = (-E, 0, 0)$, $B = (0, B_0, 0)$.

$$\begin{pmatrix} a_x \\ a_y \\ a_z \end{pmatrix} = \frac{e}{m^*}\begin{pmatrix} cE + B_0 v_z \\ 0 \\ -B_0 v_x \end{pmatrix}.\tag{3}$$

- Case (c): $E_{\text{surf}} = (-E, 0, 0)$, $B = (0, 0, -B_0)$.

$$\begin{pmatrix} a_x \\ a_y \\ a_z \end{pmatrix} = \frac{e}{m^*}\begin{pmatrix} E + B_0 v_y \\ -B_0 v_x \\ 0 \end{pmatrix}.\tag{4}$$

– Case (d): $\boldsymbol{E}_{\text{surf}} = (-E, 0, 0)$, $\boldsymbol{B} = (B_x, B_y, 0)$

$$\begin{pmatrix} a_x \\ a_y \\ a_z \end{pmatrix} = \frac{e}{m^*} \begin{pmatrix} E + B_y v_z \\ -B_x v_z \\ -B_y v_x + B_x v_y \end{pmatrix}, \tag{5}$$

where B_x and B_y in (5) are $B_0 \cos\theta$ and $B_0 \sin\theta$.

In the case of (a), radiation originating from the Lorentz force is not emitted in the x direction because the initial acceleration by $\boldsymbol{E}$ is parallel to $\boldsymbol{B}$. In the case of (b) and (c), since electrons rotate along the y- or z-axis, linearly polarized THz radiation should be observed. On the contrary, case (d) is slightly complicated due to the tilted magnetic field. In this case, one can expect that polarization of THz radiation is not linear but rather elliptical because accelerations of electrons along each axis are completely different. From the application viewpoint, linearly polarized THz radiation is preferable and desirable.

In the case of (b) and (c), E is not given explicitly. E basically originates from the complicated photo-Dember field or/and depletion field, so E should depend on various factors such as photoexcited carrier density, the surface conditions, direction of magnetic field, and so on. For simplicity, supposing E is constant (E_0) and the initial condition $v_x, v_y = 0$ at $t = 0$, the solution for the a_y component of (4) is written as

$$a_y = -\frac{eE_0}{m^*} \sin\frac{eB_0}{m^*} t. \tag{6}$$

Equation (6) is rewritten as

$$a_y = -\left(\frac{e}{m^*}\right)^2 E_0 B_0 t, \tag{7}$$

where the lowest term in the Taylor expansion is taken because of the very short time t. Since the radiated field is proportional to the carrier acceleration a_y, the radiated field is linearly dependent on the magnetic field strength [3, 8]:

$$(E_{\text{THz}})_y \propto a_y \tag{8}$$
$$\propto E_0 B_0, \tag{9}$$

where E_{THz} is the radiated field and $(E_{\text{THz}})_y$ is the y-component of E_{THz}. The THz-radiation power is written as

$$P_{\text{THz}} \propto |(E_{\text{THz}})_y|^2 \tag{10}$$
$$\propto E_0^2 B_0^2, \tag{11}$$

where P_{THz} is the THz-radiation power. The THz-radiation power depends quadratically on the magnetic field strength. Additionally, as the number of electrons is proportional to the excitation power as described in (12a) in the Chapter by *Gu* et al., the THz-radiation power P_{THz} increases quadratically as the excitation power increases and is written as,

$$P_{THz} \propto \left(\frac{n}{m^*}\right)^2 \propto P_{pump}^2, \tag{12}$$

where n is the photoexcited electron density and P_{pump} is excitation power. Equation (12) also shows that electrons with smaller effective mass are advantageous to generate intense THz radiation. Table 1 summarizes the effective mass of electrons in the Γ-valley [15].

Table 1. Comparison of effective mass, where m_0 is rest mass of an electron such as 9.1×10^{-31} kg

Semiconductor	Effective mass of electrons
GaAs	$0.067m_0$
InP	$0.073m_0$
InAs	$0.026m_0$
InSb	$0.015m_0$

The listed semiconductors are general ones used as THz-radiation emitters. InP and GaAs have almost the same effective mass and are heavier than InAs and InSb. Since the effective mass of InSb is the smallest, it is expected that the most intense THz radiation will be emitted from InSb.

These two enhancement schemes, such as applying an external magnetic field $\boldsymbol{B}$ and irradiating intense excitation laser, are basic methods to generate intense THz radiation.

2.2 Terahertz-Radiation Power and Polarization Emitted from GaAs, InP, InAs, and InSb

The results stated in the previous sections led us to investigate a new approach to generate intense THz radiation using several semiconductors such as GaAs(100), InP(100), InAs(100), and InSb(100). An 82 MHz repetition-rate mode-locked Ti:sapphire laser delivers nearly transform-limited 70 fs pulses at 800 nm, as shown in Fig. 2. The average power for excitation is about 700 mW with 3 mm diameter spot size on the sample. A liquid helium (liq. He) cooled silicon bolometer is provided for detecting the power of the total radiation and a wire-gird polarizer is placed in front of the bolometer.

The excitation-power dependence of the THz-radiation power in a 1.7 T magnetic field is shown in Fig. 3 [4]. The total radiation power from GaAs and InAs exhibits an approximately quadratic increase to the excitation power as (12) shows. Note that the power of the THz radiation from an InAs sample

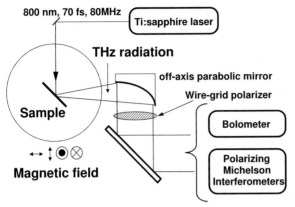

Fig. 2. Experimental setup

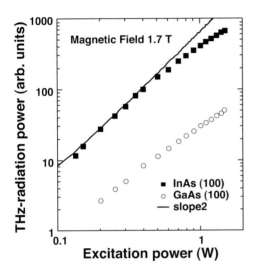

Fig. 3. Excitation power dependence of THz-radiation power from
InAs and GaAs irradiated with
femtosecond optical pulses in a
1.7 T magnetic field. *Solid line* represents slope 2

in a magnetic field is over one order of magnitude higher than that from
undoped GaAs under the same conditions, as can be seen in Fig. 3.

This difference of the power between GaAs and InAs is qualitatively understood by considering (12). Since the effective mass of GaAs is almost three
times larger than that of InAs, it is readily expected that the radiation power
from InAs is about one order stronger than that from GaAs. Figure 4 shows
the magnetic-field dependence of the THz-radiation power from InP, InAs,
and InSb irradiated with femtosecond optical pulses [16]. One can observe
anomalous behaviors such as saturation, decrease and recovery that depend
on the direction of the applied magnetic field. Contrary to our expectations,
the THz-radiation power from InAs is approximately two orders larger than
that from InP or InSb. The THz radiation from InAs is more enhanced than
the effective-mass ratio in comparison with InP. This difference has not been

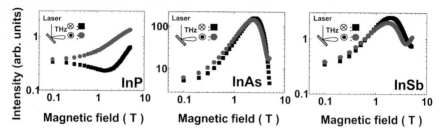

Fig. 4. Dependence of THz-radiation power from InP, InAs, and InSb on magnetic-field strength

well investigated yet. The reason for the emission from InSb being less enhanced compared with InAs in spite of the fact that the effective mass of InSb is twice as small will be attributed to the intervalley scattering of photoexcited carriers in InSb from the Γ- to the L-valley [17].

Based on the above results, among the semiconductors listed in Table 1, InAs is found to be the most efficient emitter. Therefore, it is worth investigating the properties of THz radiation from InAs in magnetic fields.

3 Terahertz Radiation from InAs in Magnetic Field

3.1 Emission of Terahertz Radiation up to 5-T Magnetic Field

The emission of THz radiation has been investigated with the use of an InAs (100) surface. The experimental setup used is almost the same as in Fig. 2. Using a specially designed magnet, THz emission from five different experimental configurations, as illustrated in Fig. 5, have been comparatively measured by changing the magnetic-field strength, magnetic-field direction, and incidence angle of the pump laser. This split-coil superconducting magnet with cross room-temperature bores can provide a magnetic field up to 5 T [6, 18].

The emitted THz radiation polarizes almost linearly or elliptically, as shown in Fig. 6. For magnetic-field strengths lower than 2 T, the THz-radiation power depends quadratically on the magnetic-field strength [4]. Moreover, in the case where the magnetic field is parallel to the surface and the laser incidence angle is 45°(G-1), the saturation at around 3 T and the reduction of the radiation power above 3 T is observed. In the G-1 and G-2 configurations (G-2: the magnetic field is parallel to the direction of the THz-wave propagation and the laser incidence angle is perpendicular to the surface), linear polarizations have been observed as expected from (2)–(5).

On the other hand, the saturation behavior is not observed clearly in the case of the G-5 (G-5: the magnetic field is parallel to the direction of THz-wave propagation and the laser incidence angle 45°) as shown in Fig. 6b.

Geometry	Magnetic field direction	Saturation field (T)	Relative power	Polarization
G-1 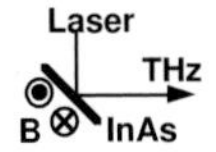 (Laser, THz, B⊗, InAs)	⊗	+3.2	1 (max)	Linear
	⊙	-3.0	0.77	
G-2 (B, InAs, THz, Laser)	N→S	+3.2	0.11	Linear
	S←N	-3.1	0.10	
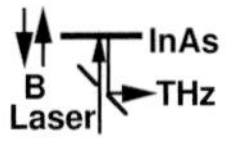 G-3 (InAs, B, THz, Laser)	No radiation was observed.			
G-4 (Laser, B, THz, InAs)	S / N ↑	+4.8	0.67	Elliptical
	N / S ↓	-4.7	0.67	
G-5 (Laser, THz, B, InAs)	N→S	+5.0 >	0.70	Elliptical
	S←N	-5.0 >	0.68	

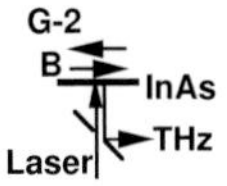

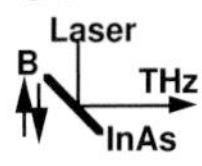

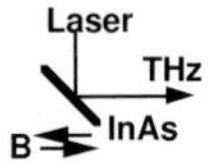

Fig. 5. Experimental setup for emission of THz radiation from InAs in the presence of an external magnetic field. The saturation of the THz-radiation power is observed in the upper two cases. The magnetic field is parallel to the $\langle 011 \rangle$ crystal axis in the G-1, G-2 configurations, and the $\langle 100 \rangle$ axis in the G-3 configuration, respectively. In the case of the G-4 and G-5 configurations, the magnetic field is $45°$ to the surface normal. Linear polarization (G-1 and G-2), no radiation (G-3), and elliptical polarization (G-4 and G-5) are observed as expected from (2)–(5)

In the case of the G-5 configuration, the polarization of the THz radiation changed dramatically with the increase of magnetic-field strength. Photoexcited electrons are accelerated on the plane perpendicular to the direction to which THz radiation propagates, as is seen in the G-5 configuration. Since the polarization of THz radiation reflects the components of the carrier accelerating direction, both components of polarization should be observed, as expected by (5). Figure 7 illustrates the time-domain measurement of the THz-emission amplitude from InAs in the G-1 configuration with a 1 T permanent magnet.

A photoconductive (PC) dipole antenna is used for detection of THz radiation. There is a very clear difference between ⊗ and ⊙. In Fig. 7a, the phase of the THz-radiation field is completely opposite, because the photoexcited electrons are accelerated to the opposite directions by the reversed magnetic field. Thus, the spectral shapes in Fig. 7b show clear differences. The power-saturation behavior of the THz radiation is observed only in the G-1 and G-2

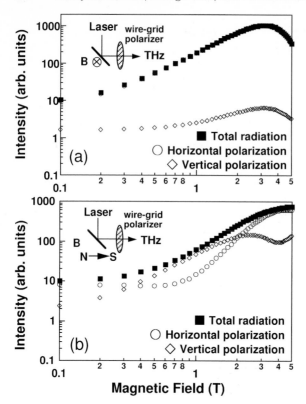

Fig. 6. Dependence of THz-radiation power on magnetic-field strength. The *inset* indicates the experimental setup for the G-1 and the G-5 configurations. (**a**) for the G-1 and (**b**) for the G-5 configurations, respectively. The saturation behavior is observed clearly at around 3 T in the case of the G-1 configuration

configurations. Additionally, in the case of the G-1 configuration, we have observed that the emitted THz-radiation power depends on the magnetic-field direction. Under this experimental condition, photoexcited electrons in the Γ-valley may be scattered to the L-valley by Lorentz acceleration of electrons due to the magnetic field, as observed in the electric-field case [19]. This process will reduce the number of electrons in the Γ-valley contributing to the generation of THz radiation.

The detailed THz-radiation spectra have been obtained by a polarizing Michelson (same as Martin–Puplett Fourier transform) interferometer. To extract the features of the THz-radiation spectra in various cases, the center frequency and the spectral width of the THz radiation are defined as the average and the standard deviation of the frequency between 0.05 THz and 5 THz by integration. Both the center frequency and the spectral width depend strongly on the magnetic-field strength that differs significantly depending on the configurations, as shown in Fig. 8. In Figs. 8b and c for

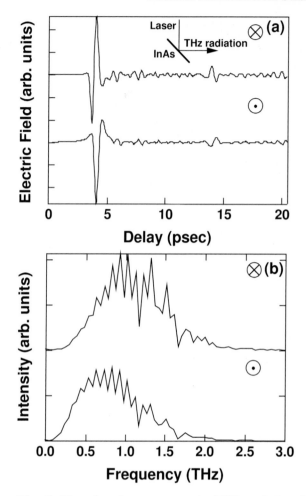

Fig. 7. Time-domain measurement of THz emission. (**a**) Temporal waveforms of emitted radiation from InAs under 1.0 T magnetic field. The direction of magnetic field is described in the *insets*. The phase is completely different from each other. (**b**) Fourier-transformed THz-radiation spectra. The interference patterns arise from the PC dipole antenna of which the substrate is transparent in the terahertz region

the G-5 configuration, the center frequency shows symmetric behavior to the magnetic-field direction, as expected from the motion of electrons. On the other hand, asymmetric behavior to the magnetic field direction is shown in Fig. 8a for the G-1 configuration. The emitted THz-radiation powers were compared for all configurations. The maximum power is obtained in the case of the G-1 configuration. As mentioned in the time-domain measurement, the electrons are accelerated in opposite directions by changing the direction of the magnetic field in the G-1 configuration. This gives rise to the change

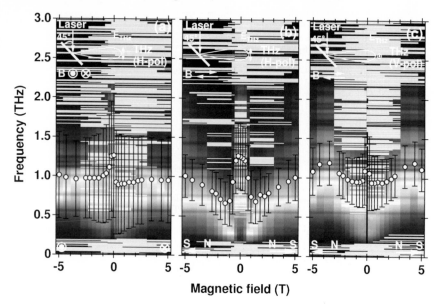

Fig. 8. Two-dimensional plots of THz-radiation spectra for different magnetic field. (a) the G-1 configuration, (b) and (c) the G-5 configurations. *Open circles* and *bars* show the center frequency and spectral bandwidth, respectively. H-pol and V-pol indicate horizontal and vertical polarization

of the center frequency. While, in the case of the G-5 configuration, since both the upward and the downward electron accelerations cause the same polarization of THz radiation, we can observe the same spectra even in the opposite magnetic field. This explanation is applied for the G-2 and G-4 configurations (G-4: the magnetic-field direction and the laser incidence angle are 45° to the surface normal).

In conclusion, we have found that the THz radiation saturates in a high magnetic field. The optimum magnetic field is 3 T, as shown in Fig. 6. We have also observed a significant dependence of the frequency and the spectral width of the radiation on the magnetic-field strength. From the practical point of view to design compact and intense THz-radiation sources, the optimum magnetic field is found to be around 3 T from these experiments. This magnetic-field strength will be easily achieved by our newly designed permanent magnet [20].

3.2 Emission of Terahertz Radiation up to 14-T Magnetic Field

In Sect. 3.1, we described enhanced generation of THz radiation from InAs up to 5 T. In order to explore more detailed physics, a magnetic field is applied to InAs up to 14 T by using a superconducting magnet [21].

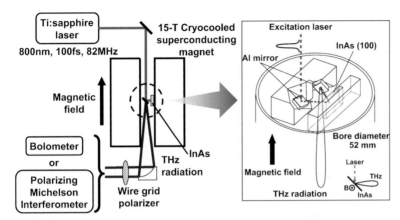

Fig. 9. Experimental setup for a THz-emitter module fitted to the 52 mm diameter bore of the superconducting magnet. The excitation laser and THz radiation should propagate parallel to the direction of the 2 m bore axis. The magnetic field was applied perpendicular to the direction of the laser incidence and parallel to the InAs surface

The experimental setup is illustrated in Fig. 9a, including the pump laser, magnet, emitter, and detection system. In the experiments up to 5 T, a 25 mm diameter bore, split-pair superconducting magnet immersed in liq. He was employed [22]. The large bore required for three orthogonal directions in the optical experiment obviously increased the heat flow to the cryostat. This heat flow limited the experimental time with the same optical alignment required for systematic measurements and restricted the practical magnetic field with the given spatial factor. To explore the dependence of the much higher magnetic-field strength, a cryogen-free superconducting magnet with sufficient bore size should be prepared. For this purpose, a specially designed, cryogen-free superconducting magnet with a 52 mm diameter room-temperature (RT) bore generates a magnetic field up to 15 T. The pump laser irradiated the sample at a 45° incidence angle. The sample was undoped bulk InAs with (100) surfaces.

A magnetic field was applied parallel to the sample surface and perpendicular to the incidence plane of the excitation laser by a specially designed setup, as shown in Fig. 9b. This configuration is consistent with the G-1 configuration, as shown in Fig. 5. A liq. He-cooled bolometer was provided to monitor the emitted power. The emitted THz-radiation power shows an anomalous behavior, as is seen in Fig. 10, when a strong magnetic field is applied.

The THz-emission amplitude demonstrated an asymmetrical dependence on the magnetic-field inversion, as observed in the low-field case [3]. The maximum intensity was obtained at approximately 3 T. For practical applications, this is an important result. The 3 T can be achieved even with a permanent

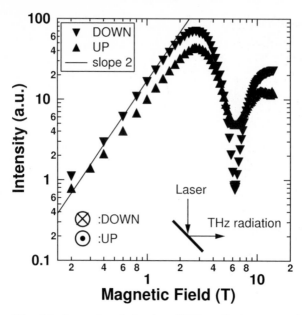

Fig. 10. Anomalous behavior of THz-radiation power from InAs pumped by a femtosecond optical pulse in a strong magnetic field. The temporal waveform reverses for the inversion of the applied magnetic field, as observed in the low-field case. At around 6 T, the radiation intensity reaches a minimum value and recovers slowly

magnet incorporating a special design. It is estimated that approximately 100 μm radiation power is emitted from InAs pumped with a femtosecond laser having the pulse width less than 100 fs and power of 1 W [4]. *Weiss* et al. have shown the quadratic dependence of THz radiation power on the magnetic-field strength up to 1 T. Our experimental result shows such dependence is valid up to 2 T, but it is no longer valid above 3 T. Analysis of anomalous behaviors such as saturation, decrease, and recover has been tried [5, 9, 23, 24, 25], however, a clear explanation has not been proposed yet. The THz-radiation spectrum was obtained by a polarizing Michelson interferometer. The radiation spectrum changes drastically in a high magnetic field, as shown in Fig. 11a. The spectrum in a high magnetic field (Fig. 11c) shows a periodic structure. The cyclotron frequency was 3.36 THz for a 3 T magnetic field, therefore the spectral structure could not be explained by this process. At around 6 T, the radiation intensity reaches a minimum and recovers slowly. This drastic variation might be attributed to the change of emission mechanism in higher magnetic fields.

We are assuming that this drastic variation may be attributed to the change of dielectric constant caused by the existence of strong magnetic fields [26]. For example, the dielectric constant of InSb is well investigated in [27] and it is reported that InSb becomes partially transparent in the THz

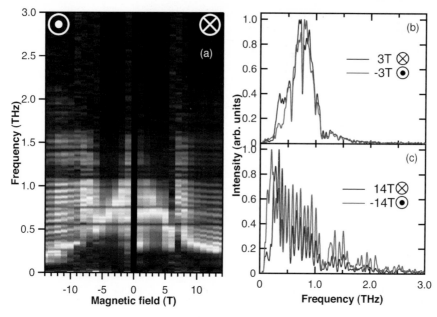

Fig. 11. The THz-radiation spectrum exhibits a significant dependence on the magnetic-field strength. (**a**) Two-dimensional plots of THz-radiation spectra. THz-radiation spectra for magnetic field strength 3 T (**b**) and 14 T (**c**)

region due to modulation of the dielectric constant in high magnetic fields. The same argument can be applied for InAs. If InAs becomes partially transparent in high magnetic fields, the radiation spectrum should exhibit a clear periodic structure, as shown in Fig. 11, due to the interference in a parallel plate.

In conclusion, we have shown experimentally that the THz-radiation power from InAs behaves anomalously, indicating saturation, decrease, and recovery up to 14 T. Furthermore, the radiation spectrum exhibited a clear periodic structure over 6 T, possibly due to the modulation of the dielectric tensor in a high magnetic field. These experimental findings imply that there still exists rich, undiscovered physics to be explored. Moreover, this new information will be helpful in designing THz emitters for new applications.

4 Compact THz-Radiation Source with 2-T Permanent Magnet and Fiber Laser

From industrial and practical points of view, a compact THz-radiation source is required. It is obvious that both a THz-radiation source with a huge laser system and a liq. He-cooled detector such as a bolometer are not useful in

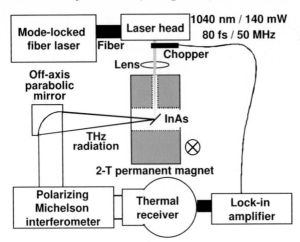

Fig. 12. Experimental setup. A 1 mm diameter laser beam is focused onto the InAs

factories. Thus, the key issue to develop a compact THz-radiation source is
a conventional small femtosecond laser and a RT-operating detector.

Regarding lasers, recently, very compact fiber lasers have been intensively
developed and the size becomes less than $(200 \times 120 \times 100)$ cm^3 [28]. One
can easily obtain femtosecond optical pulses just by turning a key switch.
Concerning room-temperature-operating detectors, a pyroelectric thermal receiver such as a deuterated triglycine sulfate (DTGS) detector is feasible.
The DTGS is used conventionally without requiring time-gating adjustment
or cryogen cooling with liq. N$_2$ or liq. He. The emitter is an InAs applied 2 T
field by a compact magnet and irradiated with femtosecond optical pulses.
The THz-radiation emitter system, including the excitation laser head, is
almost the same size as a conventional notebook computer.

4.1 Experimental Setup

An experimental setup is shown in Fig. 12. A 50 MHz repetition-rate mode-
locked fiber laser delivers nearly transform-limited 80 fs pulses at 1040 nm,
and the average power is 180 mW (IMRA Wattlite). This mode-locked fiber
laser is a complete turnkey system and is improved significantly in emitting
power, compared with the previous Femtolite operating at 1550 nm [28].

Undoped bulk InAs with a (100) surface is utilized as the THz emitter.
The carrier concentration of the InAs is less than (5×10^{16}) cm^{-3}. The average pump power is approximately 140 mW with a 1 mm diameter spot size
on the emitter.

The use of InAs as the emitter in this system is advantageous in the
following two respects. One is its approximately order of magnitude higher
efficiency in THz-radiation power compared with other semiconductors and

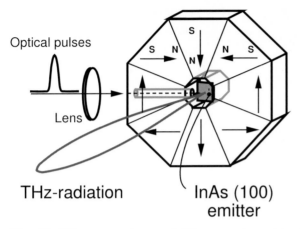

Fig. 13. 2 T permanent magnet. The magnet consists of eight pieces with different magnetic-field directions. The magnetic field reaches 2 T due to the vector sum of magnetic moments

the other is its narrower bandgap energy (0.357 eV) than the photon energy of the pumping laser (1.202 eV). The small 2 T permanent magnet consists of eight pieces of Nd-Fe-B magnet. Each piece is magnetized differently, as shown in Fig. 13. The remanence magnetic field of the Nd-Fe-B material itself is 1.3 T (NEOMAX-44H). Owing to the reasonable magnetic circuit design [29], the magnetic field in the center exceeds the remanence magnetic field. The 2 T permanent magnet is cylindrical in shape, and of 128 mm diameter and 56 mm thickness, making it smaller and much lighter than an electromagnet. It weighs only 5 kg. In the far field, the vector sum of magnetic moments is close to zero. Therefore, the magnet has very little leakage of the magnetic field owing to the above-mentioned magnetic circuit. This is highly advantageous for future system integration. Furthermore, we are planning to increase the magnetic field up to 3 T by scaling this design so as to emit the highest THz-radiation power [6].

4.2 Notebook-Computer-Size THz Emitter

A photograph of the THz-radiation system is shown in Fig. 14. The THz-radiation emitter system, including the pumping-laser head, is almost the same size as a conventional notebook computer [30]. This system is much smaller than our previous one [20].

Figure 15 shows typical THz-radiation spectra. These spectra were obtained by a polarizing Michelson interferometer with a liq. He-cooled silicon bolometer and a pyroelectric thermal detector. A typical pyroelectric thermal detector for detecting THz radiation employs DTGS. DTGS is the best material for use as a sensitive element in pyroelectric sensors due to its high pyroelectric coefficient, reasonably low dielectric constant, and best quality

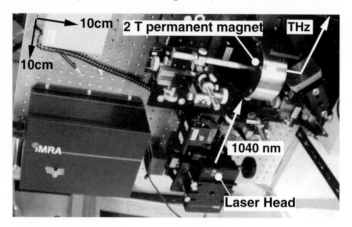

Fig. 14. Photograph of THz-radiation emitter system. Laser beam is focused onto an InAs wafer with a 2 T permanent magnet

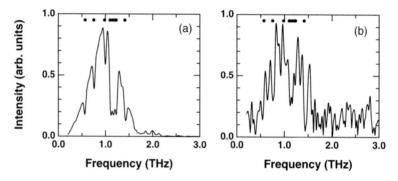

Fig. 15. THz-radiation spectra taken by different detectors. (**a**) Liquid-helium-cooled silicon bolometer; (**b**) deuterated triglycine sulfate (DTGS). In both cases, water-vapor absorption lines (*dots*) are observed [31]

factor. The DTGS is uniformly sensitive to radiation in wavelengths from ultraviolet to far infrared due to its high Curie temperature. Unlike time-gated detectors, DTGS receivers do not require time-gating adjustment and cryogen cooling for operation. We obtain the same spectrum as the THz-radiation spectrum measured by a bolometer, although DTGS has 10^{-4} less sensitivity.

Such a qualitative advancement in emitters will certainly widen the application of THz radiation [32, 33, 34, 35]. Up to now, our pumping laser power was 140 mW; however, rapid progress in high average power semiconductor lasers will enable us to generate an average power of more than 1 W in the near future. The combination of a thermal detector with a high-intensity mode-locked fiber laser will contribute to system integrations that are indispensable for industrial applications of THz radiation. These rapid improvements will enable us to realize numerous potential applications using THz radiation for industry and science.

References

[1] X.-C. Zhang, B. Hu, J. Darrow, D. Auston: Appl. Phys. Lett. **56**, 1011 (1990)
[2] X.-C. Zhang, D. Auston: J. Appl. Phys. **71**, 326 (1992)
[3] X.-C. Zhang, Y. Jin, T. Hewitt, T. Sangsiri, L. Kingsley, M. Weiner: Appl. Phys. Lett. **62**, 2003 (1993)
[4] N. Sarukura, H. Ohtake, S. Izumida, Z. Liu: J. Appl. Phys. **84**, 654 (1998) according to recalibration of the bolometer sensitivity by Sarukura *et al.*, the total radiation power from an InAs surface in a magnetic field of 1 T is corrected to be about $50\,\mu$W with pump power of 1 W
[5] R. McLaughlin, A. Corchia, M. Johnston, Q. Chen, C. Ciesla, D. Arnone, G. Jones, E. Linfield, A. Davies, M. Pepper: Appl. Phys. Lett. **76**, 2038 (2000)
[6] H. Ohtake, S. Ono, M. Sakai, Z. Liu, T. Tsukamoto, N. Sarukura: Appl. Phys. Lett. p. 05204 (2001)
[7] S. Ono, T. Tsukamoto, E. Kawahata, T. Yano, H. Ohtake, N. Sarukura: Appl. Opt. **40**, 1369 (2001)
[8] C. Weiss, R. Wallenstein, R. Beigang: Appl. Phys. Lett. **77**, 4160 (2000)
[9] J. Shan, C. Weiss, R. Beigang, T. Heinz: Opt. Lett. **26**, 849 (2001)
[10] P. Gu, M. Tani, K. Sakai, T.-R. Yang: Appl. Phys. Lett. **77**, 1798 (2000)
[11] P. Gu, M. Tani, S. Kono, K. Sakai, X.-C. Zhang: J. Appl. Phys. **91**, 5533 (2002)
[12] H. Ohtake, S. Ono, Z. Liu, N. Sarukura, M. Ohta, K. Watanabe, Y. Matsumoto: Jpn. J. Appl. Phys. **38**, L1186 (1999)
[13] R. McLaughlin, Q. Chen, A. Corchia, C. Ciesla, D. Arnone, X.-C. Zhang, G. Jones, E. Lindfield, M. Pepper: J. Mod. Opt. **47**, 1847 (2000)
[14] M. Hangyo, M. Migita, K. Nakayama: J. Appl. Phys. **90**, 3409 (2001)
[15] O. Madelung, M. Schulz, H. Weiss (Eds.): *Semiconductors*, Landolt-Börnstein, New Series III **17a–h** (Springer, Berlin, Heidelberg 1987)
[16] Y. Suzuki, H. Ohtake, S. Ono, T. Kozeki, M. Sakai, H. Murakami, N. Sarukura: in (CLEO 2002) CFI5
[17] H. Takahashi, Y. Suzuki, M. Sakai, S. Ono, N. Sarukura, T. Sugiura, T. Hirosumi, M. Yoshida: Appl. Phys. Lett. **82**, 2005 (2003)
[18] H. Ohtake, S. Ono, M. Sakai, Z. Liu, T. Tsukamoto, N. Sarukura: J. Lumin. **87–89**, 902 (2000)
[19] S. Sze: *Physics of Semiconductor Devices*, 2nd ed. (Wiley, New York 1981)
[20] S. Ono, T. Tsukamoto, M. Sakai, Z. Liu, H. Ohtake, N. Sarukura, S. Nishizawa, A. Nakanishi, M. Yoshida: Rev. Sci. **71**, 554 (2000)
[21] H. Ohtake, H. Murakami, T. Yano, S. Ono, H. Takahashi, Y. Suzuki, G. Nishijima, K. Watanabe: Appl. Phys. Lett. **82**, 1164 (2003)
[22] K. Watanabe, S. Awaji, M. Motokawa, Y. Mikami, J. Sakurabe, K. Watazawa: Jpn. J. Appl. Phys. **37**, L1148 (1998)
[23] J. Heyman, P. Neocleous, D. Hebert, P. Crowell, T. Muller, K. Unterrainer: Phys. Rev. B **64**, 085202 (2001)
[24] A. Corchia, R. McLaughlin, D. Whittaker, D. Arnone, E. Linfield, A. Davies, M. Pepper: Phys. Rev. B **64**, 205204 (2001)
[25] M. Johnston, D. Whittaker, A. Corchia, A. Davies, E. Linfield: J. Appl. Phys. **91**, 2104 (2002)

[26] H. Takahashi, Y. Suzuki, A. Quema, M. Sakai, T. Yano, S. Ono, N. Sarukura, M. Hosomizu, T. Tsukamoto, G. Nishijima, K. Watanabe: Jpn. J. Appl. Phys. **42**, L532 (2003)

[27] E. Palik, F. Furdyna: Rep. Prog. Phys. **33**, 1193 (1970)

[28] M. Jiang, G. Sucha, M. Fermann, J. Jimnez, D. Harter, M. Dageais, S. Fox, Y. Hu: Opt. Lett. **24**, 1074 (1999)

[29] H. Zijlstra: Philips J. Res. **40**, 259 (1985)

[30] H. Ohtake, Y. Suzuki, N. Sarukura, S. Ono, T. Tsukamoto, A. Nakanishi, S. Nishizawa, M. Stock, M. Yoshida, H. Endert: Jpn. J. Appl. Phys. **40**, L1223 (2001)

[31] D. Burch: J. Opt. Soc. Am. **58**, 1383 (1968)

[32] K. Saitow, K. Nishikawa, H. Ohtake, N. Sarukura, H. Miyagi, Y. Shimokawa, H. Matsuo, K. Tominaga: Rev. Sci. Instrum. **71**, 4061 (2000)

[33] K. Saitow, H. Ohtake, N. Sarukura, K. Nishikawa: Chem. Phys. Lett. **341**, 86 (2001)

[34] H. Ohtake, Y. Suzuki, S. Ono, N. Sarukura, T. Hirosumi, T. Okada: Jpn. J. Appl. Phys. **41**, L475 (2002)

[35] K. Yamamoto, K. Tominaga, H. Sasakawa, A. Tamura, H. Murakami, H. Ohtake, N. Sarukura: Bull. Chem. Soc. Jpn. **75**, 1083 (2002)

Terahertz Radiation from Bulk and Quantum Semiconductor Structures

Yutaka Kadoya[1] and Kazuhiko Hirakawa[2]

[1] Department of Quantum Matter, ADSM, Hiroshima University
   Higashi-hiroshima 739-8530, Japan
   kd@hiroshima-u.ac.jp
[2] Institute of Industrial Science, The University of Tokyo
   Komaba, Meguro-ku 153-8505, Japan
   hirakawa@iis.u-tokio.ac.jp

Abstract. An excitation of various semiconductor structures by femtosecond optical pulses leads to radiation of electromagnetic waves in the THz-frequency range. Since the waveform reflects the dynamics of the excited carriers or polarizations from which the THz wave is radiated, the time-domain studies of the emitted THz wave provide a unique way of looking directly into the temporal as well as spatial evolution of the excitation on the subpicosecond time scale. In this Chapter, we present recent studies on the emission of THz waves from nonstationary carrier transport and the coherent oscillation of carrier population in bulk and quantum semiconductor structures. Also discussed in this Chapter are the possibilities of the use of semiconductor structures for the improvement or the development of THz-wave emitters.

1 Introduction

As described in other chapters, impulsive currents in semiconductors excited by femtosecond (fs) optical pulses radiate electromagnetic (EM) waves in the THz-frequency range, serving, for example, as an EM wave source for far-infrared spectroscopy. The EM wave in the far field is proportional to the time derivative of the current, $\partial J/\partial t$. In a simplified view, J can be expressed as $J = eNv$, where e, N, and v are the elementary electric charge, total number of photoexcited carriers (electron or hole) and their velocity, respectively. If the carriers are excited by sufficiently short laser pulses and their lifetime is long enough, N can be regarded as being constant and we may approximate as $\partial J/\partial t \propto \partial v/\partial t$. Hence, in principle, the observation of such THz EM waves provides us with a way of looking into the temporal variation of v, namely the acceleration/deceleration dynamics of photoexcited carriers [1]. Actually, there is no other useful technique that can detect the carrier transport on a subpicosecond time scale. In the case of off-the-shelf semiconductor wafers, however, the acceleration is caused by the DC electric field, which is built-in near the surfaces and is out of control. Consequently, unwanted effects such as the photocarrier-induced screening make quantitative discussion quite difficult [2, 3].

K. Sakai (Ed.): Terahertz Optoelectronics, Topics Appl. Phys. **97**, 117–156 (2005)
© Springer-Verlag Berlin Heidelberg 2005

By virtue of the well-developed thin-film crystal growth techniques of the present day, various types of layered semiconductor structures can be fabricated quite precisely and reproducibly. A simple one is the undoped ($\sim$ intrinsic) layer grown on a doped substrate (i-n or i-p structures). In such structures, contrary to off-the-shelf semiconductors, the DC field can be easily controlled with the external bias. Therefore, the use of artificially structured samples is important even in the study of carrier dynamics in bulk semiconductors for a quantitative discussion. Section 2, in this Chapter, presents a rigorous elucidation of nonstationary carrier transport in bulk semiconductors.

Another important class of artificial semiconductor structures is, needless to say, the quantum structures, such as quantum wells (QWs) and superlattices (SLs). The excitation of these structures by fs optical pulses also triggers the radiation of THz EM waves, providing rich information on quantum-mechanical phenomena such as the tunneling of the electron wave packet and the coherent evolution of the excitation. The radiation of THz waves from such structures can be viewed classically in two ways. One is the current, J, flowing across the heterostructures, as in the case of bulk semiconductors. The other is the transient or oscillating polarization, P, induced by the short optical pulses. Although, under the local electric dipole approximation both views are related by $J = \partial P / \partial t$, the description based on the polarization is suited when the coherence of the excitation plays an important role. In the case of the wave-packet motion, for example, the polarization arises from the different time evolution of the wavefunctions between the electrons and holes.

Historically, the radiation of THz waves from the wave-packet motions was observed first in coupled [4] and single [5] QWs in 1992, and then in SLs (Bloch oscillation) in 1993 [6]. Since then, a good number of reports have been published. In contrast to widely used nonlinear spectroscopic techniques such as four-wave mixing, where in general the interpretation of the signal is not simple, the THz signals reflect rather directly the dynamics of the carrier motion or population, giving a big advantage to the THz-based technique. Indeed, for example, the theoretical frameworks, which are used for the analysis of the nonlinear optical response, have been critically tested on the basis of the THz-emission experiments [7]. In this Chapter, two recent THz-emission studies on semiconductor quantum structures will be presented in detail; one on the Bloch oscillation and the miniband transport in SLs in Sect. 3 and the other on the coherent nutation of the exciton population inherent to the strongly coupled exciton–photon system in Sect. 4.

On the other hand, the use of semiconductor structures is expected to open up a possibility of improving or developing THz emitters, detectors and others, which may be desired in future high-performance THz-imaging and/or spectroscopy systems. As examples, the amplification or generation of THz waves based on the negative differential resistance in SLs and the enhancement of radiation from semiconductor surfaces by the use of microcavity structures will be discussed in Sects. 3 and 5, respectively.

2 Nonstationary Carrier Transport in Strongly Biased Bulk Semiconductors

2.1 A Brief Review on Nonstationary Carrier Transport and THz Wave Radiation

With the recent progress in miniaturization of high-speed transistors, ultra-fast transistors with cutoff frequencies well above 500 GHz have been realized [8]. In such ultrashort channel transistors, carriers experience very few scattering events in the channel and drift in a very nonstationary manner, as schematically illustrated in Fig. 1. Consequently, the performance of such ultrafast transistors is not mainly determined by the steady-state properties, such as saturation velocities and mobilities, but is governed by the nonstationary carrier transport subjected to high electric fields. However, it has been difficult to characterize such very fast phenomena by using conventional electronics, such as sampling oscilloscopes, because of their limited bandwidth. Consequently, Monte Carlo calculations have been the only tool for discussing transient carrier transport.

Very recently, the pioneering work by *Leitenstorfer* et al. [9,10] has opened up a way to characterize the transient carrier velocities with sub-100 fs time resolution by measuring the THz EM radiation emitted by photoexcited carriers in semiconductors. According to Maxwell's equations, when charged carriers are accelerated by electric fields, they emit radiation whose electric-field component, E_{THz}, is proportional to the acceleration, $\partial v/\partial t$. Since the acceleration/deceleration of carriers in semiconductor devices takes place on the subpicosecond time scale, the emitted radiation is in the THz range. The basic idea is that transient carrier velocities can be recovered by integrating the traces of detected E_{THz} with respect to time. By using this scheme, *Leitenstorfer* et al. [9,10] successfully determined the transient carrier velocities in GaAs and InP, and opened up the possibility of quantitative discussion of velocity-overshoot behaviors.

What is represented by the detected THz signals is, however, not fully understood yet. Although *Leitenstorfer* et al. already pointed out that the finite sample thickness distorts the apparent transient-velocity characteristics, quantitative discussion of such an effect has not been carried out yet. Another important point is that the velocities of electrons and holes cannot be separated in the photoexcited THz measurements; in general, it is believed that the THz signal observed experimentally is dominated by electron transport because of its lighter effective mass. However, it is necessary to examine whether the hole contribution is indeed negligible or not. Therefore, it is crucial to make a quantitative comparison between the nonequilibrium

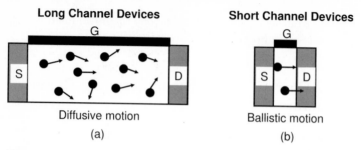

Long Channel Devices

Short Channel Devices

Diffusive motion

Ballistic motion

(a)

(b)

Fig. 1. Carrier transport in long diffusive channels (**a**) and very short ballistic channels (**b**). In conventional long-channel transistors (**a**), their performance is determined by steady-state properties, such as saturation velocities and mobilities. However, in very short channel transistors (**b**), carrier transport is quasiballistic and their switching speed is governed by how fast carriers are accelerated

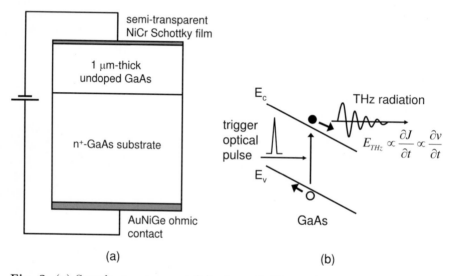

Fig. 2. (a) Sample structure and (b) schematic illustration of THz-emission experiment on biased bulk GaAs

transport data obtained by the time-domain THz spectroscopy and ensemble Monte Carlo (EMC) results to see how quantitative discussions can be made from THz data [11].

2.2 Time-Domain THz Measurements of Transient Carrier Velocities in Strongly Biased Semiconductors

The sample used in this section consisted of a $1\,\mu\text{m}$ thick undoped GaAs grown on n^+-GaAs substrate (see Fig. 2a). In this experiment, this undoped

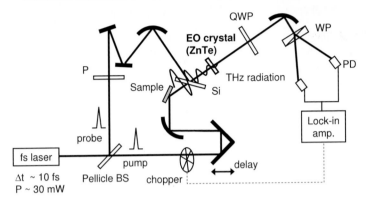

Fig. 3. Experimental setup for free-space THz electro-optic sampling measurements

GaAs layer is regarded as the channel of a transistor. An ohmic contact was fabricated by depositing and annealing AuGeNi alloy on the back surface of the sample. Then, a semitransparent NiCr Schottky film was deposited on the surface to apply a DC electric field, F, to the undoped GaAs layer. The role of the top and bottom contacts is two-fold; first, by using these electrodes, the external bias electric field, F, can be applied to the undoped GaAs regions. Secondly, these electrodes provide paths for photoexcited electrons and holes to leave the undoped regions before the next fs light pulse arrives. This is quite important in time-domain THz experiments, since if photoexcited carriers accumulate in the undoped region, the internal electric field deviates from the planned value due to a field-screening effect, as discussed in Sect. 5.

Femtosecond laser pulses from a mode-locked Ti:Al$_2$O$_3$ laser whose center wavelength was 790 nm were used to generate electron–hole pairs in the biased undoped region of the diode. The pulse width was approximately 100 fs. The excitation power was set to 30 mW and the illuminated surface area was approximately 1 mm × 2 mm. When carriers are accelerated by DC electric fields, they emit EM radiation in the THz regime, whose electric-field component, E_{THz}, is proportional to the time derivative of the electron–hole relative velocity, $\partial v_{e-h}(t)/\partial t$, as schematically shown in Fig. 2b. The free-space THz electro-optic (EO) sampling technique was used for detecting the THz radiation, as illustrated in Fig. 3. The EO sensor used in this experiment was a 100 μm thick (110) ZnTe crystal. The spectral bandwidth of this sensor was approximately 4 THz (time resolution ≈ 250 fs) [12, 13, 14, 15, 16]. The internal electric field, F, in the undoped depletion region was estimated from the voltage applied to the m-i-n diode and the surface Schottky barrier height (0.75 eV) determined from an I–V measurement. All the experiments were performed at 300 K.

The THz waves emitted from the 1 μm thick GaAs intrinsic region in the m-i-n diode was measured at various F and its temporal traces are plotted in Fig. 4a [11]. At lower fields, E_{THz} gradually increases and, then, decreases

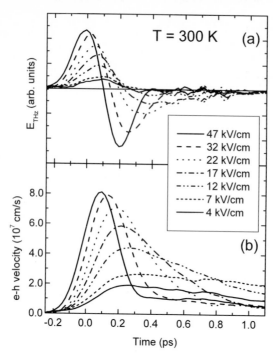

Fig. 4. Bias-field dependence of **(a)** the amplitude of THz electric field, E_{THz}, emitted from the GaAs m-i-n diode and **(b)** the transient electron–hole relative velocity, $v_{\mathrm{e-h}}(t)$, obtained by integrating E_{THz} with respect to time [11]

close to zero. At higher fields, E_{THz} increases more abruptly and its bipolar feature becomes pronounced. Figure 4b shows the transient velocity obtained by integrating E_{THz} with respect to t. When $F = 4\,\mathrm{kV/cm}$, $v_{\mathrm{e-h}}(t)$ gradually reaches its steady-state value about $0.3\,\mathrm{ps}$ after the excitation. As F is increased, the carrier acceleration increases and, furthermore, $v_{\mathrm{e-h}}(t)$ starts to exhibit a pronounced overshoot behavior due to intervalley scattering. Since the effective mass of electrons is less than that of holes, it has generally been considered that the observed THz signal is dominated by electron transport. We will come back to this point later.

Figure 5 shows the transient velocity of *electrons* in bulk GaAs calculated by using the EMC method [17, 18]. When compared with Fig. 4b, the qualitative features are similar, indicating that the THz signals are indeed dominated by electron transport. However, when we look at the data more closely, a few discrepancies are noticed. First, experimental velocity traces show broader overshoot peaks. This is due to the limited detection bandwidth ($\approx 4\,\mathrm{THz}$) of our THz measurement system. Secondly, experimentally observed velocities show a gradual decrease after the initial acceleration at low electric fields (for example, $F = 4\,\mathrm{kV/cm}$), while the EMC results do not. This point will be discussed in more detail in the next section.

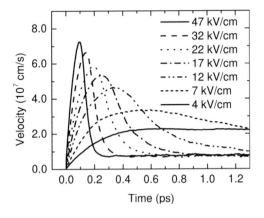

Fig. 5. Transient electron velocity in bulk GaAs calculated for various electric fields by the ensemble Monte Carlo method [11]

2.3 Effect of Sample Geometry on the THz Time-Domain Data

The observed gradual decrease in the transient velocity determined by the THz measurements shown in Fig. 4b can be explained by the effect of finite sample thickness. Considering the absorption coefficient ($\alpha \sim 10^4\,\mathrm{cm}^{-1}$) of GaAs at the photoexcitation wavelength ($\approx 790\,\mathrm{nm}$), the $1\,\mu\mathrm{m}$ thick intrinsic region of the m-i-n diode is almost uniformly illuminated. After the impulsive photoexcitation, the photoexcited electrons and holes are accelerated by the electric field and start drifting in opposite directions. However, as they drift and reach the electrodes, the number of carriers that contribute to the increase of polarization starts decreasing. Consequently, the apparent carrier velocity obtained from the THz measurements decreases.

In order to quantify such discussions, it is necessary to perform EMC calculations that take into account the actual sample geometry. In our EMC calculations, we assumed that the sample surface was illuminated by a fs laser pulse ($\lambda \approx 790\,\mathrm{nm}$) and electrons and holes with an exponential density profile expected from the absorption coefficient were instantaneously created in the $1\,\mu\mathrm{m}$ thick intrinsic bulk GaAs layer. The calculations were done by computing, iteratively, (1) the potential distribution in the biased m-i-n diode by using a drift-diffusion model and (2) transient velocities of photoexcited electrons and holes by using the EMC method. We assumed that carriers that reach the electrodes vanish and lose the contribution to the ensemble average velocity. In this calculation, two valleys (Γ: $0.067m_0$, Γ: $0.35m_0$) were assumed for electrons and one valley (Γ: $0.45m_0$) for holes. The scattering mechanisms included in the calculation were scattering by polar optical phonons, acoustic phonons, and impurities ($10^{14}\,\mathrm{cm}^{-3}$).

Figure 6a schows the comparison between the experimental and EMC results at $F = 4\,\mathrm{kV/cm}$. After an initial acceleration, the velocity trace, $v_{\mathrm{e-h}}(t)$, obtained from the THz data shows a gradual decrease for $t > 300\,\mathrm{fs}$. The calculated transient velocities are plotted by a broken line and a dotted line for electrons and holes, respectively, and the relative velocity, $v_{\mathrm{e-h}}$, is plotted

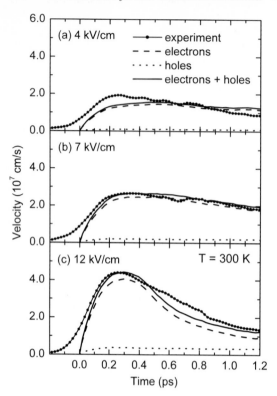

Fig. 6. Comparison between the transient carrier velocity determined from the THz data (*circles*) and the ensemble Monte Carlo results at (**a**) $E = 4\,\text{kV/cm}$, (**b**) $E = 7\,\text{kV/cm}$, and (**c**) $E = 12\,\text{kV/cm}$. The *dashed line* and the *dotted line* denote the transient velocities for electrons and holes, respectively. The *solid lines* denote the transient electron–hole relative velocities [11]

by a solid line. As seen in the figure, the calculated v_{e-h} indeed exhibits a slight decrease for $t > 400\,\text{fs}$. The agreement between the experimental results and the EMC calculations are reasonable, confirming that the apparent reduction in $v_{e-h}(t)$ originates from the carrier sweep-out effect due to finite sample thickness. This means that even when the channel region is not so thin ($\approx 1\,\mu\text{m}$), the transient velocities obtained by THz experiments are strongly affected by the geometrical effect. A similar calculation was done for higher fields ($F = 7\,\text{kV/cm}$ and $12\,\text{kV/cm}$) and the results are plotted in Figs. 6b and c, respectively. The agreement between the THz data and the EMC calculation is indeed excellent, indicating that, if actual experimental conditions (sample geometry, photoexcitation condition, etc.) are properly taken into account, the EMC calculations give a very good description of the transient carrier velocities determined from THz measurements [11]. The calculations also show that the contribution of hole transport is small (only $\approx 10\%$) for $F < 12\,\text{kV/cm}$.

2.4 Field-Dependent Carrier Velocities in Steady States

Now, let us look into the steady-state velocities. In Fig. 7, the steady-state velocities determined from the THz data in Fig. 4b are plotted, as circles,

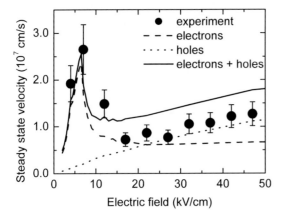

Fig. 7. Steady-state velocities determined from Fig. 1b are plotted by *circles*. Also plotted in the figure are the calculated steady-state velocities for electrons (*dashed line*), holes (*dotted line*), and the sum of the two (*solid line*) [11]

as a function of F. Because accurate determination of steady-state velocities was not possible due the carrier sweep-out effect discussed above, we simply plotted $v_{e-h}(t)$ at the time when the initial velocity transient (acceleration, overshoot, etc.) and the velocity linearly extrapolated from large t cross. The steady-state carrier velocity increases with increasing F and, then, starts to decrease above 7 kV/cm, known as negative differential velocities. In the figure, the dotted line denotes the steady-state velocity of electrons calculated by the EMC method. Comparing these two, the agreement is reasonable below 7 kV/cm. However, the discrepancy becomes significant above 20 kV/cm. In the figure, we also plotted, by a broken line, the calculated steady-state velocity for holes. The steady-state velocity of holes increases much more gradually than that of electrons because of the larger effective mass, but becomes comparable when F exceeds 20 kV/cm. The relative velocity given by the sum of both electron and hole steady-state velocities is plotted by a solid line. As seen in the figure, the agreement with the experimental and EMC results is very reasonable [11].

3 Bloch Oscillation and THz Gain in Semiconductor Superlattices

3.1 Bloch Oscillation in Semiconductor Superlattices

In an ideal periodic potential system placed in a uniform static electric field, electrons oscillate both in momentum-space and real-space. This oscillation phenomenon is called "Bloch oscillation (BO)" and corresponds to the Bragg reflection of electron wave packets at the edge of the Brillouin zone, which was predicted by *Bloch* [19] and *Zener* [20]. Bloch oscillating electrons move back and forth in the energy band at the Bloch frequency ω_B (= $eFa/\hbar$; e: elementary charge, $\hbar$: reduced Planck constant, F: applied DC electric

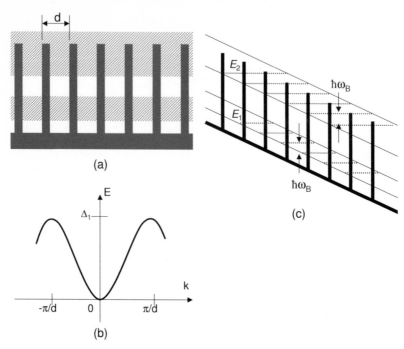

Fig. 8. (a) Superlattice (SL) structure, (b) semiclassical E-k dispersion in unbiased SLs, and (c) formation of Wannier–Stark ladders in a biased semiconductor SL

field, a: lattice constant). In actual bulk crystals, however, since the energy bandwidth is of the order of a few eV, the carriers are scattered far before reaching the edge of the Brillouin zone and, therefore, BOs cannot be seen.

Owing to recent progress in crystal growth technologies, it has become possible to grow periodic stacks of ultrathin semiconducting materials with an atomic-level precision. When two semiconducting materials with different electron affinities are periodically stacked, as shown in Fig. 8a, the size of the Brillouin zone is reduced to $\pm\pi/d$ (d: the thickness of one period of the structure; typically $\approx 10\,\mathrm{nm}$) (Fig. 8b). In such structures, called semiconductor superlattices (SLs), cosine-like modified energy bands, called "minibands", with a width of the order of a few tens of meV are formed and, when a DC bias field is applied, electrons are expected to perform BOs, as predicted by *Esaki* and *Tsu* [21]. In a quantum-mechanical picture, when an electric field is applied to an ideal SL, the miniband is split into equally spaced Wannier–Stark (WS) ladder states with an energy separation of $\hbar\omega_B$, as schematically illustrated in Fig. 8c.

Since the Bloch frequency $edF/\hbar$ can be easily tuned up into the THz range, the idea of using BOs for realizing ultrahigh frequency oscillators (often called "Bloch oscillators") has attracted much attention. Since the first proposal by *Esaki* and *Tsu* [21], considerable effort has been made to

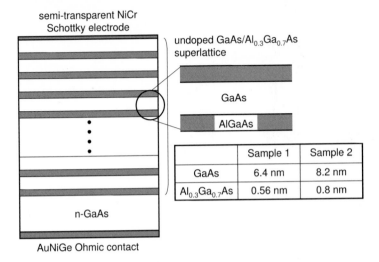

Fig. 9. Superlattice sample structures

search for BOs and obtain THz emission. Ultrafast time-domain experiments on high-quality samples unambiguously demonstrated that electrons in SL minibands perform at least a few cycle Bloch oscillations [6, 22, 23, 24, 25].

However, more recently, it has been recognized that a more fundamental question is whether BO is useful in generating/amplifying EM waves [26]. In this section, we would like to discuss this problem and present a strong experimental support for the THz gain due to Bloch oscillating electrons in SLs. For this purpose, we will show that time-domain THz spectroscopy is very useful in determining gain/loss spectra in the THz regime [27].

3.2 Superlattice Samples and Time-Domain THz Autocorrelation Spectroscopy

The SL samples studied in this section were $GaAs/Al_{0.3}Ga_{0.7}As$ SL m-i-n (metal-intrinsic-n$^+$-type) diode structures (see Fig. 9). The samples were prepared by growing 500 nm thick undoped $GaAs/Al_{0.3}Ga_{0.7}As$ SL layers on n$^+$-GaAs substrates by molecular beam epitaxy. We designed the GaAs wells (6.4 nm) and $Al_{0.3}Ga_{0.7}As$ barriers (0.56 nm) so as to set the first miniband width, Δ_1, to be 100 meV for sample 1. The other sample (sample 2) had 8.2 nm thick GaAs wells and 0.8 nm thick $Al_{0.3}Ga_{0.7}As$ barriers, and Δ_1 was designed to be 50 meV. In both samples, the second miniband was separated by a 40 meV wide minigap. The top contacts were formed by depositing a semitransparent 4 nm thick NiCr Schottky film and the bottom ohmic contacts were formed by alloying AuGeNi/Au.

When a fs laser pulse excites the sample, electron–hole pairs are optically created in the miniband. Due to an applied electric field, F, the carriers

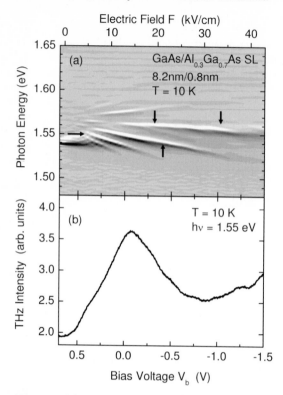

Fig. 10. (a) The contour plot of the derivative spectra of the interband photocurrent, $\partial I_{ph}/\partial(h\nu)$, measured on sample 2 as a function of the applied bias voltage. The *horizontal arrow* denotes the photon energy of the fs laser pulses used for the time-domain THz measurements. The *vertical arrows* indicate the anticrossing points due to Zener tunneling into the second miniband. (b) The integrated intensity of the THz emission is plotted as a function of the bias voltage [27]

start drifting and a THz EM wave that is proportional to the carrier acceleration is emitted into free space. Since the miniband width for heavy holes is much narrower than that for electrons, heavy holes are almost localized. Furthermore, absorption due to light holes is 1/3 that due to heavy holes. Consequently, the motion of electrons dominates the emitted THz signal.

To verify the formation of the miniband and the WS ladder in our samples, we first took weak-excitation interband photocurrent spectra, I_{ph}. Figure 10a shows the contour plot of the derivative of the interband photocurrent spectrum, $\partial I_{ph}/\partial(h\nu)$, measured on sample 2 ($\Delta = 50\,\text{meV}$), where $h\nu$ is the incident photon energy. For $+0.8\,\text{V} > V_b > +0.5\,\text{V}$, only an excitonic absorption at the bottom of the miniband is observed. However, when V_b is reduced below $+0.5\,\text{V}$, the spectrum starts showing rich structures that originate from the formation of the WS ladder. Six ladder levels are clearly seen in the figure, indicating a good quality sample. Even at $T = 300$ K (not shown here), although the structures become weaker and broader, the sign of the WS ladder was still discernible.

For time-resolved THz spectroscopy, an autocorrelation-type interferometric set up was used [28], i.e., the power of the THz radiation generated by a pair of time-correlated fs laser pulses was measured as a function of the time interval, τ, between the two laser pulses imposed by a Michelson interferometer (Fig. 11). When the laser pulses excite the sample, two almost identical THz EM waves are emitted consecutively. Experiments were performed by using 100 fs laser pulses delivered from a mode-locked Ti:Al$_2$O$_3$ laser. The laser pulses were loosely focused onto the sample surface at a polar angle of 70 °C. Under a loosely focused condition, the mechanism for the THz emission is dominated by acceleration/deceleration of photoexcited carriers in the SL miniband. The typical pump power used in the experiment was 10 mW to avoid field screening. The generated THz emission was collimated by off-axis parabolic mirrors and focused onto a Si bolometer operated at 4.2 K. The optics and the sample were kept in an enclosure purged with nitrogen to minimize absorption by atmospheric moisture. In our quasi-autocorrelation measurements the detected signal is a convolution of the two THz electric fields consecutively emitted from the sample.

In Fig. 10b, the intensity of emitted THz radiation is plotted by open circles as a function of applied electric field, F. The THz intensity starts rising at around $V_b = +0.65$ V, indicating that this voltage is the flatband voltage, V_{b0}, in this particular sample. At low fields, the emitted THz intensity increases with increasing F, which is consistent with an ordinary band transport picture. However, when F exceeds 15 kV/cm, the THz intensity starts decreasing with increasing F, indicating that the dipole moment of oscillating electrons decreases with F. In the WS regime, where edF becomes larger than the scattering broadening, electrons start to be localized on each well-resolved quantized level in the WS ladder (field-induced Stark localization); the spatial amplitude of each level is given by Δ_1/edF and is inversely proportional to F. Therefore, the observed reduction of emitted THz intensity is clear evidence for the formation of a WS ladder above a critical electric field.

Figure 12a shows the time-domain autocorrelation traces of the THz emission measured at 10 K for various F. The pump photon energy was set to be 1.55 eV, which is close to the bottom of the miniband, as indicated by a horizontal arrow in Fig. 10a [25]. The reason why the bottom of the miniband was pumped will be discussed in the next paragraph. At very low fields, a hump is seen at around $\tau = 0$. However, as we increase F, a few cyclic oscillations start to show up. From the F-dependence of the oscillation frequency, the observed oscillations were identified as Bloch oscillations. The Fourier spectra of the time-domain traces are plotted in Fig. 12b. As seen in the figure, the frequency of BO shifts to higher frequency as F is increased. However, for $F > 17$ kV/cm, the spectral shape drastically changes and becomes

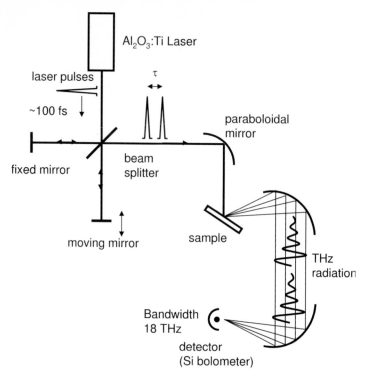

Fig. 11. Experimental setup for autocorrelation-type THz time-domain measurements

very broad. The behavior in the large reverse bias region will be discussed in Sect. 3.4.

Figure 13 shows the THz autocorrelation traces measured at $F \sim 18\,\mathrm{kV/cm}$ for various pump photon energies, $\hbar\omega$, at 300 K [25]. The pump photon energy was varied from 1.442 eV (bottom of the miniband) to 1.498 eV (top of the miniband) so as to cover the whole miniband. It is noted from the figure that clear oscillations are observed only when the electronic states close to the bottom of the miniband are photoexcited. On the contrary, the oscillations are washed out for $\hbar\omega > 1.462\,\mathrm{eV}$. This behavior can be understood in the following way; when the SL is pumped at the bottom of the miniband, the photoexcited electrons occupy the initial momentum states with $k_z \sim k_{xy} \sim 0$. In this case, when electrons are accelerated by F, they move almost in phase from $k_z \sim 0$. However, when the top of the miniband is photoexcited, not only electrons with $k_z \sim \pi/d$ and $k_{xy} \sim 0$ but also those with small k_z and large k_{xy} are photoexcited. Hence, the initial momentum of electrons in the acceleration direction, k_z, is not uniquely determined. Consequently, the phase of BOs is not well defined and clear oscillations are not visible. Thus, the visibility of BOs is strongly dependent on the preparation

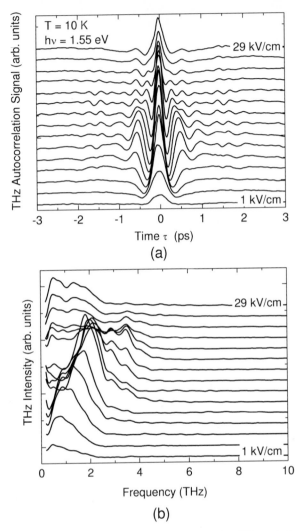

Fig. 12. (a) Autocorrelation traces of the THz emission and (b) their Fourier spectra measured on sample 2 for various bias electric fields from 1 kV/cm to 29 kV/cm in 2 kV/cm steps. The traces are shifted for clarity [27]

of the initial wave packets by fs laser pulses [25]. It is also noted that a few cyclic BOs are observed at 300 K even when the miniband width is larger than the LO-phonon energy, indicating that BO is a rather robust phenomenon.

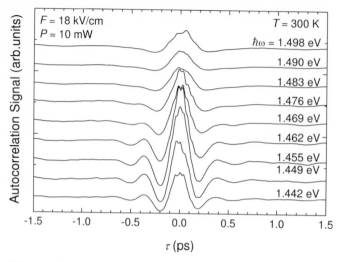

Fig. 13. Autocorrelation traces of the THz emission from the SL sample measured at $F = 18\,\text{kV/cm}$ for various pump photon energies [25]

3.3 Time-Domain Determination of High-Frequency Carrier Conductivities in Superlattices

Owing to the progress in fine epitaxy techniques and fs laser spectroscopy, ultrafast time-domain experiments on high-quality SL samples unambiguously demonstrate that electrons in SL minibands perform at least a few cyclic Bloch oscillations [6, 22, 23, 24, 25]. However, the idea of a Bloch oscillator was challenged in a more fundamental way by asking whether a BO is useful in generating/amplifying EM waves [26]. When an electric field is applied to an ideal SL, the miniband is split into equally spaced WS ladder states with an energy separation of $\hbar\omega_B$, as schematically illustrated in Fig. 8c. Because of the translational symmetry of the structure, the electron population on each level is identical, too. Since the emission and absorption of EM waves take place between the adjacent two WS levels, they occur at the same frequency with the same intensity and, hence, perfectly cancel each other. Consequently, an ideal SL has no net gain or loss for EM waves [26].

However, when scattering exists in the system, the situation changes; in the presence of scatterers, new transition channels, i.e., scattering-assisted transitions, become available, as shown in Fig. 14. Since the scattering-assisted emission (absorption) occurs at frequencies slightly lower (higher) than ω_B, gain (loss) for EM waves is expected below (above) ω_B. This means that a Bloch oscillator is not a device that generates an EM wave at ω_B, but is a medium that has a broad gain band *below* ω_B [29, 30].

The direct experimental proof of such broad THz gain is to show that the real part of the carrier conductivity, $\text{Re}[\sigma(\omega)]$, becomes negative in the THz range. The conventional method for determining $\text{Re}[\sigma(\omega)]$ in the THz range

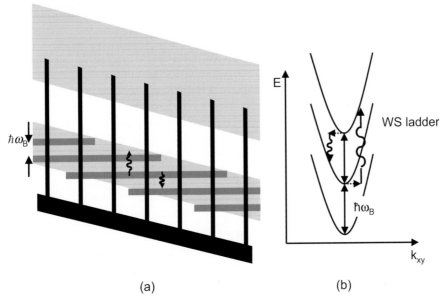

(a)   (b)

Fig. 14. (a) Schematic illustration of the WS ladder in a biased SL with scatterers. (b) Inplane dispersion of WS levels and scattering-assisted emission/absorption processes

is the measurement of free-carrier absorption in doped SLs under bias electric fields by using, for example, Fourier transform spectrometers [31, 32]. If we want to do so, we immediately face the following two problems; 1. For such measurements, a large sample size, typically a few mm^2, is necessary. Then, an extremely large current flows through the sample under biased conditions. 2. The profile of internal electric field in doped SLs becomes very nonuniform due to the formation of high-field domains. These two inevitable problems have made the conventional measurements practically impossible so far.

We have proposed a new experimental method to determine carrier conductivities in the THz region [27]. For determining the high-frequency electron conductivity in SL minibands, we took a totally different approach; we used undoped SL samples in order to avoid a large current density and high-field domain formation, and took all the necessary information within a few ps, by noting that the time-domain THz spectroscopy inherently measures the step response of the electron system to the applied bias electric field and that the Fourier spectra of the THz emission is closely related to $\sigma(\omega)$.

Let us discuss what the THz emission spectra shown in Fig. 12b represent. In the THz-emission experiment, we first set a DC electric field F in the SL and, then, shoot a fs laser pulse at the sample at $t = 0$ to create a step-function-like carrier density, $n_0\Theta(t)$, in the sample. Subsequently, THz waves are emitted from the accelerated photoexcited electrons. Now, by using our imagination, we will view the experiment in a different way. Let us perform the following thought experiment; we first set electrons in the conduction miniband under a flatband condition and, at $t = 0$, suddenly apply a step-function-like bias electric field, $F\Theta(t)$. We note that electrons in the thought experiment emit the same THz radiation as in the actual experiment [27]. This fact implies that the time-domain THz experiment inherently measures the step response of the electron system to the applied electric field.

By noting this important conclusion, we formulate the THz-emission process as follows; the creation of a step-function-like carrier density by fs laser pulses in the actual process is replaced with the application of a step-function-like electric field in the thought experiment as,

$$F(t) = F\Theta(t), \quad \tilde{F}(s) = \frac{F}{s}. \tag{1}$$

Here, the variables with $\tilde{}$ denote their Laplace transformation. The transient current by the photoexcited electrons in the miniband is given by

$$\tilde{J}(s) = \tilde{\sigma}(s)\tilde{F}(s). \tag{2}$$

Then, the emitted THz electric field, E_{THz}, can be obtained as,

$$E_{\text{THz}} \propto \frac{\partial J(t)}{\partial t} = \frac{1}{2\pi i} \int_{c-j\infty}^{c+j\infty} s\tilde{\sigma}(s)e^{ts}\frac{F}{s} = \sigma(t)F, \tag{3}$$

where $\sigma(t)$ is the electron conductivity in the time domain. Equation (3) tells us that since the spectra shown in Fig. 12b are the Fourier spectra of the THz autocorrelation traces they are proportional to $|\sigma(\omega)|^2 F^2$.

In the following, we will compare the observed THz spectra with the conductivity theoretically predicted for the Bloch oscillating electrons by *Ktitrov* et al. [33] by solving the semiclassical Boltzmann transport equation

$$\sigma(\omega) = \sigma_0 \frac{1 - \omega_{\text{B}}^2 \tau_{\text{m}} \tau_{\text{e}} - i\omega\tau_{\text{e}}}{(\omega_{\text{B}}^2 - \omega^2)\tau_{\text{m}}\tau_{\text{e}} + 1 - i\omega(\tau_{\text{m}} + \tau_{\text{e}})} \tag{4}$$

with

$$\sigma_0 = \frac{\sigma_{00}}{1 + \omega_{\text{B}}^2 \tau_{\text{m}} \tau_{\text{e}}},$$

where τ_{m} and τ_{e} are the momentum and energy relaxation times of electrons, respectively, and σ_{00} is a constant. The top panels of Fig. 15 show the calculated real and imaginary parts of the miniband conductivity, and the bottom panels show the measured THz spectra and calculated $|\sigma(\omega)|^2$ for two representative bias conditions. τ_{m} and τ_{e} in the calculation were determined as $0.15\,\text{ps}$ and $1.5\,\text{ps}$, respectively, from numerical fitting. ω_{B} at $F = 1.0\,\text{kV/cm}$ was simply set to be $edF/\hbar$. However, we found that the peak position of the THz spectrum at $11.1\,\text{kV/cm}$ is slightly lower than the

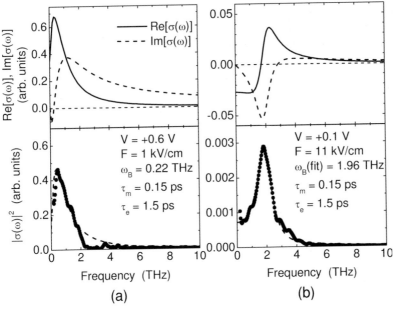

Fig. 15. The *top panels* show the calculated real (*solid line*) and imaginary (*dashed line*) parts of $\sigma(\omega)$ for the SL miniband. The *bottom panels* show the calculated $|\sigma(\omega)|^2$-spectra (*dashed line*) and the THz emission spectra (*dots*) measured for sample 2 by time-domain THz spectroscopy; (a) $F = 1.0\,\mathrm{kV/cm}$ and (b) $F = 11\,\mathrm{kV/cm}$ [27]

value simply expected from $edF/\hbar$ (2.41 THz), which is due to anticrossing with the WS ladder states of the second miniband, as will be discussed later. We set ω_B at $F = 11.1\,\mathrm{kV/cm}$ to be 1.96 THz, which gives the best fit to the experimental data.

At very low field [$F = 1.0\,\mathrm{kV/cm}$, $\omega_B(\tau_m\tau_e)^{0.5} = 0.66$], the conductivity spectra and the THz emission show more or less a Drude-like behavior (see Fig. 15a). However, as F is increased, (4) predicts that a negative $\mathrm{Re}[\sigma(\omega)]$ below ω_B develops and that a clear dip in $\mathrm{Im}[\sigma(\omega)]$ appears at ω_B. Consequently, the expected $|\sigma(\omega)|^2$ spectrum becomes asymmetric with an enhanced low-frequency spectral component. It should be noted in Fig. 15b that the THz emission at $F = 11.1\,\mathrm{kV/cm}$ [$\omega_B(\tau_m\tau_e)^{0.5} = 5.8$] indeed has a characteristic asymmetric spectral shape and that the agreement between the observed THz emission and the calculated $|\sigma(\omega)|^2$ is excellent, indicating that E_{THz} is indeed proportional to $\sigma(\omega)$ of Bloch oscillating electrons. Furthermore, the calculated $\mathrm{Re}[\sigma(\omega)]$ strongly suggests that the THz gain persists at least up to 1.7 THz [27]. Similar experiments were performed for another sample ($\Delta_1 = 100\,\mathrm{meV}$). Figure 16 summarizes the THz spectra and the calculated $|\sigma(\omega)|^2$ for various F. Except for the case at $F = 2.6\,\mathrm{kV/cm}$, where the THz

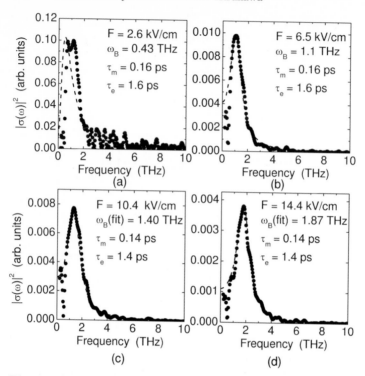

Fig. 16. The calculated $|\sigma(\omega)|^2$ spectra (*dashed line*) and the THz emission spectra (*dots*) measured for sample 1 are shown for four bias electric fields; (**a**) 2.6 kV/cm, (**b**) 6.5 kV/cm, (**c**) 10.4 kV/cm, and (**d**) 14.4 kV/cm. The sharp dips at 0.5 THz in the measured data are due to water-vapor absorption [27]

emission is quite weak and the signal/noise ratio is low, the excellent agreement between experiment and theory further confirms the above conclusion.

3.4 Zener Tunneling into Higher Minibands and High-Frequency Limit of the Bloch Gain

Now, we would like to discuss what limits the maximum frequencies of the Bloch gain in SLs. The good agreement between theory and experiment discussed above was obtained only up to 13 kV/cm for the THz data shown in Fig. 12b. As F is further increased, the period of the Bloch oscillation becomes shorter but, at the same time, the amplitude gradually diminishes. Concomitantly, the Fourier spectra become very broad (see Fig. 12b). We attribute this behavior to the anticrossing of the WS states of the first miniband with those of the second miniband, i.e., Zener tunneling into the second miniband. Let us take a look at Fig. 10b again. It is noted that the THz emission becomes brighter with increasing field for $V_b > -0.8\,\mathrm{V}(F > 29\,\mathrm{kV/cm})$. This contradicts what is expected from simple field-induced Stark localization in a single miniband and indicates that electrons in the first miniband find a new conduction channel. This new channel is the Zener tunneling into the second miniband. Further support can be seen in the interband photocurrent spectra shown in Fig. 10a, where an anticrossing feature can be observed in the high-bias region, as indicated by vertical arrows [34]. From these data, it is concluded that the Zener tunneling into the higher miniband sets the high-frequency limit to the Bloch gain.

Hence, it is of prime importance to clarify the electronic structures of higher-lying minibands/minigaps in SLs and understand the electron transport in a strong bias field range. In this section, we will discuss the Zener tunneling effect in biased wide-miniband $GaAs/Al_{0.3}Ga_{0.7}As$ SLs in more detail.

The samples were the same $GaAs/Al_{0.3}Ga_{0.7}As$ SL m-i-n diode structures as those used in the previous section. We simply measured the intensity of the THz radiation by using a Si bolometer. Experiments were performed by using 100 fs laser pulses delivered from a mode-locked $Ti:Al_2O_3$ laser. The laser pulses were loosely focused ($\phi = 1$ mm) onto the sample surface at a polar angle of 70 °C. The pump photon energy was set to excite carriers at the bottom of the first miniband. The generated THz emission was collimated by off-axis parabolic mirrors and focused onto a Si bolometer.

Figure 17a shows the bolometer output for the THz radiation emitted from sample 1 as a function of the bias voltage, V_b, applied to the SL. The THz intensity starts rising at around $V_b = +0.83$ V, indicating that this voltage is the flatband voltage, V_{b0}, in this particular sample. The applied electric field F was estimated by subtracting V_{b0} from V_b and dividing by the total thickness of the SL region of 500 nm.

Two distinct regions can be seen in the figure; one region (I) is the THz emission for 0 kV/cm to 30 kV/cm. The THz emission in region I is due to intraminiband electron transport [25]. When $F \lesssim 8$ kV/cm, strong interwell coupling leads to the formation of energy minibands. Furthermore, the Bloch energy ($\equiv eFd$; d: the SL period) is less than the scattering broadening of the WS levels. Therefore, electrons drift in a quasicontinuous miniband. From a closer look at Fig. 17a, a weak kink in the THz emission curve is noticed at ≈ 8 kV/cm. For $F \gtrsim 8$ kV/cm, the miniband splits into equally spaced WS states and the wave functions are gradually localized (WS localization). In this intermediate field region, clear Bloch oscillations are observed in the time-domain THz traces [25, 27]. When F exceeds ≈ 18 kV/cm, the THz intensity starts rolling off due to stronger WS localization. The other region (II) is a broad THz-emission band observed for 30 kV/cm to 110 kV/cm. When F exceeds 30 kV/cm, the reduction of THz intensity saturates and the emission becomes brighter again. Furthermore, quasiperiodic structures are observed on top of the broad peak [35]. The dashed curve is a blowup of the THz intensity in the high field region. Since the THz intensity is proportional to the carrier acceleration, the THz emission becomes brighter whenever carriers

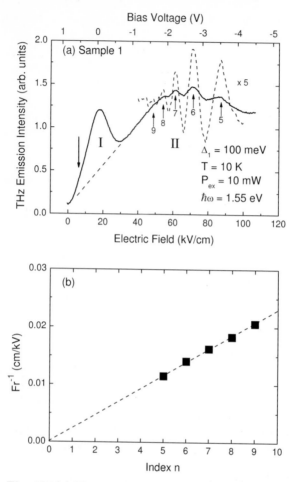

Fig. 17. (a) The bias-field dependence of the THz intensity measured for sample 1. The *dashed curve* shows a blowup of the THz intensity in the high-field region. The *arrow* at ≈ 8 kV/cm shows the position of a weak kink in the THz-intensity spectrum. (b) The inverse of the resonance electric fields plotted as a function of the index number (see Fig. 18) [35]

find new transport channels. We attribute the broad feature and the fine structure to the nonresonant and resonant interminiband Zener tunneling effects, respectively [34, 36, 37]. In the following, we will focus our attention on the fine structure in the THz-emission traces.

Let us consider the band diagram under a strong bias field, where the interminiband Zener tunneling takes place. Figure 18 shows a schematic illustration of the SL band diagram in a high-field condition. Equally spaced WS levels with an energy separation of eFd are formed for both the first and the second minibands. It is also noted that the energy separation between

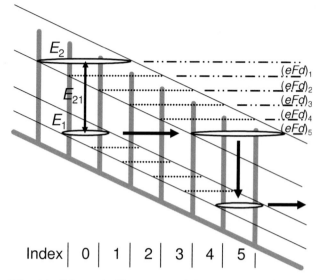

E_2

E_{21}

E_1

$(eFd)_1$
$(eFd)_2$
$(eFd)_3$
$(eFd)_4$
$(eFd)_5$

Index | 0 | 1 | 2 | 3 | 4 | 5 |

Fig. 18. Schematic illustration of the conduction-band diagram of the SL under a strongly biased condition [35]

the first and the second quantized energy levels in each quantum well (QW), E_{21} ($= E_2 - E_1$), is almost field independent, since the quantum-confined Stark effect is weak in narrow QWs. Consequently, the energy level of the ground state E_1 in the 0-th well is aligned with the excited state E_2 in the n-th well when F satisfies the condition; $E_{21} = neFd$, where the index n denotes the position of the QW numbered with respect to the QW where the wave function of the ground state is localized (see Fig. 18). Under such a condition, electrons in the first miniband can resonantly tunnel into the excited state in the n-th well and the THz emission is enhanced. In Fig. 17b, the inverse of the resonance electric field, F_r, indicated by arrows in Fig. 17a is plotted as a function of n. The linear dependence between F_r^{-1} and the index n indicates that the quasiperiodic peak structure is indeed due to the interminiband resonant Zener tunneling between the WS ladders associated with the first and the second minibands. Resonant Zener tunneling up into the 8th-nearest neighbor QW was clearly resolved, indicating that the wave function in the SL coherently extends over $\approx 60\,\mathrm{nm}$ (8 periods $\times$ 7 nm) [35].

4 THz Radiation From Semiconductor Microcavities in Strong Exciton–Photon Coupling Regime

4.1 Cavity-Polaritons and the Idea of THz Wave Radiation

When quantum wells (QWs) are embedded in a planar wavelength-scale optical cavity (microcavity), the light–matter interactions are largely modified owing to the confinement of the light field (a review is given, for example, in [38]). In particular, in high-Q semiconductor microcavities (SMCs) at low temperatures, where the damping of the cavity-mode photons and the exciton coherence are sufficiently slow, coherent coupling between the excitons and photons brings about the formation of two-dimensional exciton–polariton states called cavity-polaritons [39]. The coupling is often called "strong coupling". In such a system, neither the exciton nor the cavity-mode photon, but the two polariton states become the normal modes of the system. Because of the two-dimensional nature of the system, where the motions in the direction normal to the QW and cavity mirror planes are quantized, the cavity and exciton states, from which the polaritons are formed, are uniquely determined once an inplane wavevector $K_{\parallel}$ is given. Experimentally, $K_{\parallel}$ is determined by selecting the angle θ_{in} of the incident or emitted light with respect to the normal of the layer plane. Hence, the cavity-polariton modes are observed spectrally as the splitting (anticrossing) in the reflection spectra taken at various θ_{in} [39], and the dispersion curve can be drawn directly, as illustrated in Fig. 19.

The angular frequency of the two (upper and lower) polariton modes, ω_+ and ω_- are expressed as, neglecting the damping terms,

$$
\omega_{\pm} = \frac{\omega_{\mathrm{ex}} + \omega_{\mathrm{c}}}{2} \pm \frac{\sqrt{(\omega_{\mathrm{ex}} - \omega_{\mathrm{c}})^2 + \Omega_0^2}}{2} , \tag{5}
$$

where $\hbar\omega_{\mathrm{ex}}$ and $\hbar\omega_{\mathrm{c}}$, both functions of $K_{\parallel}$, are the energy of the exciton and the cavity mode, respectively. Practically, the cavity mode depends strongly on $K_{\parallel}$, whereas the exciton energy is effectively dispersionless for $K_{\parallel}$, as illustrated in Fig. 19. The splitting Ω_0 at zero detuning, $\delta = \omega_{\mathrm{ex}} - \omega_{\mathrm{c}} = 0$, which is sometimes called the "Rabi" frequency, is determined by the coupling strength between the exciton and the cavity mode, and is proportional to the square-root of the exciton oscillator strength, $\sqrt{f_{\mathrm{osc}}}$. In the case of GaAs-based SMCs, the splitting energy $\hbar\Omega_0$ is of the order of meV.

An excitation of such a system by a short laser pulse can trigger a complementary oscillation at the frequency $\Omega/2\pi = (\omega_+ - \omega_-)/2\pi$, in the exciton and photon populations. The oscillation, called normal mode oscillation (NMO), corresponds to the intuitive time-domain picture of polaritons, where the energy is exchanged periodically between the excitons and photons. In a sense that neither the exciton nor the photon state is the eigenstate of the system, the occurrence of NMO is natural, because the laser pulse is a photonic state. In GaAs-based SMCs, the frequency of NMO, $\Omega/2\pi$, lies in the THz regime. Alternatively, NMO can be understood as the beating in the polarization or the light field carried by the two polariton states, at ω_+ and ω_-. For example, at low excitation, the exciton density can be expressed as the square of the polarization, namely $n_{\mathrm{ex}} = |P_{\mathrm{ex}}|^2 = |P(\omega_+) + P(\omega_-)|^2$, which oscillates at the frequency $\Omega = \omega_+ - \omega_-$. Hence, one can note that, to initiate NMO, it is necessary to excite both the polariton modes coherently. NMO has been

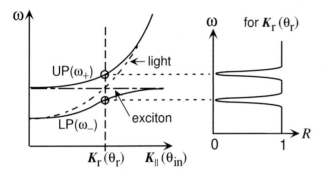

Fig. 19. Illustration of the dispersion curve of cavity-polariton modes and the surface reflection spectra for the incidence angle that corresponds to the inplane wave vector $K_{\parallel}$ producing the exciton-cavity resonance

observed in resonant luminescence [40, 41, 42] or pump-probe type measurements including four-wave-mixing [40, 43, 44, 45, 46]. In these experiments, however, all the relevant light beams are in the frequency range that is confined in the microcavity, and the signal could suffer from the higher-order interference effect.

On the other hand, it has been pointed out that THz EM waves are radiated from cavity-polaritons undergoing NMO [47]. The idea is as follows. In the presence of a static electric filed in the QW, the electron and hole wavefunctions are displaced in the opposite direction, as illustrated in Fig. 20. Since such a polarized exciton carries a static dipole, $\mu_{ex} = e(z_h - z_e)$, where $z_h = \langle h|z|h \rangle$ and $z_e = \langle e|z|e \rangle$ are the average position of a hole and electron, respectively, the oscillation in the exciton population N_{ex} due to NMO results in an oscillation of macroscopic dipole $D \propto N_{ex}\mu_{ex}$, which is expected to radiate EM waves, according to the well-known electromagnetic theory, at the frequency $\Omega/2\pi \sim$ THz. Therefore the radiation of the THz waves expressed as

$$E_{THz} \propto \frac{\partial^2}{\partial t^2}[N_{ex}\mu_{ex}] = \frac{\partial^2 N_{ex}}{\partial t^2}e(z_h - z_e),\tag{6}$$

can manifest the oscillation of N_{ex} associated with the NMO. In contrast to the pump-probe-type measurements mentioned above, the THz-emission experiments, where the signals are not confined in the cavity, allow us to observe the signal free from the unwanted interference effects.

The radiation of THz waves from cavity-polaritons can be also understood as a kind of optical rectification in the second-order nonlinear process, since it comes from the oscillation n_{ex}, which is proportional to $|P_{ex}|^2$ as mentioned above. From the viewpoint of optical nonlinearity, to obtain second-order polarization, spatial inversion symmetry of the system must be broken, which is indeed the case ($\Delta z \neq 0$) in the QW excitons under a static electric field, as shown in Fig. 20.

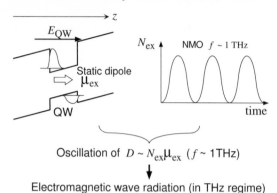

Oscillation of $D \sim N_{ex} \mu_{ex}$ ($f \sim 1$THz)
$\downarrow$
Electromagnetic wave radiation (in THz regime)

Fig. 20. The mechanism of THz wave radiation from cavity-polaritons undergoing NMO

4.2 The Microcavity Samples and the Cavity-Polariton Modes under a Static Electric Field

The samples used here are the SMC devices [48, 49], as illustrated in Fig. 21a, where two GaAs QWs separated by an AlAs barrier are embedded near the center of the half-wavelength cavity made of highly reflective distributed Bragg reflectors (DBRs), which is composed of alternating quarter-wavelength AlAs and $Al_{0.2}Ga_{0.8}As$ layers. The number of layers in the air- and substrate-side DBR are 33 and 58, respectively, for which the Q-value of the empty cavity was evaluated to be ≈ 3700 from the surface reflection spectrum of the sample in which the QW can be regarded to be transparent ($\omega_c \ll \omega_{ex}$). The substrate-side DBR is doped with Be while in the air-side DBR, only a part (12 nm) of an AlGaAs layer is doped with Si, to reduce the absorption for the THz wave.

A DC voltage, V_{DC} was applied between the Si-doped conducting Al-GaAs layer, and the p-type substrate. In Fig. 21b, the energy levels of the exciton-cavity coupled system is schematically illustrated as functions of the DC electric field in the QW, E_{QW}. Here, only a lowest heavy-hole exciton state is considered. With the increase of E_{QW}, the exciton transition frequency ω_{ex} is redshifted through the quantum-confined Stark effect (QCSE), while the cavity resonance ω_c is determined by the incidence angle of the excitation light θ_{in}, independently of E_{QW}. Around the exciton-cavity resonance $\omega_{ex} \sim \omega_c$, the anticrossing shows up, corresponding to that in the polariton dispersion curve as shown in Fig. 19. By changing E_{QW} (V_{dc}), while keeping ω_c (θ_{in}), the polariton modes with various detuning $\delta = \omega_{ex} - \omega_c$ are realized. On the other hand, by the combined change of V_{dc} and θ_{in}, $\delta = 0$ is achieved at various values of E_{QW}. In the tuning of E_{QW}, f_{osc} varies also due to QCSE so that Ω_0 can be also modified. Hence, in the experiment, Ω was varied through the change of δ and Ω_0.

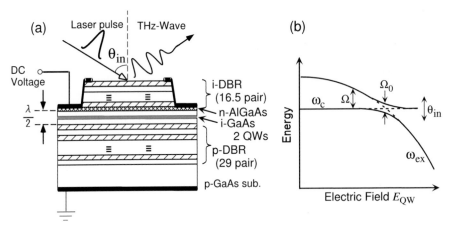

Fig. 21. (a) Schematic drawing of the high-Q SMC device used for THz wave radiation. (b) Schematic drawing of the polariton modes under a static electric field

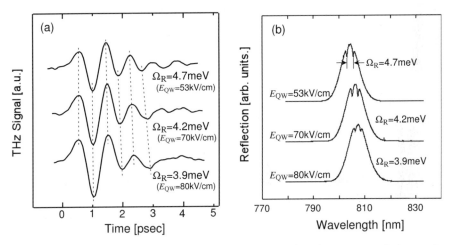

Fig. 22. (a) Examples of the detected THz signal for three cases of the mode splitting, all at $\delta = 0$ realized by the combined tuning of E_{QW} and θ_{in}. (b) Reflection spectra of the excitation laser pulses, for the three cases shown in (a). The two dips in each spectrum correspond to the polariton modes

4.3 Observation of THz Wave Emissions from Cavity-Polaritons

In the THz-wave-generation experiment, the sample was cooled to 10 K in a cryostat, and the time-domain traces of the THz waves were measured in a standard THz time-domain spectroscopy (TDS) setup using a mode-locked Ti-sapphire laser and a photoconductive antenna.

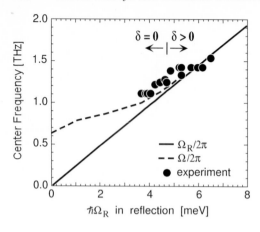

Fig. 23. Frequency of the THz wave is plotted as a function of the mode splitting Ω_R observed in the reflection spectra. The *solid line* shows the frequency Ω_R/π. The *dashed curve* shows the frequency Ω/π, where the internal value of the mode splitting Ω was calculated from Ω_R, taking into account the damping effect

Figure 22a shows the examples of the detected signals for three cases of the mode splitting at the exciton-cavity resonance, $\delta = 0$, realized by the combined tuning of V_{dc} and θ_{in}, for which the splitting depends on E_{QW} through the change of Ω_0. The values of Ω_R indicated in the figure are the frequency difference between the two polariton modes seen in the reflection spectra as shown in Fig. 22b. The Ω_R value is very close to, but can be slightly different from the true mode splitting $\Omega = \omega_+ - \omega_-$ ($= \Omega_0$ for $\delta = 0$) calculated from (5) as mentioned below. In all the cases, several-cycle oscillations are clearly seen, manifesting the oscillation of exciton population associated with the NMO. The dashed lines connecting the corresponding maxima and minima were drawn to guide the reader's eye. Obviously, for the smaller splitting $\Omega_R(\sim \Omega)$, the oscillation gets slower, as expected for the NMO. Larger damping of the oscillation at higher E_{QW} corresponds to the shallower polariton-mode dips seen in Fig. 22b. In the present case, the increase of the damping rate is due to the increase of the dephasing rate of the exciton.

The frequency of the THz signals evaluated as the inverse of the oscillation period is summarized in Fig. 23 as a function of Ω_R. As mentioned above, Ω (and Ω_R) can be also modified with the change of δ. The data points in the figure include both the cases of exciton-cavity resonance ($\delta = 0$, Ω_0: varied) and offresonance (δ: varied, $\Omega_0 \sim$ constant). For the offresonant cases, the exciton was detuned to the higher-energy side of the cavity mode ($\omega_c < \omega_{ex}$), to avoid the influence by the light-hole exciton and continuum states. Basically, the experimental data points are fitted quite well by the solid line representing the values of $\Omega_R/2\pi$.

The observed frequency, however, deviates from the straight line, particularly in small Ω_R cases. Recalling that the decrease of Ω_R is realized at $\delta = 0$ by the decrease of Ω_0 with the increase of E_{QW} (see Fig. 22), we can draw an important conclusion. In general, Ω_R is different from Ω, which is the intrinsic frequency of NMO, because of the finite linewidths (dampings) of the cavity and exciton modes, particularly when the linewidth is not sufficiently smaller than Ω [50]. In the present sample, the linewidths of the excitons and the cavity resonance are about $2\,\mathrm{meV}$ and $0.6\,\mathrm{meV}$, respectively, for which the arithmetic average, $1.3\,\mathrm{meV}$, is not very small compared with $\Omega \approx 4\,\mathrm{meV}$. The exciton linewidth increases gradually with the increase of E_{QW} [47]. Taking into account the linewidth, we calculated Ω for each Ω_{R} and plotted it by the broken curve in Fig. 23. Though the fitting is not perfect, the difference between Ω_{R} and Ω accounts qualitatively for the observed deviation of the data points from the Ω_{R} line. In other words, the frequency of the emitted THz wave is not that of the beating between the optical fields, whose frequency is given by Ω_{R}, but is equal to the frequency of the NMO. Hence, we can say that, in the THz-emission technique, the oscillation of the exciton population due to NMO is directly observed.

In order to confirm the prediction, (6), the E_{QW} and excitation power dependencies of the EM wave amplitude were studied. In the experiment, the exciton-cavity resonance ($\delta = 0$) was kept constant by the tuning of the incidence angle, or E_{QW}, and the center frequency of the excitation pulses was adjusted to the polariton doublets. Figure 24a shows the EM wave amplitude as a function of E_{QW}. The amplitude increases monotonically with the increase of E_{QW}, demonstrating the significance of the application of a static electric field. In the figure, the relative change of the displacement between the electrons and holes $\Delta z = \langle 1hh|z|1hh \rangle - \langle 1e|z|1e \rangle$, which is proportional to the exciton self-dipole μ_{ex}, calculated for the $12\,\mathrm{nm}$ thick GaAs is also shown by the solid curve. The observed dependence of the THz amplitude on E_{QW} corresponds quite well to the increase of Δz, supporting the prediction, (6). Shown in Fig. 24b is the dependence of the amplitude on the excitation laser power P_{ex}, to which the exciton population N_{ex} is proportional. The data points clearly fall on the line $E_{\mathrm{THz}} \propto P_{\mathrm{ex}}$ manifesting again the validity of (6). Then, combining the two dependencies, we can summarize that the THz waves are radiated as a result of the second-order nonlinear response caused by the field-induced break of the inversion symmetry.

4.4 Transition from Strong- to Weak-Coupling Regime

As demonstrated clearly in the preceding sections, the THz waves are radiated from the coherent nutation of the exciton population associated with the NMO of the cavity-polariton system. Such coherent nutation decays due to the dephasing of the exciton coherence or the leakage of the light field from the cavity. The effect of the increase in the exciton dephasing has been shown, though not clearly, in Fig. 22, as the faster damping of the oscillation for higher E_{QW}. Here, we discuss the influence of the cavity-mode damping. For this purpose, we prepared SMC samples where the number of layers in the air-side DBR, N_{air} was reduced to 23 and 13 (originally 33), for which the Q-values of the empty cavities are about 1100 and 240, respectively (originally 3700). In the lowest Q-value sample, an anticrossing is not observed

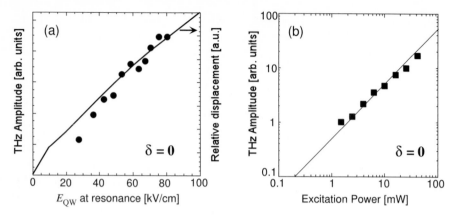

Fig. 24. Amplitude of the THz wave is plotted as a function of (a) the static electric field in the QW, E_{QW}, and (b) the excitation power P_{ex}. The experiments were performed keeping the exciton-cavity resonance ($\delta = 0$). In (a) the calculated variation of the relative displacement Δz between the electron and hole is shown by the *solid curve* and in (b) the *solid line* is drawn to show the dependence $E_{THz} \propto P_{ex}$

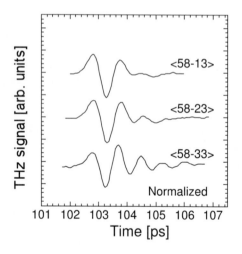

Fig. 25. THz signal obtained with various Q-value samples. The number n in the labels $\langle 58\text{-}n\rangle$ indicates the number of layers in the air-side DBR. The signals were normalized by the first half-cycle

in the reflection spectrum (not shown). Shown in Fig. 25 are the observed THz signals for these samples. To see clearly the decay of the oscillation, the signals were normalized by the amplitude of the first half-cycle. Obviously the decay is fast in low-Q samples, and nearly one cycle is left in the lowest Q-value sample. The first one cycle results from the instantaneous creation of the excitons by the laser pulses, which is observed even in the excitation of a single exciton or continuum states in bare QWs [51]. In other words, it is the oscillation following the first cycle that manifests the coherent nutation of the exciton population.

5 THz Radiation from Bulk Semiconductor Microcavities

5.1 THz Wave Radiation from Semiconductor Surfaces: A Revisit

The excitation of unbiased semiconductor surfaces by short laser pulses is an attractive method for obtaining THz EM waves [52], because of the simplicity of the setup and the robustness against the excitation-induced damage of the emitter. In the case of rather low excitation density, where the optical rectification effect [53] is negligible, the source of the THz wave is the surge current carried by impulsively excited carriers near the surface [52]. In wide-gap semiconductors such as GaAs and InP where the built-in field is rather high, the current stems from the drift motion of the photoexcited carriers by the built-in surface electric field. Depending on the excitation pulse width with respect to the period of carrier acceleration, the time derivative of the current, to which the far-field THz wave is proportional, is dominated either by the variation of the velocity [9] or of the density [52] of the photoexcited carriers. In both cases, the strength of the static electric field and the total number of carriers excited in the field region are the important parameters determining the amplitude of the THz waves.

The built-in surface electric field is caused by the difference in the Fermi levels between the surface and the bulk region. In most III–V semiconductors, the Fermi level at the surface is pinned at a certain energetic position, while it lies around the conduction band edge in n-doped bulk region. In off-the-shelf bulk semiconductors, only the built-in potential difference can be chosen. For a given built-in potential difference, a thinner field region is beneficial for obtaining a higher field, while it is unfavorable for exciting larger number of carriers, as illustrated in Figs. 26a and b. Hence, there exists a sort of tradeoff relation between the field strength and the photocarrier number.

In an optical cavity, in contrast, the excitation pulses can be absorbed efficiently during the round trips, even though a thin absorption layer is involved, allowing us to realize a rather high static field with only the built-in potential, as illustrated in Fig. 26c. As a result, one may expect larger THz wave amplitude for a given excitation light intensity. In this section, we show that the microcavity structures, which have been described in the previous section, can be used also for the enhancement of the THz waves, by replacing the QWs by a bulk GaAs [54]. In addition, we discuss an influence of the carrier accumulation on the radiation of THz waves that has not been considered yet.

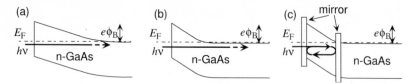

Fig. 26. Illustration of the potential profile of n-GaAs and SMC near the surface at the thermal equilibrium, displaying that (**a**) the light absorption is efficient but the field is low in a thick depletion region, (**b**) the field is high but the absorption is inefficient in a thin depletion region, (**c**) both a high field and efficient absorption is realized in the SMC

5.2 Enhancement of THz Wave Generation Efficiency

The cross-sectional structure and the energy-band diagram around the THz-wave generating region of the SMC samples considered here are schematically drawn in Fig. 27a. A 182.6 nm thick undoped GaAs and a 50 nm thick Si-doped ($\sim 2 \times 10^{18}$ cm^{-3}) GaAs are sandwiched by a pair of DBRs, forming a one-wavelength cavity. The substrate-side DBR is doped with Si, while the air-side one is undoped. The THz waves are radiated from the surge current created by the laser pulses in the undoped GaAs layer, where a static electric field is built-in, as in the case of bare n-type GaAs substrate. In order to obtain a higher field in the undoped GaAs region, the Fermi-level pinning is desired to take place not at the surface but at the interface between the undoped GaAs layer and the air-side DBR. Such an intentional pinning was introduced in the sample preparation by interrupting the molecular beam epitaxy growth and exposing the GaAs surface to the room air.

Assuming that the energetic position of the Fermi level at the interface is the same as that at the bare GaAs surface (≈ 0.7 eV below the conduction-band edge), and that in the n-GaAs region is located at the band edge, the built-in electric field E_B at thermal equilibrium is expected to be 38 kV/cm. For the present purpose, not a very high but a low-Q cavity is better, as shown later. Hence, in the experiments, two types of cavity were tested; one having 11.5 and 4 AlAs/Al$_{0.2}$Ga$_{0.8}$As pairs in the substrate- and air-side DBRs, respectively, referred to as a (23-8) cavity, and the other having 11.5 and 2 pairs in each side, referred to as a (23-4) cavity. The results were compared with those obtained with a reference sample, whose structure is shown in Fig. 27b, and an InAs sample, in which an undoped 500 nm thick InAs has been grown on semi-insulating GaAs substrate by molecular beam epitaxy. Figure 28 shows the surface reflection spectra of the cavity samples for an incidence angle of 45°. The cavity resonance is clearly seen around 820 nm, where the reflectivity is very low, as expected. In particular, in a (23-4) cavity, nearly zero reflectance is realized.

The THz-wave generation experiments were performed with a standard THz TDS setup where a mode-locked Ti-sapphire laser and a photocon-

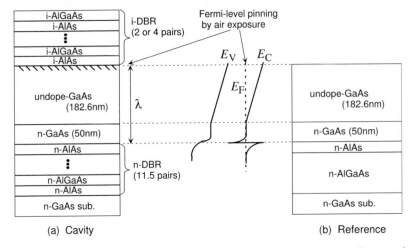

Fig. 27. Schematic drawing of the bulk SMC and reference samples together with the potential profile at the thermal equilibrium

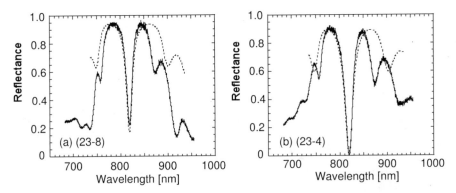

Fig. 28. Surface reflection spectra of bulk SMC devices for 45° incidence; (**a**) (23-8) cavity, (**b**) (23-4) cavity. The *solid* and *dotted curves* are the experimental results and theoretical predictions, respectively

ductive antenna were used. For the excitation, the laser pulses centered at ≈ 820 nm were incident on the samples with an angle of 45° for which the cavities satisfy the resonant condition. The power reflectance of a (23-8) cavity, a (23-4) cavity, the reference, and InAs samples for this condition were 41%, 17%, 40%, and 44%, respectively. The amplitude of the THz wave was evaluated from the difference between the positive and negative peaks in the time-domain signals.

Shown in Fig. 29 are the amplitudes of the THz waves obtained with (23-8) and (23-4) cavities, the reference, and InAs samples as functions of the excitation power. The excitation spot is about 0.8 mm in diameter. In

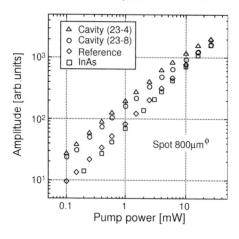

Fig. 29. The amplitude of the THz waves obtained with a (23-8) cavity, a (23-4) cavity, a reference, and InAs samples as functions of the excitation power [54]

the low-excitation-power regime, the amplitudes of the cavity samples are obviously larger than that obtained with the reference and InAs samples. With the increase of the excitation intensity, the amplitude obtained with InAs increases linearly, while the increase is obviously sublinear in the case of cavities, even in the low-excitation regime, and the deviation from the linear relation becomes larger in the high-excitation regime. The sublinear relation can be explained by the screening of the static field by the photoexcited carriers as discussed in the next section. By taking into account the difference of the reflectance between the samples, and comparing the results at the same net excitation power, which penetrate into the samples, one can find that the difference between the cavities becomes negligible and that the amplitude of the cavities are 2.6 times larger than that of the reference sample, and 3.1 times larger than that of InAs.

Assuming the absorption coefficient in GaAs to be $\alpha = 1.2 \times 10^4 \, \text{cm}^{-1}$ [55], the absorbed power in the undoped GaAs region in the reference sample is evaluated to be 20% of the net excitation. In the cavity samples, assuming that all the net power is absorbed inside the cavity (no transmission) and taking the standing-wave pattern into account, the absorption in the undoped GaAs region can be evaluated roughly to be 75%. From these values, one may expect an enhancement factor of 3.8, if the amplitude of the THz wave is proportional to the total number of the carriers excited in the undoped GaAs region. The smaller experimental value may be due to the screening effect, which exists even in the lowest-excitation case, as indicated by the sublinear dependence of the amplitude on the excitation power.

One point that should be mentioned is that the THz amplitude obtained with the reference sample is the same as or even stronger than that with InAs. On the other hand, the THz amplitude obtained with n-type GaAs doped with Si to $2 \times 10^{18} \, \text{cm}^{-3}$ is about a factor of three smaller than the undoped InAs (not shown here). Hence, the results indicate clearly that the

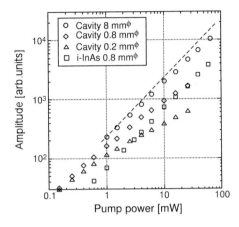

Fig. 30. The amplitude of THz waves obtained with various excitation spot size, as functions of the excitation power. The *dashed line* indicates the linear relationship [54]

THz intensity depends on the doping profile, and that a comparison between different materials needs care due to the difference of the potential as well as the electronic structures.

From an application point of view, it is important to realize the enhancement even in the high-excitation-power regime. If the photocarrier-induced screening is the origin of the sublinear dependence of the THz wave amplitude on the excitation intensity shown in Fig. 29, one simple way to increase the efficiency at a high excitation power is to use a larger excitation spot. Figure 30 shows the excitation-power dependence of the THz amplitude for various spot sizes, obtained with a (23-8) cavity, in comparison with the case of InAs. As anticipated, the deviation from the linear relationship between the excitation power and the amplitude is smaller for a larger excitation spot. In particular, for the 8 mm spot case, the amplitude of the THz wave obtained with the cavity increases almost linearly with the excitation intensity, and is still a factor of 3 (factor of 9 in intensity) larger than that with InAs, at the excitation of 65 mW. However, as can be seen by a comparison with the linear relationship indicated by the broken line in the figure, the dependence of the amplitude on the excitation intensity is still slightly sublinear. Hence, to utilize the cavity-induced enhancement even at higher excitation intensities, a larger spot should be used.

5.3 Field Screening by Carrier Accumulation

In the 8 mm spot case shown in Fig. 30, even at the highest excitation, 65 mW, the carrier density created by a single pulse is about 8×10^9 cm^{-2}, for which the screening field, ≈ 1.2 kV/cm, is much smaller than the equilibrium field, $E_B \approx 38$ kV/cm. Hence, the screening associated with each excitation pulse cannot explain the sublinear dependence. Rather, for such a case, one may need to take into account the accumulation of the photoexcited carriers. In the microcavity samples, the photoexcited electrons and holes accumulate

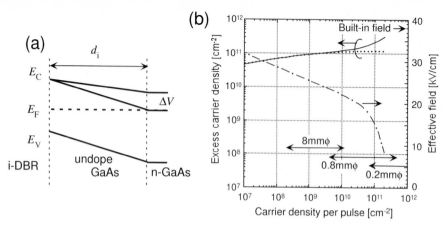

Fig. 31. (a) The calculation model for the recovery process. (b) The excess carrier density, n_0, just before the arrival of each pulse (*dotted curve*) and that, $n_0 + n_P$, just after the arrival (*solid curve*), and the effective electric field E_{eff} for the THz wave radiation as functions of the carrier density n_P created by each pulse [54]

in the n-doped GaAs region and the interface between the undoped GaAs and the air-side DBR, respectively. They may recombine inside the device, through the transportation across the undoped GaAs region. However, if the recombination does not complete between the pulses, an amount of the excess carriers may be left.

Here, to make the analysis simple, we treat the transportation as the thermionic emissions of electrons from the n-GaAs layer to the DBR/GaAs interface, considering the simplified potential profile shown in Fig. 31a. The nonequilibrium voltage ΔV induced by the excess carriers of sheet density $n_s(t)$ is expressed as $\Delta V(t) = [en_s(t)/\epsilon]d_i$, where e, ϵ and d_i are the elementary charge, dielectric constant, and the thickness of the undoped GaAs layer, respectively. The thermionic current density is expressed as [56] $J_{te}(t) = J_s[\exp(e\Delta V(t)/k_BT)-1]$, where J_s and k_BT/e are the saturation current density and the thermal energy, respectively. Neglecting the carrier transport during each excitation pulse, the time dependence of n_s after an impulse excitation can be obtained by solving the equation, $\partial n_s(t)/\partial t = -J_{te}(t)/e$. Then, for the steady-state situation where the carriers of density n_P are created periodically by each pulse with the interval, $\tau_{ML} = 13$ ns, of the mode-locking laser, the carrier density n_0 remaining just before the arrival of each pulse is obtained. Using thus-calculated values of n_0, the effective static electric field, by which the photoexcited carriers are accelerated, was evaluated as $E_{eff} = E_B - e(n_0 + n_P/2)/\epsilon$, where a half of n_P is included as a rough estimation of the average field during an excitation pulse.

Figure 31b shows the estimated values of n_0, $n_0 + n_P$, and E_{eff} as functions of n_P, using the values $J_s = 1.5 \times 10^{-6}$ A/cm^2 [56], $T = 300$ K, and the dielectric constant of GaAs, 13.1. One can notice that, even for a weak excitation, $n_P \ll 10^{11}\,\mathrm{cm}^{-2}$, the accumulated excess carrier density n_0 is quite high, $n_0 \sim 10^{11}\,\mathrm{cm}^{-2}$, and the effective field E_{eff} is apparently smaller than the equilibrium value, $38\,\mathrm{kV/cm}$. This can be understood as follows. In the early stage of the recovery process, the thermionic current is high because of large ΔV. However, with the decrease of ΔV, the current decreases exponentially and the recovery process slows down drastically, remaining rather high n_0 at the arrival of the next excitation pulse. Hence, the value of ΔV, and hence the current density, at a time long after each excitation pulse, is not dominated by the initial carrier density, n_0 and ΔV depend weakly on the carrier density per pulse, n_P. The gradual decrease of E_{eff} with increasing n_P results in the weakly sublinear dependence of the THz amplitude on the excitation power. As indicated by the arrows in the figure, the experiment with the 8 mm spot and that with the 0.8 mm spot under the weak excitation in Fig. 30 correspond to such a regime. Therefore, in the low-excitation-density regime, the screening of the built-in surface electric field by the accumulated photocarriers explains qualitatively the observed sublinear dependence of the THz amplitude on the excitation intensity.

In contrast, in the high-excitation-density condition, $n_P > 10^{11}\,\mathrm{cm}^{-2}$, which corresponds to the 0.2 mm spot case and the excitation $> 6\,\mathrm{mW}$ in the 0.8 mm spot case in Fig. 30, $n_0 + n_P$ is dominated by n_P and increases rapidly, and E_{eff} drops steeply, resulting in a strong saturation of THz wave amplitude. In such a case, however, the recovery current during each pulse can not be neglected, since ΔV becomes large (the potential barrier for the thermionic emission becomes small), and the present simple calculation does not predict correctly the experimental results. Possibly, as observed experimentally, due to the instantaneous recovery, the THz amplitude may increases gradually with the increase of the excitation intensity.

6 Summary

In this Chapter, recent topics on the time-domain THz-emission studies of bulk and quantum semiconductor structures were presented, including the discussion on the applications of such structures as the emitter of THz waves.

In Sect. 2, the nonequilibrium carrier transport in DC-biased bulk GaAs were discussed quantitatively on the basis of the time-domain THz-emission experiment compared with the ensemble Monte Carlo calculations. As a result, it was proven that the finiteness of sample thickness has to be taken into account in the analysis of the THz data. Furthermore, from the discussion on the steady-state carrier velocity, we showed that the contribution of hole transport becomes significant for a DC-bias higher than $20\,\mathrm{kV/cm}$. It was then concluded that, when actual experimental conditions (sample geometry, photoexcitation condition, etc.) are properly taken into account, Monte Carlo calculations give a very good description of the transient carrier transport determined by the THz spectroscopy.

Section 3 was devoted to the THz wave emissions from semiconductor superlattices. In particular, we focused on a long-standing question, i.e., Bloch gain in semiconductor superlattices. To answer this question experimentally, we have showed that the time-domain THz spectroscopy is very useful in determining gain/loss spectra in the THz regime. By noting that the time-domain THz emission spectroscopy inherently measures the step response of the electron system to the bias electric field, the obtained THz spectra were compared with the high-frequency conductivities predicted for miniband transport. Excellent agreement between theory and experiment strongly supports that the THz gain due to Bloch-oscillating electrons persists at least up to 1.7 THz. It was also shown that Zener tunneling into the second miniband sets the high-frequency limit to the THz gain for the samples studied here.

Radiation of THz waves from high-Q semiconductor microcavities, where the quantum-well excitons and the cavity-mode photons are strongly coupled, was presented in Sect. 4. A systematic investigation on the frequency and amplitude of the THz waves showed clearly that the observed THz waves are radiated from the coherent oscillation of the exciton population, and that the radiation can be viewed as a kind of second-order nonlinear response. In addition, the transition from strong- to weak-coupling regime with the decrease of the cavity Q-value was demonstrated as the decay of the oscillation in the THz wave.

In Section 5, we discussed the radiation of THz waves from low-Q bulk semiconductor microcavities containing a bulk GaAs layer. It was shown that, in the low-excitation-density regime, $< 10^{10}$ cm^{-2} photons per pulse, the use of an optical microcavity enhances the THz intensity in comparison with the bare semiconductor surfaces, owing to the efficient use of the excitation pulses as well as to the reduction of the reflection of the excitation pulses at the surface. In addition, some results of a numerical estimation were shown to point out that the screening of the surface DC field by the accumulated photocarriers can cause sublinear dependence of the THz amplitude on the excitation intensity, even in the low-excitation-density regime.

As demonstrated in this Chapter, the time-domain THz-emission studies provide rich information on the temporal as well as spatial dynamics of photoexcited carriers in semiconductors on the subpicosecond time scale, where the nonstationary and/or coherent effects play a significant role. In addition to the excellent time resolution, the technique has a feature that the obtained waveform is related directly to the photocarrier dynamics. Hence, the technique is unique and useful, and expected to be used in the future even more widely. On the other hand, the use of various semiconductor structures is expected to lead to the improvement or the development of the devices for the emission, detection, and control of THz waves, which may be required in the future high-performance, low-cost, compact, and functional THz systems. Further investigations in such a direction are strongly desired.

References

[1] W. Sha, J. Rhee, T. Norris, W. Schaff: IEEE J. Quantum Electron. **28**, 2445 (1992)
[2] A. Iverson, G. Wysin, D. Smith, A. Redondo: Appl. Phys. Lett. **52**, 2148 (1988)
[3] B. Hu, E. de Souza, W. Knox, J. Cunningham, M. Nuss, A. Kuznetsov, S. Chuang: Phys. Rev. Lett. **74**, 1689 (1995)
[4] H. Roskos, M. Nuss, J. Shah, K. Leo, D. Miller, A. Fox, S. Schmitt-Rink, K. Köhler: Phys. Rev. Lett. **68**, 2216 (1992)
[5] P. Planken, M. Nuss, I. Brener, K. Goossen, M. Luo, S. Chuang, L. Pfeiffer: Phys. Rev. Lett. **69**, 3800 (1992)
[6] C. Waschke, H. Roskos, R. Schwedler, K. Leo, H. Kurz, K. Kohler: Phys. Rev. Lett. **70**, 3319 (1993)
[7] P. Bolivar, F. Wolter, A. Müller, H. Roskos, H. Kurz, K. Köhler: Phys. Rev. Lett. **78**, 2232 (1997)
[8] Y. Yamashita, A. Endoh, K. Shinohara, K. Hikosaka, T. Matsui, S. Hiyamizu, T. Mimura: IEEE Electron Dev. Lett. **23**, 573 (2002)
[9] A. Leitenstorfer, S. Hunsche, J. Shah, M. Nuss, W. Knox: Phys. Rev. Lett. **82**, 5140 (1999)
[10] A. Leitenstorfer, S. Hunsche, J. Shah, M. Nuss, W. Knox: Phys. Rev. B **61**, 16 642 (2000)
[11] M. Abe, S. Madhavi, Y. Shimada, Y. Otsuka, K. Hirakawa: Appl. Phys. Lett. **81**, 679 (2002)
[12] Q. Wu, X.-C. Zhang: IEEE J. Sel. Top. Quantum Electron. **2**, 693 (1996)
[13] Q. Wu, X.-C. Zhang: Appl. Phys. Lett. **70**, 1784 (1997)
[14] A. Leitenstorfer, S. Hunsche, J. Shah, M. Nuss, W. Knox: Appl. Phys. Lett. **74**, 1516 (1999)
[15] G. Gallot, J. Zhang, R. McGowan, T.-I. Jeon, D. Grischkowsky: Appl. Phys. Lett. **74**, 3450 (1999)
[16] E. Palik: *Handbook of Optical Constants of Solids 2* (Academic, New York 1991)
[17] K. Tomizawa: *Numerical Simulation of Submicron Semiconductor Devices* (Artech House, Boston 1993)
[18] C. Jacoboni, P. Lugli: *The Monte Carlo Method for Semiconductor Device Simulation* (Springer, Berlin, Heidelberg 1989)
[19] F. Bloch: Z. Phys. **52**, 555 (1928)
[20] C. Zener: Proc. R. Soc. London Ser. A **145**, 523 (1934)
[21] L. Esaki, R. Tsu: IBM J. Res. Dev. **14**, 61 (1970)
[22] J. Feldmann, K. Leo, J. Shah, D. Miller, J. Cunningham, T. Meier, G. von Plessen, A. Schulze, P. Thomas, S. Schmitt-Rink: Phys. Rev. B **46**, 7252 (1992)
[23] T. Dekorsy, P. Leisching, K. Kohler, H. Kurz: Phys. Rev. B. **50**, 8106 (1994)
[24] F. Loser, Y. Kosevich, K. Kohler, K. Leo: Phys. Rev. B **61**, 13373 (2000)
[25] Y. Shimada, K. Hirakawa, S.-W. Lee: Appl. Phys. Lett. **81**, 1642 (2002)
[26] G. Bastard, R. Ferreira: Cr. Acad. Sci. II **312**, 971 (1991)
[27] Y. Shimada, K. Hirakawa, M. Odnoblioudov, K. Chao: Phys. Rev. Lett. **90**, 46806 (2003)
[28] R. Kersting, K. Unterrainer, G. Strasser, H. Kauffmann, E. Gornik: Phys. Rev. Lett. **79**, 3038 (1997)

[29] H. Kroemer: On the nature of the negative-conductivity resonance in a super-lattice bloch oscillator, arXiv: cond-mat/0007428 (2000)

[30] H. Willenberg, G. H. Dohler, J. Faist: Phys. Rev. B **67**, 085315 (2003)

[31] S. Allen, et. al.: Semicond. Sci. Technol. **7**, B1 (1992)

[32] M. Helm: Semicond. Sci. Technol. **10**, 557 (1995)

[33] S. Ktitorov, G. Simin, V. Sindalovskii: Sov. Phys. Solid State **13**, 1872 (1971)

[34] B. Rosam, K. Leo, M. Gluck, F. Keck, H. Korsch, F. Zimmer, K. Kohler: Phys. Rev. B **68**, 125301 (2003)

[35] Y. Shimada, N. Sekine, K. Hirakawa: Appl. Phys. Lett. (2004) in press

[36] A. Sibille, J. Palmier, F. Laruelle: Phys. Rev. Lett. **80**, 4506 (1998)

[37] M. Helm, W. Hilber, G. Strasser, R. D. Meester, F. Peeters, A. Wacker: Phys. Rev. Lett. **82**, 3120 (1999)

[38] Y. Kadoya: *Optical Properties of Low-Dimensional Materials 2* (World Scientific, Singapore 1998) Chap. 7

[39] C. Weisbuch, M. Nishioka, A. Ishikawa, Y. Arakawa: Phys. Rev. Lett. **69**, 3314 (1992)

[40] T. Norris, J. Rhee, C. Sung, Y. Arakawa, M. Nishioka, C. Weisbuch: Phys. Rev. B **50**, 14663 (1995)

[41] H. Cao, J. Jacobson, G. Björk, S. Pau, Y. Yamamoto: Appl. Phys. Lett. **66**, 1107 (1995)

[42] J. Jacobson, S. Pau, H. Cao, G. Björk, Y. Yamamoto: Phys. Rev. A **51**, 2542 (1995)

[43] H. Wang, J. Shah, T. Damen, W. Jan, J. Cunningham, M. Hong, J. Mannaerts: Phys. Rev. B **51**, 14713 (1995)

[44] G. Bongiovanni, A. Mura, F. Quochi, S. Gurtler, J. Staehli, F. Tassone, R. Stanley, U. Oesterle, R. Houdre: Phys. Rev. B **97**, 7084 (1997)

[45] M. Koch, J. Shah, T. Meier: Phys. Rev. B **57**, R2049 (1998)

[46] H. Wang, H. Hou, B. Hammons: Phys. Rev. Lett. **81**, 3255 (1998)

[47] Y. Kadoya, K. Kameda, M. Yamanishi, T. Nishikawa, T. Kannari, T. Ishihara, I. Ogura: Appl. Phys. Lett. **68**, 281 (1996)

[48] Y. Hokomoto, Y. Kadoya, M. Yamanishi: Appl. Phys. Lett. **74**, 3839 (1999)

[49] M. Kusuda, S. Tokai, Y. Hokomoto, Y. Kadoya, M. Yamanishi: Physica B **272**, 467 (1999)

[50] V. Savona, F. Tassone: Solid State Commun. **95**, 673 (1995)

[51] P. Planken, M. Nuss, W. Knox, D. Miller, K. Goossen: Appl. Phys. Lett. **61**, 2009 (1992)

[52] X.-C. Zhang, B. Hu, J. Darrow, D. Auston: Appl. Phys. Lett. **56**, 1011 (1990)

[53] S. Chuang, S. Schmitt-Rink, B. Greene, P. Aseta, A. Levi: Phys. Rev. Lett. **68**, 102 (1992)

[54] T. Sakurada, Y. Kadoya, M. Yamanishi: Jpn. J. Appl. Phys. **41**, L256 (2002)

[55] H. Casey, M. Panish: *Heterostructure Lasers* (Academic, New York 1978) p. Pt. A 153

[56] S. Sze: *Physics of Semiconductor Devices*, 2nd ed. (Wiley, New York 1981) Chap. 5

Generation of CW Terahertz Radiation with Photomixing

Shuji Matsuura[1,2,3] and Hiroshi Ito[4]

[1] National Institute of Information and Communications Technology
588-2 Iwaoka, Nishi-ku, Kobe 651-2492, Japan
[2] California Institute of Technology
1200 E. California Blvd., Pasadena, CA 91125, USA
[3] *Present address:* Institute of Space and Astronautical Science,
Japan Aerospace Exploration Agency
3-1-1 Yoshinodai, Sagamihara, Kanagawa 229-8510, Japan
matsuura@ir.isas.jaxa.jp
[4] NTT Photonics Laboratories
3-1 Morinosato Wakamiya, Atsugi-shi, Kanagawa 243-0198, Japan
hiro@aecl.ntt.co.jp

Abstract. The generation of continuous-wave (CW) terahertz radiation by the photomixing (optical heterodyne downconversion) with the ultrafast optoelectronic devices is reviewed. The principle, design, and characteristics of conventional photomixers made of low-temperature-grown GaAs are introduced. Current limits on the terahertz output power for the conventional design of the photomixer are stated, and the recent designs to increase the terahertz output power are described in detail, showing experimental results. In addition to the output power, the performance of the photomixer is discussed in terms of the optical-to-terahertz conversion efficiency and the noise of the terahertz output. To explain the importance of the CW terahertz source, some examples of the terahertz systems for spectroscopic applications are shown. Finally, future trends of the design and application of the CW terahertz source are discussed.

1 Introduction

Coherent, tunable continuous-wave (CW) THz sources are strongly needed in many applications such as high-resolution spectroscopy, heterodyne receiver systems, local area networks and so on, and various methods have hitherto been investigated. The state-of-the-art performance of conventional coherent tunable sources is shown in Fig. 1 with their covering ranges [1, 2, 3, 4, 5, 6, 7, 8, 9, 10].

Photomixing (or optical heterodyne downconversion) is one of them and it has been a well-known technique for many years [11]. However, for generating coherent radiation in the THz region (gray zone in Fig. 1), useful levels of output power have not been obtained because of the lack of robust photomixing devices and high-quality tunable lasers. With recent advances in high-speed III-V optoelectronic devices and solid-state lasers, interest in

K. Sakai (Ed.): Terahertz Optoelectronics, Topics Appl. Phys. **97**, 157–202 (2005)
© Springer-Verlag Berlin Heidelberg 2005

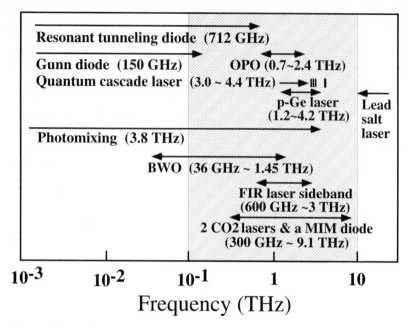

Fig. 1. State-of-the-art performance of conventional coherent tunable sources [1, 2, 3, 4, 5, 6, 7, 8, 9, 10]

power generation by photomixing has been revived. The photomixing technique using the photoconductive (PC) antenna, initiated by *Brown* et al., has spread and developed [12, 13, 14, 15, 16, 17, 18, 19]. Following *Brown* et al., we call the PC-antenna device a "photomixer" intentionally when the device is used for the CW THz-wave generation.

In many cases of coherent spectroscopy, the CW operation of the radiation source is crucially required. Among the sources illustrated in Fig. 1, an optical parametric oscillator (OPO) and p-Ge emit in pulse mode, and quantum cascade lasers emit in pulse mode and one example in CW [7] quite recently, while the photomixer emits real CW radiation. It is competitive with the traditional solid-state CW devices such as Gunn diodes and multiplier, or the quantum cascade laser that will be developed further. In principle, the photomixer as the frequency downconverter should have a much wider tunable range than these fundamental oscillators. This is true only when a highly tunable two-color laser is available as the photomixing source. An example of a specifically designed tunable laser for the photomixng is shown in the description of a THz spectrometer.

The maximum THz output power of the photomixer is currently limited by a thermal damage threshold to the level that is insufficient for some practical applications. Many efforts of optimizing the antenna design and the substrate material of the photomixer were rewarded with success in the improvement of the THz output power. However, there is a shortfall to the required output power levels, and there is still much room for output-power improvement by changing the device design. The past evolution of the device design and recent progress in the output-power improvement with them are reviewed in this Chapter.

Photomixing is deeply relating to the photoexcitation of the substrate semiconductor, hence the incident photon energy must be higher than the bandgap energy. Photomixing has been made using the device fabricated on the low-temperature-grown GaAs (LT-GaAs) substrate and using the laser source of which the emitting wavelength is around 800 nm [20, 21]. However, the technology of key components necessary for photomixing, such as the diode laser, optical fiber and related components, has progressed remarkably in the optical communication band. This situation requires the development of ultrafast photoconductors having such excellent qualities as the LT-GaAs in the communication band. The bandgap energy of such materials as Ge and GaSb is lower than the photon energy of the communication band. Ion-implanted Ge has been proposed, but it has not reached the quality comparable with or superior to the LT-GaAs [22]. The photomixer made of self-assembled GaAs:Er/GaAs superlattice material has been demonstrated [23, 24]. Although the maximum THz output power of this device is currently lower than the LT-GaAs devices, it has the advantage that the carrier lifetime can be precisely controlled by changing the lattice spacing in the MBE process. The general concept of an ultrafast superlattice device could be extendable to the communication band and would induce future new device ideas. Uni-traveling-carrier photodiodes (UTC-PDs) developed recently at NTT Co. in Japan are quite promising, and are described in this Chapter.

In general, the photomixing with two independently operated lasers is well known and widespread. In addition, it has been proposed and experimentally verified that the excitation of a photomixer with a multimode laser diode (LD) gives rise to a CW sub-THz radiation comb. The radiation comb covers a broad (sub-THz) spectral range simultaneously, like the THz pulse emitter. This source is especially suitable for spectroscopy of solid-state materials, hence a full description of this is given in the Chapter by *Nishizawa* and *Sakai*.

Finally, it is noted that a review paper on two-beam photomixing has been written by *Verghese* et al. [25] and a review chapter has been written by *Duffy* et al. [26] in a book on THz-radiation sensing.

2 Principles of Generation of CW Terahertz Radiation

The photomixing process is based on optical heterodyne downconversion and it occurs by illuminating the photomixer with two single-mode CW laser beams having average powers P_1 and P_2 and angular frequencies ω_1 and ω_2, respectively. Figure 2a schematically shows the two-beam photomixing

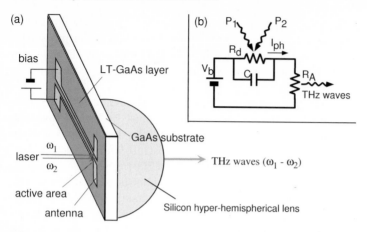

Fig. 2. Principle of THz generation. **(a)** Schematic view of two-beam photomixing with a photomixer. **(b)** Equivalent circuit of the photomixer

with a photomixer coupled to a hemispherical substrate lens. This scheme of photomixing is expressed by an equivalent circuit such as is shown in Fig. 2b.

The instantaneous optical power incident on the photomixer is given by [11]

$$P_i = P_0 + 2\sqrt{mP_1P_2}\left[\cos(\omega_1 - \omega_2)t + \cos(\omega_1 + \omega_2)t\right], \tag{1}$$

where $P_0 = P_1 + P_2$ is the total incident power averaged over a long time period and m the mixing efficiency that ranges in value between 0 and 1 depending on the spatial overlap of the two laser beams. The first cosine term modulates the photoconductance at the difference frequency $w(= \omega_1 - \omega_2)$ but the second term, approximately twice the optical frequency, varies on a time scale much shorter than the carrier lifetime τ, and thus does not modulate the photoconductance significantly.

The dynamic current equation for the equivalent circuit of the photomixer is given by

$$C\frac{dV}{dt} = \frac{V_b - V}{R_A} - G(t)V, \tag{2}$$

where $G(t)$ is the photoconductance modulated at angular frequency w, C the capacitance across the antenna gap, R_A the antenna-radiation resistance and V_b is the DC bias voltage. The time-dependent photoconductance is expressed as [12]

$$G(t) = G_0 \left(1 + \frac{2\sqrt{mP_1P_2}\sin(\omega t + \phi)}{P_0\sqrt{1 + (\omega\tau)^2}}\right), \tag{3}$$

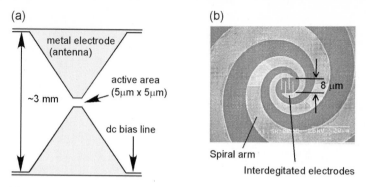

Fig. 3. Photomixers with self-complementary antennas, (**a**) bow-tie antenna device and (**b**) log-spiral antenna device [25]

where G_0 is the DC photoconductance for an average incident power P_0 and $\phi = \tan^{-1}(1/\omega\tau)$. Solving (2), we can write the output THz-radiation power P_{THz} as [12]

$$P_{\text{THz}}(\omega) = \frac{I_{\text{ph}}^2 R_{\text{A}}}{2\left[1 + (\omega\tau)^2\right]\left[1 + (\omega R_{\text{A}} C)^2\right]}, \tag{4}$$

where $I_{\text{ph}}(= G_0 V_{\text{b}})$ is the DC photocurrent. This expression is valid for moderate optical power such that $G_0 R_{\text{A}} \ll 1$. Equation (4) shows that the output power of THz radiation increases in proportion to the squares of the bias voltage V_{b} and the photoconductance G_0, hence, total incident power P_0. It also shows that the output power decreases, on the higher-frequency side, influenced by whether τ or $R_{\text{A}} C$ is dominant, or by both when they are comparable.

3 Basic Characteristics of CW Terahertz Radiation with Photomixing

The photomixer is usually fabricated on ultrafast photoconductive (PC) materials to achieve the current modulation at THz frequencies. The most widely used PC material is the LT-GaAs. Annealed LT-GaAs has unique properties, such as short carrier lifetime ($\approx 0.5\,\text{ps}$), large breakdown-field threshold ($> 300\,\text{kV/cm}$), and relatively high carrier mobility ($\approx 200\,\text{cm}^2/(\text{V}\cdot\text{s})$), each of which is important for the efficiency of photomixing. The photomixer is fabricated on typically $1.5\,\mu\text{m}$ thick LT-GaAs layer grown on a semi-insulating (SI) GaAs wafer. Figures 3a and b show examples of the device structure with broad-band self-complementary antennas; a bow-tie antenna (schematic view) and a spiral antenna with an interdigitated electrode (photograph), respectively.

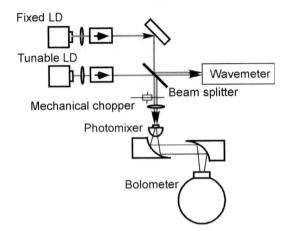

Fig. 4. An example of experimental setup for photomixing

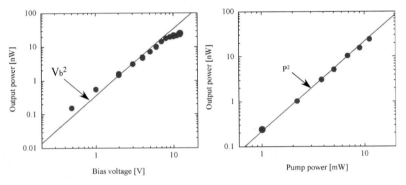

Fig. 5. THz output-power dependence on (**a**) applied bias voltage and (**b**) pump-laser power for an LT-GaAs photomixer

In general, two single-mode lasers are used for photomixing. In order to make photomixing efficient, it is necessary to collimate and align the two laser beams precisely to achieve a good spatial mode matching of the laser beam. Figure 4 is an example of an optical setup of photomixing with two laser diodes (LDs). Commercially available diode lasers in visible and near-infrared ranges of 800 nm to 850 nm have been usually used for the photomixing in LT-GaAs. The tunable range of the diode lasers is typically 10 nm in wavelength or 5 THz in frequency by changing the operational temperature. The beams from the two lasers are combined with a 50/50 beam splitter and then focused to the antenna gap of the photomixer. The incident laser power on the photomixer of such a fiber-coupled diode laser system is typically 30 mW.

The generated THz radiation is emitted into free space through a silicon hyperhemispherical substrate lens with a typical diameter of 10 mm. The output radiation is focused onto a power detector, e.g., a 4.2 K silicon composite

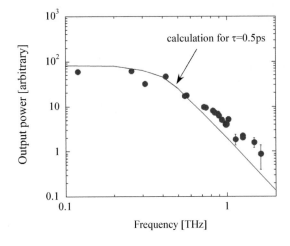

Fig. 6. THz output spectrum for a bow-tie antenna device

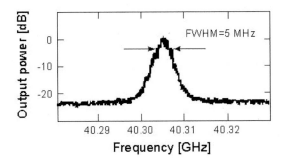

Fig. 7. Linewidth of microwaves generated by photomixing [18]

bolometer. The difference frequency of the two lasers, the frequency of the output radiation, is monitored by using an optical spectrum analyzer or a wavemeter. The full width at half maximum (FWHM) linewidth of the output radiation is directly measured with an RF spectrum analyzer by tuning the difference frequency to the microwave range.

Figure 5a shows the output-power dependence of a bow-tie antenna device on the bias voltage and Fig. 5b is the dependence on the pump-laser power. Both of them are plotted at the frequency of 1.4 THz. As expected from (4), both plots obey simply the square-law dependences, V_b^2 and P_0^2, respectively.

Figure 6 shows the THz output power of the spiral device with interdigitated electrodes as a function of frequency. The THz spectrum shows a flat-top region at the low-frequency limit and high-frequency rolloff behavior due to the carrier lifetime and the RC time constant, following (4). The maximum THz output power is close to 1 μW at lower frequencies. On cooling the device to 77 K, the heat capability increases and pump-laser power can be increased, which leads to an output power of more than 1 μW [16].

The frequency stability of the emitted THz radiation is basically determined by the stability of the pump lasers. Since no good tool to measure the THz linewidth directly is available, it has usually been examined by down-converting the THz frequency to the microwave range. Shown in Fig. 7 is the spectrum of microwaves generated by a LT-GaAs photomixer with free-running external-cavity diode lasers. The FWHM linewidth of this system was about 5 MHz, which is wider than the value expected from a typical value for this type of laser, ≈ 1 MHz, due to the contribution of acoustic noise. In addition to the reduction of acoustic noise, the active frequency control would narrow the linewidth to less than 1 MHz, as described later.

4 Increased Terahertz Radiation Power

4.1 Optimization of Antenna Design

From (4) the THz output power of the photomixer in the small-signal limit is expected to be proportional to the antenna impedance. The impedance of a planar antenna on dielectric substrate is at most a few $100\,\Omega$ on GaAs ($\varepsilon = 12.8$) and much lower than the resistance of the LT-GaAs photoconductor of typically $> 10\,\mathrm{k\Omega}$ under the illumination at modest laser power. Thus, the photomixer works as a current source loaded by the antenna under such a highly mismatched condition. In the initial stage of the photomixer development, broadband self-complementary antenna design, such as bow-tie, log-spiral and log-periodic antenna, had been used to enjoy the wide bandwidth of the photomixer [14, 25]. Since the impedance of broadband antenna is relatively low ($72\,\Omega$), there is still room to improve the THz output power by optimizing the antenna design. In fact, remarkable improvement of the THz output power has been achieved with high-impedance antenna designs as described below.

Resonant antennas, e.g., dipole and slot, have an advantage of high impedance against broadband antennas for generating high output power at a certain frequency. In fact, H-shaped quasidipole designs, which have commonly been used for the THz-pulse generation, show the resonance features in the THz-output characteristics depending on the dipole length [17]. For these devices, however, the Q-factor of the resonance was not so high as to improve the THz output power drastically, because their electrode structure including the DC bias line were not specifically tuned.

For a well-tuned dipole antenna device, the capacitive part of the impedance associated with the electrode is designed to be canceled with the inductive part of the impedance of the antenna and the DC bias line [15, 27]. Figure 8b shows the THz output characteristics of a full-wave dipole antenna device depicted in Fig. 8a. The THz output power takes the maximum at the resonance and shows a clear enhancement in output power by a factor of 2–3 compared with the result for the spiral antenna device at the same

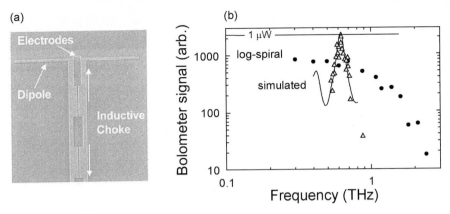

Fig. 8. (a) A full-wave dipole antenna device and (b) its output-power spectrum [25]

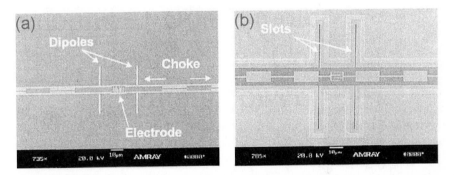

Fig. 9. (a) A dual-dipole antenna device, (b) a dual-slot antenna device [27]

frequency. The bandwidth of the resonance feature is about 10 % of the center frequency, in good agreement with the calculation by applying an equivalent circuit model [12]. At the resonance frequency, the antenna impedance is a real number (pure resistance), and the THz output power is not limited by the RC time constant, while the broadband antenna device has the RC-time-limited bandwidth as described in Sect. 2. The improvement of the output power at the resonance is not only due to high impedance but also due to the RC-time-free property.

The dual-dipole antenna design shown in Fig. 9a has the advantage of the flexibility in circuit design for tuning at higher frequencies compared with the single-dipole design [25, 26, 27, 28]. As seen in the figure, the photoconductor part is connected to twin-dipole antennas with a coplanar strip (CPS) transmission line. The length of the transmission line is tuned to cancel out the electrode capacitance of the photoconductor part. Although the max-

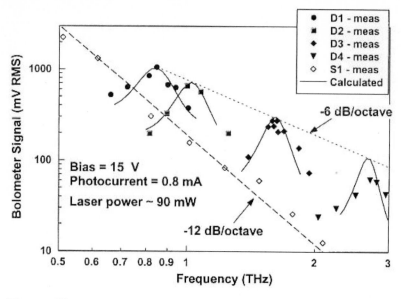

Fig. 10. The output-power spectra of several dual-dipole antenna devices tuned at frequencies ranging from 0.8 THz to 2.7 THz [27]

imum attainable impedance for the dual-dipole design is half that for the single-dipole design, the required inductive tuning is also half that for the single-dipole, and this property allows the design for much higher-frequency operation. A dual-slot device with coplanar waveguide (CPW) transmission line was designed as shown in Fig. 9b [27]. In contrast to the case for the dual-dipole design, the dual-slot design has the potential to have very high impedance, which leads to very high THz output power, but this is at the cost of flexibility of circuit tuning especially at higher frequency. Another advantage of dual-antenna designs is a symmetric near-Gaussian beam pattern [27], while the single-dipole antenna device shows relatively large side lobes [29, 30]. High directional gain prevents the reflection loss at the silicon lens surface and results in high THz output power.

Figure 10 shows THz output power spectra of several dual-dipole devices tuned in a range from 0.8 to 2.7 THz compared with the result for a spiral antenna device [27]. The photomixers used here have the AlAs-layered structure described later in Sect. 4.3. In contrast to the $-12\,\mathrm{dB/Oct}$ rolloff behavior for the spiral device, the dual-dipoles show the $-6\,\mathrm{dB/Oct}$ rolloff behavior without the bandwidth limitation by the RC time constant. The THz-output characteristics calculated by applying equivalent-circuit models for the dual-dipole devices show again good agreement with the measurements. The slow rolloff behavior is a great advantage in the high-frequency operation above 1 THz. The maximum THz output power of the dual-dipole

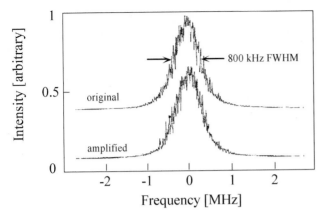

Fig. 11. Spectral shape of the beat signal at 12 GHz before and after the amplification with two-frequency MOPA [31]

devices is 3–10 times higher than that of the spiral device, owing to their high impedance, slow rolloff behavior, and high directional gain.

4.2 High-Power Laser Sources

The THz output power of photomixers is currently limited to sub-μW levels [25], which is not sufficient for practical use in advanced scientific measurements, such as nonlinear spectroscopy, local oscillator application for radio astronomy, etc. One of the simplest ways to improve the THz output power is the use of high-power laser sources, because the THz output power has a quadratic dependence on the optical power, as described in Sect. 3. However, high-power tunable lasers in the wavelength required for pumping the LT-GaAs photomixer ($\lambda < 850$ nm) have recently been used mainly for basic research, while the 1.55 μm band laser technology has been energetically developed for fiber optics communication.

A continuous-wave Ti:sapphire laser has been used for the photomixing as well as THz-pulse generation as described in other chapters, because of its wide tunable range, high spectral purity, and high optical output power. However, Ti:sapphire laser may not make the best match for the photomixer, because the total system of Ti:sapphire laser including the pump laser is relatively large and power consuming. One of the most important advantages of the photomixer against the other THz sources is the potential for constructing a very compact system. Low power consumption is a practical requirement especially for space-based applications. A wide tunable range is crucial to achieve new scientific results in spectroscopic applications as discussed in Sect. 8. The diode laser is currently the best candidate to satisfy all of these requirements.

The linewidth of a diode laser is inherently wide due only not to its small cavity size but also the complex interplay between the refractive index and the carrier density in the laser medium. In order to improve the spectral purity and the frequency stability of the diode laser, much effort has been applied [32]. The most popular frequency-stabilization technique is the combination of the optical feedback with tunable external cavity and the electrical feedback by applying the injection-current modulation [33]. With this technique, the linewidth of a few tens of kHz, sufficient for many spectroscopic applications, was achieved. By applying these stabilization methods to two lasers with individual cavities or to a two-frequency laser with different modes of a single cavity as described in Sect. 8, the difference frequency for the THz generation can be stabilized [34].

The optical power of frequency-stabilized diode lasers is typically a few tens of milliwatt, which is too low to generate the terahertz wave efficiently. In order to increase the laser power to acceptable levels, a two-frequency MOPA (master oscillator power amplifier) technique was exploited for the photomixer source system [31, 34]. The two-frequency laser beams are simultaneously amplified with a tapered semiconductor amplifier, while the MOPA scheme has usually been used for the single-frequency laser. This system provides coaxial two-frequency beams with almost the same power and the total output power of approximately 500 mW. The gain bandwidth of this type of amplifier is typically wider than 20 nm, and therefore the two-frequency amplification up to 10 THz is possible.

As shown in Fig. 11, the spectral shape of the two-frequency beat signal at the amplifier output is identical to that of the master laser as long as the difference frequency is greater than 10 GHz. At a difference frequency of less than 10 GHz, unbalanced amplification of the two frequencies and sideband generation occur due to nonlinear gain effects [31].

According to the quadratic dependence, the THz output power for the 500 mW laser power is expected to be 1–2 orders of magnitude higher than the best results for the maximum laser power that conventional LT-GaAs photomixers can handle (< 100 mW). In order to enjoy high optical power of the MOPA system, a new design of the photomixer with high thermal damage threshold is required, as described in the following sections.

4.3 Thermal Conductivity of Substrate

Even though high-power lasers are available as described above, the damage threshold of conventional LT-GaAs photomixers with a small active area of $50\,\mu m^2$ to $100\,\mu m^2$ is 50 mW to 100 mW, i.e., an incident laser flux of $\approx 10^5$ W/cm^2. The thermal failure arises from the device heating by the combination of optical power and ohmic power [25]. The maximum THz output power is then limited to $1\,\mu W$ levels. The thermal-damage threshold of conventional LT-GaAs photomixer is basically determined by the thermal conductivity of the GaAs substrate.

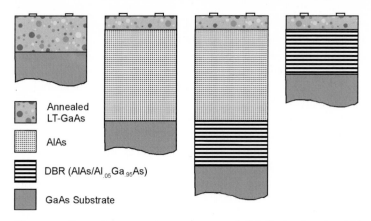

Annealed
LT-GaAs

AlAs

DBR (AlAs/Al$_{.05}$Ga$_{.95}$As)

GaAs Substrate

Fig. 12. From *left* to *right*, conventional LT-GaAs devide, AlAs buffer-layered device, AlAs buffer-layered device with disributed-Bragg-reflector (DBR), and LT-GaAs device with DBR are shown [28]

The actual temperature of the active area of the LT-GaAs photomixer in operation was measured by a clever method described in[25]. The DC photo-current of the photomixer illuminated by a weak laser beam at 850 nm, whose photon energy is just above the energy gap of GaAs at room temperature, can be a good measure of the energy gap shift by the temperature change. The calibration of the photocurrent for the probe beam is carried out by installing the photomixer in a cryostat and changing the bath temperature in a range from 37 K to 77 K. The temperature measurement of the active area was done by pumping the photomixer with another powerful laser at 810 nm and measuring the photocurrent for the probe beam. The result shows that the thermal failure occurs when the temperature difference between the active area and the substrate is roughly 110 K and that the absolute temperature of the active area is not important for the failure.

Some attempts to increase the damage threshold have been made by changing the substrate material. At room temperature, the ther-mal conductivity of silicon $(1.1\,\mathrm{W/(cm \cdot K)})$ is three times that of GaAs $(\approx 0.4\,\mathrm{W/(cm \cdot K)})$. The LT-GaAs grown on silicon substrate has been used for the photomixer, and it produced about twice the THz output power com-pared to the GaAs substrate device, according to the increase of the available laser power [30].

It has been pointed out that a thin buffer layer of high thermal conduc-tivity material under the LT-GaAs film should be effective to improve the thermal property of the device, because the temperature of the active area is estimated from the thermal model to drop within roughly 10 μm from the surface [25, 28]. In fact, photomixers with AlAs buffer layers, which have about twice the higher thermal conductivity of GaAs, showed remarkable improvement of the maximum THz output power [28]. A schematic view of the photomixer with AlAs layer is given in Fig. 12. For conventional photomixer designs the thickness of the LT-GaAs films is set to be greater than the absorption length of $\approx 1\,\mu$m, in order to avoid the bandwidth reduction caused by long-lived carriers generated in the substrate of normal GaAs. The bandgap energy of AlAs is higher than that of GaAs, and it is transparent at 850 nm. If the AlAs layer is thicker than the penetration depth of the bias electric field, the photocarrier generated in the GaAs substrate below the AlAs layer does not contribute to the photocurrent. Then, the LT-GaAs can be thinner than the absorption length, and the thermal conductivity near the active area increases considerably.

The dual-dipole devices, whose THz output power spectra were already shown in Fig. 10, are fabricated on 0.35 μm thick LT-GaAs / 2.5 μm thick AlAs layers. The maximum THz output power at the resonance is 3 μW, 2 μW, 0.8 μW, and 0.2 μW at 0.85 THz, 1.05 THz, 1.6 THz, and 2.7 THz, respectively [27,28], and 3–5 times higher than that of the device without AlAs layers. This is the best result ever attained with LT-GaAs photomixers at a frequency above 1 THz, relying on their high damage threshold.

4.4 Quantum Efficiency

According to (4) derived from a simple photoconductor theory, the THz output power is proportional to the square of the photocurrent that is proportional to the quantum efficiency, η, and the photoconductive gain, g, given by the ratio between the carrier lifetime, τ, and the carrier transit time, T, as

$$g = \frac{\tau}{T} = \frac{\tau \mu E}{L}, \tag{5}$$

where L is the electrode gap of the photoconductor, μ is the effective carrier mobility (mainly contributed by the electron mobility in LT-GaAs), and E is the bias electric field strength. The photoconductive gain is calculated to be $g = 0.03$ by substituting typical physical parameters for LT-GaAs, $\tau = 0.3$ ps, $L = 2\,\mu$m, $\mu = 200\,\mathrm{cm}^2/(\mathrm{V \cdot s})$, and $E = 10^5\,\mathrm{V/cm}$ for (5). For conventional LT-GaAs photomixers a ratio of the DC photocurrent to the pump-laser power gives $\eta g \approx 0.005$ [35]. In this case, the quantum efficiency for $g = 0.03$ is $\eta \approx 0.2$, which is lower than that expected from a simple plane-parallel photoconductor model by a factor of 2–3.

The reason for low ηG can be explained by using a more realistic photoconductor model with an inhomogeneous electric field for the MSM (metal-semiconductor-metal) planar electrode structure [35]. As shown in Fig. 13, the electric field in the device is strong near the electrode and the top surface but decreases rapidly with the depth in the medium. Since photocarriers are mostly generated in the bulk region where the electric field is weak, the total photoconductive gain (or the photocurrent) is consequently reduced. It is

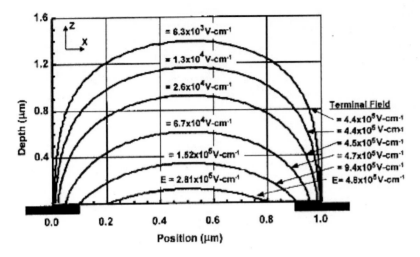

Fig. 13. Electric-field distribution in the photomixer [35]

noteworthy that the reduction of the photoconductive gain by the inhomogeneous field effect may be attributed to lowering of the quantum efficiency
in the expression of (5). Hereafter, the term 'quantum efficiency' is used with
the meaning of the quantum efficiency including the field effect.

The quantum efficiency can be improved by restricting the incident light
to the LT-GaAs layer near the top surface, where the bias field is relatively
uniform and strong. Accordingly, the device with a built-in optical cavity
structure has been proposed [28, 35]. The cavity structure can be fabricated
by growing an AlGaAs/GaAs distributed Bragg reflector (DBR) under the
LT-GaAs layer, which is thinner than the absorption length, as depicted in
Fig. 12. The incident light is reflected back and forth between the top surface
and DBR, and the total light path can be long enough to attain high quantum
efficiency at the cavity mode frequencies. The photomixer with 10 periods of
$Al_{0.05}Ga_{0.95}As/AlAs$ quarter-wave DBRs provided the quantum efficiency
of approximately twice that of the device without the DBR structure [28].
Such a resonant device has a narrow optical bandwidth corresponding to a
Q-value of the cavity. The measured 1.5 dB bandwidth for the DBR device
is greater than 5 nm, which corresponds to ≈ 2.5 THz and is sufficiently wide
for many spectroscopic applications. However, the thermal conductivity of
DBR is lower than that of GaAs, and the resultant maximum THz output
power at the thermal failure remains at similar levels to those obtained by
using devices with the AlAs heat sink.

Recently, THz-wave generation with a vertically integrated LT-GaAs photomixer was demonstrated [36]. The vertical configuration that provides a
uniform electric field in the active region was realized by sandwiching the
LT-GaAs in a plane-parallel arrangement with a semitransparent contact and

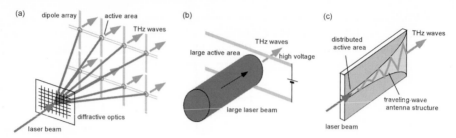

Fig. 14. Variety of large active area design, (**a**) array antenna, (**b**) single large active area, and (**c**) traveling-wave design

a buried contact under the LT-GaAs layer. According to the above discussions, this device provides very high quantum efficiency that corresponds to the responsivity of $0.04\,A/W$, where a conventional photomixer with MSM structure has the responsivity of approximately $0.005\,A/W$. Although the current maximum THz output power of the demonstrated device does not exceed that of conventional MSM-type photomixers, the vertical configuration has the potential to surpass the MSM device in the output power as well as the quantum efficiency.

The reflection loss at the top surface of the LT-GaAs and the metal electrode in the active area reduces the external quantum efficiency. According to the refractive index of GaAs of 3.6, the Fresnel coefficient for the reflection is 32%. The antireflection (AR) coating onto the photomixer surface with Si_3N_4 was attempted, and the AR coating raises the optical–THz conversion efficiency [25]. Again, the increase of the quantum efficiency by AR coating does not help to increase the maximum THz output power limited by the thermal-damage threshold.

4.5 Large Active Area Design

As described in Sect. 4.3, the thermal failure of conventional photomixers occurs at the power density in the active area of $\approx 10^5\,W/cm^2$. A simple way to avoid the thermal failure of the photomixer is by reducing the optical and ohmic heat power density by increasing the active area. One can easily imagine a large variety of designs to increase the active area as depicted in Figs. 14a–c. However, there is an important restriction in designing large active area devices. This is the requirement of coherent (inphase) superposition of THz waves generated in the distributed area of the device, while a conventional photomixer has an active area smaller than the wavelength and guarantees the coherent operation automatically.

Figure 14a shows an array-antenna design. Each array element consists of a single small-area photomixer, and THz output beams from all array elements are coherently combined together and form a synthesized beam of a phased-array antenna. In order to perform coherent superposition of the THz waves in phase, a single laser beam may be divided into multiple laser beams with the same optical path lengths for each array element. Such power division can be done with optical-fiber or holographic optics. When the laser power for an individual array element is given, e.g., just below the thermal limit, the total THz output power should be proportional to the square of the number of array elements because of coherent overlap of the waves. A narrow beam pattern of the array-antenna devices depending on the array dimension is another advantage against single-antenna devices. Improvement of the maximum THz output power with such an array device has never been reported.

Figure 14b shows a single large active area design. An active area with a large aperture is biased with high voltage and illuminated by a high-power laser with a large beam size. If the wavefront of the laser beam is parallel to the device surface, the coherent overlap of THz waves is automatically guaranteed, and a diffraction-limited beam pattern determined by the laser spot size is generated. If the active area is larger than the wavelength of the THz wave, the antenna is no longer necessary for efficient radiation. This type of large-aperture device has been used for pulsed THz-wave generation, and increased THz power was reported [37]. However, the increase of the electrode gap results in lowering of the photoconductive gain that is more important for efficient generation of the CW THz waves than for the pulsed THz generation. If the electrode gap is decreased to obtain a high photoconductive gain while keeping the active area size constant, the device bandwidth decreases due to the increase of the electrode capacitance. Such a limitation of the gain–bandwidth product is a common problem for the high-speed operation of photodetectors, and it can be solved by operating the device in a traveling-wave mode, as described in the following section.

4.6 Traveling-Wave Photomixer

A promising design to have a large active area while keeping a wide bandwidth is the traveling-wave design depicted in Fig. 14c. The THz waves generated in a distributed active area propagate along a microwave transmission line associated with the active area keeping the coherent superposition condition, and they are radiated by an antenna terminating the transmission line. The active area can be very large corresponding to the transmission line length. The traveling-wave design can provide an extremely wide bandwidth, because the transmission line is a distributed circuit and has basically no bandwidth limitation by the electrode capacitance, while conventional small-area designs are regarded as lump-element circuits goverred by the capacitance-limited bandwidth.

To date, essentially two types of traveling-wave photomixers with LT-GaAs have been proposed. One is an optical-waveguide type, as shown in Fig. 15 [38]. An optical waveguide is fabricated along the LT-GaAs absorbing layer with a DC-biased microstrip transmission line. The optical waveguide

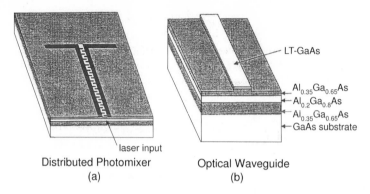

laser input

Distributed Photomixer   Optical Waveguide
            (a)   (b)

Fig. 15. Optical-waveguide-type traveling-wave photomixer [38], (a) schematic view of an LT-GaAs photomixer with a periodically loaded CPS waveguide, (b) detail of the optical waveguide with an LT-GaAs absorber

is edge illuminated, and the light is gradually absorbed in the distributed area of the LT-GaAs layer during the propagation. The photocarrier density waves are induced by moving optical fringes due to the beat between the two-color optical waves along the microwave transmission line. The THz waves accompanied with the photocarrier density waves travel along the microwave transmission line and emerge to the free space from a planar antenna. For the inphase superposition of the THz waves along the transmission line, velocity matching between the optical waves and the THz waves is required. This can be done by tuning the structure of the microwave transmission line, the same as the velocity matching between microwaves and electron bunches with slow-wave structure in the traveling-wave tube. The frequency dispersion of the microwave transmission line may reduce the device bandwidth, but it has been shown that the velocity matching in a wide frequency range over a few terahertz is possible [38].

The improvement of the THz output power with the optical waveguide traveling-wave photomixer has not been reported, but it was shown that a low-frequency ($\approx 100\,\mathrm{GHz}$) version of a similar device provides the highest gain–bandwidth product ever for high-speed photodetectors [39]. Although the measurement for the low-frequency version has been done in the pulsed operation, the measured capacitance-free wide bandwidth promises high performance of the THz version. Since the thermal-failure limit of the optical-waveguide-type device in the CW operation has not been well studied, the maximum THz output power of such devices is not predictable at present.

Recently, a high optical–THz conversion efficiency of an advanced design of the optical-waveguide traveling-wave device, 20 times higher than the highest value described in Sect. 4.4, was reported [40]. However, since the THz output power was measured in the pulsed operation, the reported efficiency converted to the value in the CW operation could be similar to that of the

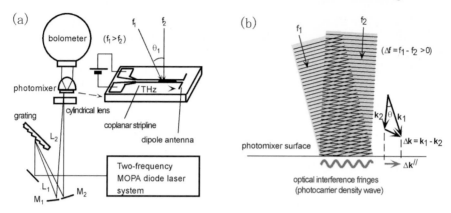

Fig. 16. (a) Free-space traveling-wave photomixer showing the optical setup. (b) The principle of the optical–THz phase matching for the free-space traveling-wave photomixer

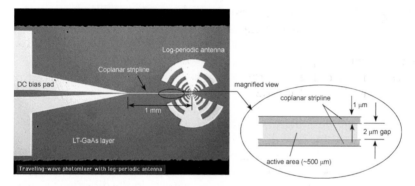

Fig. 17. Traveling-wave photomixer with log-periodic antenna showing schematic of the coplanar stripline structure

other devices. Because the optical–THz conversion efficiency is proportional to the laser power [12], while the peak power of the pulse laser is much higher than the average laser power, e.g., $\approx 10^4$ times higher for a 82 MHz repetition rate and sub-ps pulse width of a conventional mode-locked Ti:sapphire laser. Besides the conversion efficiency, this tentative result shows that the traveling-wave design is potentially one of the best solutions to overcome the gain–bandwidth limitation.

An alternative design of the traveling-wave photomixer is a free-space coupled device [41,42] as shown in Fig. 16a. A DC-biased coplanar stripline (CPS) microwave transmission line fabricated on LT-GaAs is vertically illuminated by a linear-shaped laser beam formed with a cylindrical lens. The active area of this device, determined by the laser-beam size (≈ 0.5 mm) and the gap of the CPS line ($\approx 2\,\mu m$), is one order of magnitude larger than that of conventional photomixer design. The generated THz waves are propagated along the CPS line and radiated by a planar antenna at the terminal. A hyperhemispherical silicon lens is attached onto the back side of the antenna to shape the beam pattern. The velocity matching between optical and THz waves is done by angle tuning of two nonaxial laser beams described as follows.

Two nonaxial laser beams with a small offset angle are focused on the same area of the device. The combined two beams generate a spatial interference fringe pattern accompanied with the carrier-density modulation in the LT-GaAs, as depicted in Fig. 16b. The interference fringe moves along the device surface with a velocity given by

$$V_{op} = \frac{\omega}{k_p}, \tag{6}$$

where ω is the difference (angular) frequency between the two laser frequencies, and k_p is the projection of the difference wavevector onto the device surface. The velocity of the moving fringe can be tuned by changing the offset angle so as to be equal to the velocity of the generated THz wave in the CPS line. The smaller the offset angle, the greater the velocity of the interference fringe. If the CPS line has no frequency dispersion, the THz wave velocity is a constant, given by

$$V_{THz} = \frac{c}{\sqrt{\varepsilon_{eff}}}, \tag{7}$$

where c is the speed of light, and ε_{eff} is the effective dielectric constant, $(\varepsilon + 1)/2$ [43]. For GaAs ($\varepsilon = 12.8$), the offset angle required for the velocity matching is ≈ 1 deg, which is within an accessible range with normal optical alignment tools. The angle-tuning bandwidth for the velocity matching is inversely-proportional to the square of the active area in the direction parallel to the CPS line. The angle-tuning bandwidth of a few tens of percent for a sub-mm-long laser spot size is reasonable for practical operation [41].

For the traveling-wave device, the THz output spectrum is determined not only by the antenna impedance but also the impedance-matching condition between the CPS line and the antenna. The CPS line is designed to have a real impedance of $\approx 100\,\Omega$ for the GaAs substrate, similar to that of the broadband antenna. In this case, the antenna impedance in (4) may be replaced by the CPS line impedance. The best result has been obtained with a log-periodic antenna design, while the dipole-antenna design [41] and endfire-antenna design [44] have also been used. The THz output spectrum for the dipole device was broad and had no clear resonance feature because of the impedance mismatch between the CPS line and the antenna at the resonance [41]. The endfire-antenna device seems to suffer from low output power due to an asymmetrical beam pattern and the power loss caused by the substrate effects, such as multireflection between top and bottom surfaces of the substrate [43].

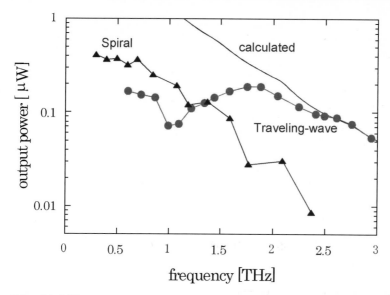

Fig. 18. THz output power spectrum of a traveling-wave device with a log-periodic antenna

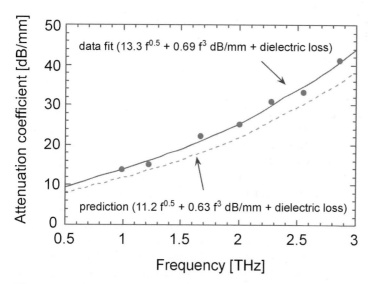

Fig. 19. Frequency-dependent loss of the coplanar stripline of the free-space traveling-wave photomixer [45]

The measured THz output power spectrum of the log-periodic antenna device with the $2\,\mu$m gap CPS is shown in Fig. 18. For comparison, the best result attained with a conventional spiral antenna device is also shown in

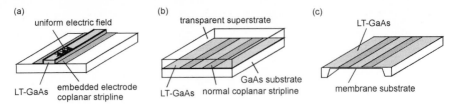

(a)
uniform electric field

LT-GaAs   embedded electrode
            coplanar stripline

(b)
transparent superstrate

                    GaAs substrate
LT-GaAs normal coplanar stripline

(c)
LT-GaAs

membrane substrate

Fig. 20. Advanced design of free-space coupled traveling-wave device, (**a**) embedded electrode, (**b**) superstrate design, and (**c**) membrane structure

Fig. 18. The output spectrum of the traveling-wave photomixer is extremely broad in comparison with that of the spiral device whose bandwidth is limited by both the carrier lifetime and the RC time constant. As described above, the bandwidth of the traveling-wave photomixer is free from the RC time constant and is limited by the carrier lifetime of LT-GaAs and the frequency-dependent loss of the CPS line, such as ohmic loss, dielectric loss, and radiation loss [42, 45]. The measured loss as a function of the frequency is compared with the calculated loss in Fig. 19 [45]. At very high frequencies of $> 2\,\mathrm{THz}$, the attenuation coefficient of the CPS line was measured to be $> 30\,\mathrm{dB/mm}$, i.e., most of the generated THz power is lost during the travel in the 0.5 mm long active area. The frequency dependence of the measured loss is explained by the combination of various frequency-dependent loss mechanisms.

The CPS line loss is mainly due to the radiation loss ($\propto f^3$) and the dielectric loss ($\propto f$), and these loss factors originate from the existence of the substrate in one half of the free space. An almost flat spectrum over a range from 0.5 THz to 3 THz was observed, and the rolloff behavior at higher frequencies, above 2 THz, is roughly consistent with the calculation based on the carrier lifetime of LT-GaAs and the CPS line loss. It is noteworthy that the flat spectrum in the low-frequency region is quite different from the calculated spectrum, which is shown in Fig. 18 by the solid line. This may be related to the interference from the backward wave reflected back to the active area by the DC bias pad structure at the terminal of the CPS line. Further circuit analysis is required to make the cause clear.

The maximum THz output power of the free-space traveling-wave photomixer was greater than $0.1\,\mu\mathrm{W}$ in a wide range extending to 3 THz. This power level has never been achieved with the other photomixers with broadband antennas, though only dual-dipole devices provided similar power levels at the resonance frequencies, as described in Sect. 4.3. The pump-laser power for the present measurement of the traveling-wave photomixer is 380 mW, which is well below the thermal-damage threshold of the device. Since the maximum THz output power of this device is still limited by the maximum available optical power of the pump laser, further increase of the THz output power with a more powerful laser source can be expected. According to the square dependence on the laser power, the expected THz output power is $\approx 1\,\mu W$ at 3 THz for the 1 W laser power, which corresponds to the thermal-damage threshold. The quantum efficiency of this device estimated from the DC photocurrent is similar to that of the small-area devices. If the quantum efficiency is increased by using some methods described in Sect. 4.4, $\approx 1\,\mu W$ THz output power is attainable with the present laser setup.

Some advanced traveling-wave designs to increase the THz output power are shown in Figs. 20a–c. As described in Sect. 4.4, a nonuniform electric field in LT-GaAs reduces the quantum efficiency. Shown in Fig. 20a is a free-space traveling-wave device with an embedded-electrode structure that provides a more uniform electric field in the photoconductor [46]. In fact, this device showed improvement of the quantum efficiency as expected, but the maximum THz output power of this device at present is similar to that of the previous design. In order to increase the THz output power from the traveling-wave devices, the reduction of the CPS line loss is crucial. A superstrate made of high-dielectric material sandwiching the CPS line with the LT-GaAs substrate (Fig. 20b) would be effective to reduce the radiation loss [43]. AR-coated sapphire is suitable for the superstrate material, because it is transparent at the laser wavelength and its dielectric constant is comparable to that of GaAs. A drawback of the superstrate design is that the impedance of the sandwiched CPS line is lower than that of the half-air (normal) CPS line with the same dimension. Another way to reduce the transmission line losses is by removing the substrate completely or making it thinner than the wavelength of the THz wave. In contrast to the case for the superstrate design, high impedance of the CPS line on the membrane should also contribute to increase the THz output power. Figure 20c shows a design of the free-space traveling-wave device with a very thin (a few micrometer) membrane substrate. The measurements for these devices are ongoing work [47].

5 Uni-Traveling-Carrier Photodiode – Novel Photomixer

Light signals in the 1.55 μm range are beneficial to generate THz waves because we can use the various devices and materials developed for optical communications systems, while the LT-GaAs photomixer uses the 850 nm laser. To date, coherent wave generation up to 1 THz by a long-wavelength pin-photodiode (pin-PD) [48] has been investigated. However, the obtained output power at around 1 THz is only a few μW [48, 49], which has to be increased for practical applications.

The uni-traveling-carrier photodiode (UTC-PD) [50] is a novel photomixer that enables us to achieve a wide bandwidth and a high output power simultaneously [51, 52, 53]. Figure 21 shows the band diagram of a UTC-PD. The UTC-PD has a p-type narrow-gap absorption layer (InGaAs) and an

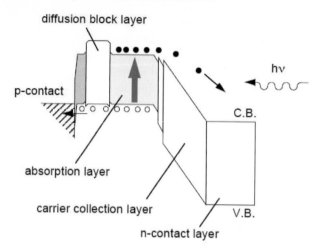

Fig. 21. Band diagram of a UTC-PD

undoped (or a lightly n-type doped) wide-gap collection layer (InP). Because the absorption layer is quasi-neutral, majority holes respond very fast, within the dielectric relaxation time, by their collective motion. Therefore, only electrons are the active carriers, and their transport determines the total delay time. This is in contrast to the conventional pin-PD, in which both electrons and holes are the active carriers [54]. In addition, we can effectively utilize the electron-velocity overshoot [55] in the depletion layer in order to improve the device performance. The fact that the velocity and mobility of electrons in the materials used in an InP/InGaAs UTC-PD are an order of magnitude higher than those of holes simultaneously provides two major merits: 1. higher device operation speed and 2. higher output saturation current due to the lower space charge effect in the depleted collection layer. In addition, one can independently design the collection layer and the absorption layer thicknesses in this structure. Thus, a very thin absorption layer can be used to attain an extremely high 3 dB down bandwidth (f_{3dB}) without sacrificing the CR charging time. This is an essential difference from the conventional pin-PD, in which the CR charging time increases significantly when the absorption layer thickness is reduced in order to decrease the carrier transit time in the device [56]. Another important point is that high speed with a high saturation output is maintained even at a low bias voltage, because the high electron velocity in the depletion layer can be maintained at a relatively low electric field. Thus, the UTC-PD is suitable for low-power operation.

To increase the bandwidth of the UTC-PD, reduction of the CR charging time, as well as the carrier transit time, is essential. For this purpose, a small junction area (S), a small load resistance (R_L), and a thin absorption layer (W_A) with a relatively thick collection layer are effective. Figure 22 shows a pulse photoresponse of a UTC-PD with $S = 5\,\mu m^2$, $R_L = 12.5\,\Omega$, and $W_A = 30\,nm$ [57]. With these parameters, the CR charging time is con-

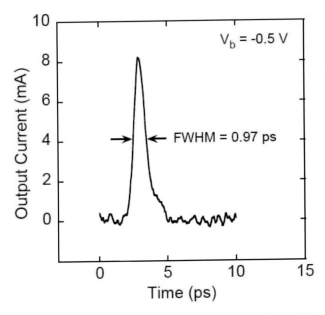

Fig. 22. Pulse photoresponse of a UTC-PD with an absorption layer thickness of 30 nm

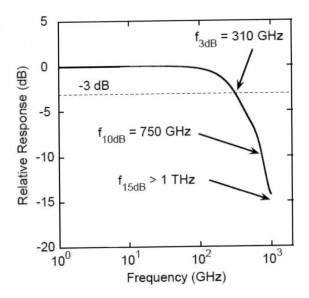

Fig. 23. Fourier transform of the pulse photoresponse shown in Fig. 22

sidered to be a minor portion of the total delay time. The photoresponse was measured by pump-probe electro-optic (EO) sampling [58] with a 1.55 μm incident pulse (full width at half maximum (FWHM): 280 fs). The FWHM of

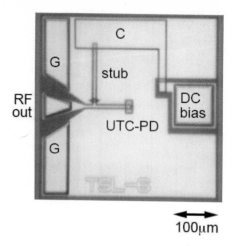

Fig. 24. Micrograph of a UTC-PD integrated with a matching circuit

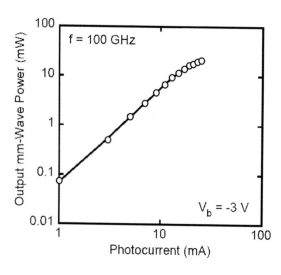

Fig. 25. Relationships between the measured millimeter-wave output power and diode photocurrent at 100 GHz for the device shown in Fig. 24

this pulse response is as narrow as 0.97 ps even at a small reverse bias voltage (V_b) of -0.5 V. The Fourier transform of this pulse waveform is shown in Fig. 23. The $f_{3\,\mathrm{dB}}$ is estimated to be 310 GHz, which is the highest value ever reported for PDs operating at 1.55 µm. This device also exhibited a 10 dB down bandwidth of 750 GHz, and a 15 dB down bandwidth of over 1 THz. These results indicate the superior potential of the UTC-PD for generating coherent waves up to the THz range.

An effective way to improve the efficiency of optoelectronic (O/E) conversion at desired frequencies is to integrate a UTC-PD and a resonant matching circuit [59, 60]. This circuit works to compensate the imaginary part of the internal impedance in the UTC-PD at a designed frequency and increase

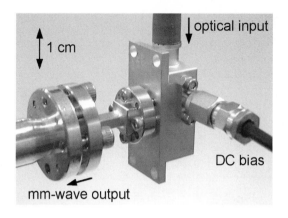

1 cm

optical input

DC bias

mm-wave output

Fig. 26. Photograph of an F-band waveguide-output UTC-PD module

the output power by increasing the effective load resistance. In addition, the depletion layer of the UTC-PD should be designed to be relatively thick so that the junction capacitance becomes relatively small. The small capacitance is important if we want to be able to easily attain the required impedance transform at resonance using the short-stub circuit. A thick depletion layer can only be used in UTC-PDs, because, in pin-PDs, a thick depletion layer significantly reduces the bandwidth due to the low hole velocity. Figure 24 is a micrograph of a device integrating a UTC-PD and a short-stub circuit consisting of a coplanar waveguide and a metal-insulator-metal capacitor. This short-stub circuit simultaneously acts as a matching circuit and a bias-tee circuit. Considering the device parameters of the UTC-PD, the circuit was designed to provide increased output power at 100 GHz with an effective load resistance of 100 Ω. Figure 25 shows the relationship between the millimeter-wave output power and the diode photocurrent at 100 GHz for a bias voltage of −3 V. As seen in this figure, a wide linearity is maintained up to a high millimeter-wave output power of over 10 mW, and the highest output power was 20.8 mW at a photocurrent of 25 mA. To our knowledge, this is the highest output power directly generated from a PD at frequencies above the millimeter-wave range. The 1 dB compression bandwidth is evaluated to be as wide as 40 GHz [61].

For practical use, we need to have the device in a module. However, the reported photodiode modules with a waveguide output port [62, 63] are usually bulky and incompatible with the standard O/E device assembly technology. Figure 26 shows a photograph of a newly developed butterfly-type module for operation in the F-band (90 GHz to 140 GHz) [64]. Its size and configuration are equivalent to those of conventional semiconductor O/E devices, so that it is compatible with the standard assembly/testing equipment for O/E device modules. On the bottom side of the module is a WR-8 rectangular waveguide output port. The size of the module is 12.7 mm × 30 mm × 10 mm (excluding the optical fiber and the leads). The responsivity of the module is about 0.35 A/W. The maximum output power at 120 GHz was measured

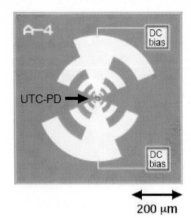

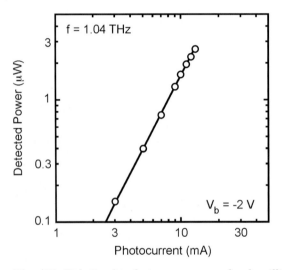

Fig. 27. Micrograph of a UTC-PD integrated with a log-periodic antenna

Fig. 28. Relationship between measured sub-millimeter-wave output power and diode photocurrent at 1.04 THz for the device shown in Fig. 27

to be 17 mW (at a photocurrent of 25 mA), with a very wide linearity range. The output 3 dB bandwidth is about 55 GHz, which fully covers the F-band. This bandwidth is mainly determined by the frequency characteristics of the matching circuit.

In the frequency range above the F-band, monolithic integration of a photodiode and a miniaturized antenna [65] is advantageous because it eliminates the loss and reflection in electrical transmission lines. It may also be necessary to employ a quasi-optical approach to configure a system that can handle sub-millimeter-wave signals. Figure 27 shows a micrograph of a device integrating a UTC-PD and a self-complementary log-periodic toothed

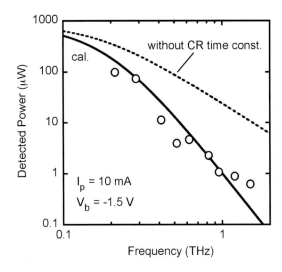

Fig. 29. Relationship between measured THz output power and frequency for a quasi-optical module using the device shown in Fig. 27 at a diode photocurrent of 10 mA

planar antenna, whose teeth correspond to frequencies from 150 GHz to 2.4 THz [66,67]. The millimeter wave was mainly emitted toward the back side of the wafer. The device was placed on a Si hyper-hemispherical lens and the output power was detected using a Martin–Puplett-type Fourier transform spectrometer (FTS) and an InSb hot-electron bolometer. The absolute value was calibrated against a blackbody. Figure 28 shows the relationship between the measured THz output power and the diode photocurrent at 1.04 THz for a bias voltage of $-2\,$V. The output power increased linearly in proportion to the square of the photocurrent, and the maximum output power obtained was $2.6\,\mu$W at a photocurrent of 13 mA [68]. To our knowledge, this is the highest output power directly generated from a PD in the THz band. It is even higher than the highest output power reported for the LT-GaAs photoconductive switch [69], and is more than two orders of magnitude higher than that obtained by a pin-PD [49]. More importantly, the bias voltage applied was about an order smaller than that required for the LT-GaAs photoconductive switch [69]. Thus, total power dissipation in the device is much smaller, which should be advantageous in regard to device reliability.

Figure 29 shows the frequency characteristics of the device shown in Fig. 27. Here, the output power was measured in a form of a quasi-optical module at a constant photocurrent of 10 mA [68] with a bias voltage of $-1.5\,$V. The output power decreased gradually first as the frequency increased, and then dropped more steeply by approximately $-12\,$dB/Oct above 600 GHz. The solid line in Fig. 29 shows the calculated frequency characteristics, assuming a CR-time-constant-limited bandwidth of 210 GHz for a $75\,\Omega$ load and a carrier-transport-limited bandwidth of 170 GHz. The good correlation between the experiment and the calculation indicates that the frequency characteristics of the photodiode response itself explain the entire behavior. Thus, if one can eliminate the influence of the CR time constant by employing a resonant narrow-band matching circuit [60], it would be possible to obtain an output power of more than $10\,\mu\mathrm{W}$ at $1\,\mathrm{THz}$, as indicated by the broken curve in Fig. 29. These results clearly indicate that the UTC-PD is a promising and realistic device for generating a continuous THz wave with a practical output power, which is required in various applications.

6 Optical–Terahertz Conversion Efficiency

As described in Sect. 4 and Sect. 5, the THz output power of the photomixer has been greatly improved in several ways. However, the optical–THz conversion efficiency defined by a ratio of the THz output power to the pump laser power is quite low; $P_{\mathrm{THz}}/P_{\mathrm{opt}} = 10^{-6} - 10^{-5}$ in the frequency range of $1\,\mathrm{THz}$ to $3\,\mathrm{THz}$. This value is one order of magnitude lower than that from the measured DC photocurrent, which indicates the net laser power coupled to the photomixer. Therefore, only a small fraction of the photocurrent contributes to the THz generation, and/or there exists a large amount of power loss in the THz optics. The field-dependent carrier lifetime described in Sect. 4.4 could explain the former reason for low efficiency. The latter could be due to the power loss in the THz optics, e.g., reflection and absorption in the silicon lens and the substrate.

The optical–THz conversion efficiency of the traveling-wave device is $\approx 10^{-6}$ at frequencies around $2\,\mathrm{THz}$. It is slightly lower than small-area devices due to its lower quantum efficiency, which is caused by the spherical aberration of the cylindrical lens that focuses the laser beam onto the device. Various types of loss in the CPS line also decrease severely the THz output power, as described in Sect. 4.6. Since the optical–THz conversion efficiency of photomixers is proportional to the laser power, the traveling-wave device pumped by a high-power laser promises high efficiency. If advanced traveling-wave designs to reduce the CPS line loss work properly, as described in Sect. 4.6, they could provide very high levels of optical–THz conversion efficiency.

The optical–THz conversion efficiency of photomixers can, in principle, exceed the Manley–Rowe limit for the three-wave mixing process in nonlinear optical materials, by the order of 10^{-3}. In order to attain such levels, better understanding of the power-loss mechanism in the device and drastic change of the photomixer design would be necessary.

The optical–THz conversion efficiency of UTC-PD has already reached the Manley–Rowe limit at $300\,\mathrm{GHz}$. The efficiency of UTC-PD at $1\,\mathrm{THz}$ is already one order of magnitude higher than the LT-GaAs photomixers, though this frequency is above the device bandwidth of UTC-PD. As discussed in the previous section, the optical–THz conversion efficiency and the maximum output power of the photomixer depend linearly and quadratically on the quantum efficiency, respectively. Excellent performance of UTC-PD relies on its high quantum efficiency. The challenge for UTC-PD is the extension of the bandwidth to the THz range keeping high quantum efficiency. Since the average carrier traveling distance in UTC-PD is much greater than that in LT-GaAs photomixers, the challenge for UTC-PD is the further improvement of the quantum efficiency rather than the considerable extension of the bandwidth to the THz range.

7 Noise Behavior

The performance of the spectroscopic source is determined not only by the maximum output power but also the stability of the output power. The noise of photomixer is caused by various physical processes depending on operational conditions. Here, we describe briefly the noise behavior of the LT-GaAs photomixer.

A fundamental noise limit for the THz output power is the photocurrent noise (shot noise), which arises from the quantum nature of the photocarrier. When the mean square noise of the photocurrent is represented by I_n^2, the noise power of THz waves, P_n, is simply given by

$$P_n = I_n^2 R_A \,. \tag{8}$$

Note that this expression is the same as (4) when $I_n^2 = I_{ph}^2/2$, because I_n is the effective amplitude of the AC signal, while I_{ph} is its peak amplitude.

For the photodiode-type photomixer such as UTC-PD, the dominant noise component is the shot noise (quantum noise) in the photogeneration process, where the bias field is usually very low and the recombination process is absent. The mean square noise current of the shot noise is given by

$$I_n^2 = 2e(I_{ph} + I_{dark})B \,, \tag{9}$$

where e is the elementary electric charge, I_{ph} is the DC photocurrent, I_{dark} is the dark current (usually very small), and B is the device bandwidth [70]. In most cases, the device bandwidth is determined by the carrier transit time (τ_e) and the RC time constant, i.e., $B = 2.8/(2\pi\tau_e)$ or $1/(2\pi R_A C)$ [67]. Because a much higher photocurrent level is available in the photodiode-type photomixer, the signal-to-noise ratio, I_{ph}^2/I_{dark}^2, can be much larger than that of photoconductor-type photomixers such as the LT-GaAs photomixer.

For the photoconductor-type photomixer, the shot noise is given by an equation different from (9). If the pump-laser power is high or/and the bias field is high, the dominant noise component of photoconductors is so-called g–r noise, which is caused by the random process of generation and recombination (g–r) of carriers in photoconductors [71]. The amplitude of the g–r noise in current unit is given by

$$I_{\mathrm{gr}}^2 = 4e(I_{\mathrm{ph}} + I_{\mathrm{dark}})gB\,, \tag{10}$$

where g is the photoconductive gain. The expression of the g–r noise is the same as that of the shot noise in photodiodes, except for a factor of 2, but depends linearly on the photoconductive gain. Note that higher photoconductive gain results in higher noise because of fewer carriers being required for a given photocurrent. By substituting a typical photocurrent of $I_{\mathrm{ph}} = 0.5\,\mathrm{mA}$ for the pump-laser power of $\approx 100\,\mathrm{mW}$ and $g = 0.03$ for (10), a typical value of the g–r noise of the LT-GaAs photomixer is estimated to be $I_{\mathrm{gr}} = 3(B/\mathrm{THz})^{1/2}\mu\mathrm{A}$.

Under very low illumination condition or/and at low bias filed, the Johnson (thermal) noise corresponding to the photomixer resistance could dominate in the noise current. The photomixer resistance, R_{d}, and the temperature, T, give the Johnson noise in current units as

$$I_{\mathrm{J}} = \frac{4kTB}{R_{\mathrm{d}}}\,, \tag{11}$$

where k is the Boltzmann constant [71]. As an example, the Johnson noise at room temperature for a typical resistance of the LT-GaAs photomixer, $\approx 10\,\mathrm{k\Omega}$, is $I_{\mathrm{J}} \approx 1(B/\mathrm{THz})^{1/2}\mu\mathrm{A}$, which is the same order as the g–r noise. Of course, the Johnson noise can be considerably reduced by cooling the device, but the reduction is not very significant because of the square-root dependence on the temperature, unless it is cooled down to, say, liquid-helium temperature.

If the electrode contact of the photoconductor is not perfect, the current flows intermittently and generates the photocurrent noise [72]. However, such contact noise shows a $1/f$ type spectrum, and it would not contribute to the frequency range of interest.

The laser-intensity fluctuation, the so-called relative intensity noise (RIN) defined by

$$\mathrm{RIN} = \frac{I_{\mathrm{n}}^2}{I_{\mathrm{ph}}^2 B} \tag{12}$$

also contributes to the photocurrent noise. The RIN of typical diode laser for the laser power of $> 10\,\mathrm{mW}$ is of the order of $< -120\,\mathrm{dB/Hz}$, which is a similar order to the g–r noise and Johnson noise [73]. Since these laser-induced noise components are highly dependent on the individual laser and independent of the photomixer property.

The source-noise-limited signal-to-noise ratio of the measurement is given by the ratio of the THz output power to the noise power,

$$\frac{P_{\mathrm{THz}}}{P_{\mathrm{n}}} = \frac{I_{\mathrm{ph}}^2}{2I_{\mathrm{n}}^2}\,. \tag{13}$$

For a bandwidth of 1 THz, the DC photocurrent of 0.5 mA and the photocurrent noise estimated from (10)–(12), the source-noise-limited signal-to-noise ratio is $P_{\mathrm{THz}}/P_{\mathrm{n}} \approx 10^5$. Since the total system bandwidth for the THz power measurement with a slow detector can be ≈ 10 Hz, and the random noise presented here should be negligible. In this case, the system-noise performance would be limited by the other factors including the optical alignment change due to acoustic vibration.

In practical use of the photomixer for spectroscopic applications, the amplitude noise in the detected THz signal is not always the photocurrent noise as described above. If there is a spectral line in the transfer function of the total system, the frequency jitter (FM noise) of the laser is converted to the amplitude noise depending on the line profile. This type of noise is serious in the spectroscopy of narrow molecular lines as described in the following section. In this case, the frequency stabilization is obviously important to achieve high signal-to-noise ratio.

8 Spectroscopic Applications

8.1 Terahertz Spectroscopy in Laboratory

As described in other chapters, the THz pulse source is quite useful in solid-state and liquid-state spectroscopy with relatively low spectral resolution. On the other hand, a CW THz source is suitable for atomic and molecular

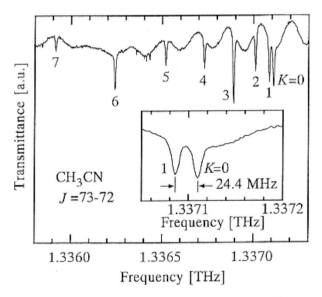

Fig. 30. Absorption spectrum of CH_3CN measured by using the photomixer source with free-running diode lasers

spectroscopy, which requires very high spectral resolution, e.g., $\delta\nu/\nu < 10^{-5}$. Wide spectral coverage is also crucial to search for new molecular lines. The photomixer as a spectroscopic source satisfies these requirements. Fourier transform spectroscopy (FTS) with a thermal (incoherent) source can also provide wide spectral coverage, but the photomixer has many advantages over the FTS instrument, at least in the THz region. The spectral power density of thermal sources commonly used for FTS, such as Hg lamps and electric heaters, is at most $\approx 1\,nW/MHz$ in the THz region, while currently available photomixers provide $\approx 100\,nW/MHz$. In order to obtain a high spectral resolution of $\delta\nu/\nu \approx 10^{-5}$ at 1 THz with FTS, a $\approx 10\,m$ long path difference of the interferometer is required. In contrast, the photomixer spectrometer takes up only a small space, say, $\approx 0.1\,m$, to provide the same spectral resolution.

Applications of CW THz photomixers to the spectroscopy of simple molecules, such as SO_2, CH_3CN, CO and H_2O, have been demonstrated [18, 34, 74, 75, 76, 77, 78], and it has been shown that the photomixer has sufficiently high performance to produce new scientific results [77, 78]. Figure 30 shows an example of the THz absorption spectrum of acetonitrile CH_3CN gas obtained by using a photomixer [18]. The setup for the spectroscopy is the same as that shown in Fig. 4, but a sample gas cell is placed in the THz beam. A quasidipole photomixer is pumped by two free-running external-cavity diode lasers. The THz radiation from the photomixer goes through a $\approx 50\,cm$ long gas cell with polyethylene windows, and the transmitted power is measured with a 4.2 K silicon-composite bolometer followed by a lock-in amplifier. The gas cell is filled with the gas sample at a pressure of 0.2 Torr. The spectrum shown in Fig. 30 was taken by sweeping the frequency of one of the two lasers. A sinusoidal pattern seen in the spectrum is due to the interference fringes between the photomixer and the bolometer. This slow baseline change does not affect the accuracy of the high-resolution measurement. The spectral resolution was limited to $\approx 5\,MHz$ by instrumental linewidth, while the Doppler width and the pressure broadening width at room temperature are 1.5 MHz and 1 MHz, respectively. Series spectra of the K-structure for symmetric-top CH_3CN molecule are seen, and the nearest two adjacent lines are clearly separated, i.e., consistent with the instrumental linewidth.

The noise of photomixer does not dominate in the total noise of spectroscopic system, because the THz output power of the photomixer is presently not so high. Owing to a lowpass filter in front of the bolometer to reduce thermal infrared radiation, the background photon noise is negligible compared with the detector noise. Hence, the detection limit for spectral lines is limited by the detector noise. The noise equivalent power (NEP) of the bolometer used here is $\approx 10\,pW/Hz^{1/2}$. For a THz output power of the photomixer of $\approx 0.1\,\mu W$, the minimum detectable absorption of this system is estimated to be $\approx 10^{-4}$. The measured noise performance was actually one order of magnitude worse than this estimation [16]. The excess noise would be due to the alignment change by acoustic vibration and the laser jitter

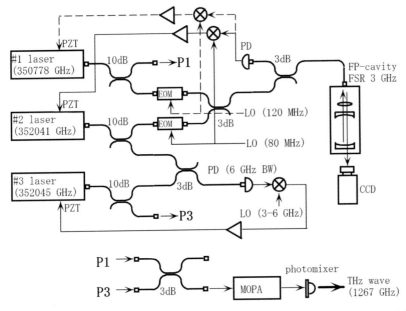

Fig. 31. Frequency-stabilized difference-frequency synthesizer system with three diode lasers

around the spectral line. The laser stabilization is obviously required to improve the performance. The fast data sampling and the fast frequency sweep, which shift the signal frequency out of the noise-dominating frequency range, would also help to reduce the noise. Once the detector-limited noise performance is achieved, it is meaningful to use a detector with better NEP, e.g., ultrasensitive detectors developed for infrared and sub-mm astronomy.

8.2 Frequency-stabilized Systems for Molecular Spectroscopy

For detailed study of molecular lines such as the line profile and the absorption strength, the instrumental linewidth and the frequency stability of the source should be sufficiently high compared with the molecular linewidth, which is typically a few MHz at room temperature. The frequency-stabilized photomixer systems with such a very narrow linewidth have already been built. A photomixer pumped by tunable dye lasers was used to study the self-broadening of SO_2 [74]. The dye lasers were stabilized to a confocal Fabry–Perot cavity by using edge-lock servocontrol, but the instrumental linewidth of the THz radiation is limited to 2 MHz by the laser jittering. The instrumental linewidth of $\approx 50\,\mathrm{kHz}$ was achieved by using a pair of distributed-Bragg-reflector (DBR) diode lasers stabilized by an optical feedback method using the reflected fringe of a tilted high-finesse confocal Fabry–Perot cavity [75]. The second overtone THz molecular spectroscopy with the best ever

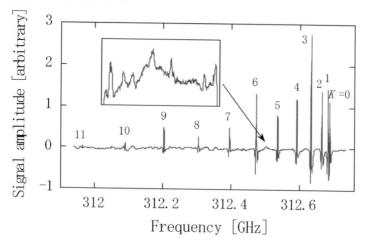

Fig. 32. Absorption spectrum of CH_3CN measured by using the photomixer source with frequency-stabilized systems

spectral resolution was performed by using this THz-source system. However, this system was simply designed to achieve very narrow linewidth by compromising on the frequency tunability.

Shown in Fig. 31 is the most progressive photomixer system that is designed to provide both excellent tunability and stability [34, 76]. Three fiber-coupled external-cavity diode lasers are used to synthesize the difference frequency. Two of the three lasers (indicated as #1 and #2) are locked to different longitudinal modes of an ultralow-expansion (ULE) confocal Fabry–Perot cavity with the Pound–Driver–Hall method; the phase-sensitive detection of the laser fringe reflected from the reference cavity provides a discriminator signal that crosses zero at the cavity resonance, and the discriminator signal is fed back to the laser-frequency control loop. For coarse tuning of the THz difference frequency, one of the two cavity-locked lasers is discretely changed by a step of the free spectral range (FSR) of the cavity (3 GHz). The third laser (#3) is phase-locked to one of the cavity-locked lasers (#2) with a tunable offset frequency generated by a microwave synthesizer. The difference frequency between lasers #2 and #3 is the sum of integral multiples of FSR of the reference cavity and the tunable microwave offset frequency. For fine tuning of the difference frequency, the offset frequency is changed up to the cavity FSR. The accuracy of the difference frequency is basically limited by the stability of the reference cavity and the determination accuracy of the FSR value.

The instrumental linewidth of the three-diode-laser system is $\approx 1\,\text{MHz}$, 5 times better than that of the free-running diode-laser system described in Sect. 8.1. The second-derivative absorption spectrum of CH_3CN measured by using the photomixer source with the three-laser system is shown in Fig. 32.

The $K = 0, 1$ lines, which are separated by $\approx 6\,\text{MHz}$, are clearly resolved, and this spectral resolution is consistent with the instrumental linewidth. In the present system, fast laser jitter out of the feedback-loop bandwidth limits the linewidth. If a higher-frequency control circuit and/or the optical feedback to the laser with higher-finesse cavities are used, much narrower linewidths could be achieved with the same frequency synthesizing strategy. The absolute calibration of the THz frequency can be done by measuring the electrical offset/drift in the cavity lock loops and monitoring very slow temporal change of cavity mode spacing. The measurement of molecular lines as described above is suitable for the absolute frequency calibration; any electrical and optical DC offsets in the system can be determined and corrected with high precision. Measurements of the rotational transitions of carbon monoxide (CO) are especially suitable for frequency calibration in the THz region, because their frequencies have been precisely determined in the millimeter-wave region and they are strong in the THz region. The absolute frequency calibration of the present system with an accuracy of $\delta\nu/\nu \approx 10^{-7}$ ($\delta\nu \approx 100\,\text{kHz}$ at $\nu \approx 1\,\text{THz}$), was achieved by performing the absorption spectroscopy for CO lines [34]. This accuracy has never been obtained with the other THz sources. The frequency-stabilized photomixer source has already provided the performance required for general molecular spectroscopy, and it has actually been producing many scientific outputs [77, 78].

8.3 Local-Oscillator Application for Heterodyne Detection

The photomixer is expected to be a local oscillator (LO) in a heterodyne receiver system for astronomy. The LO system for astronomy is required to be compact for ease in mounting on the telescope, and this is especially pronounced for space astronomy. As already described in the last section, wide spectral coverage and high tunability are crucial to search for new molecular species in the interstellar space and observe molecular lines in distant objects at cosmological redshift. The wide spectral coverage is especially important for space astronomy that is free from the restriction of the observable frequency range such as the atmospheric window. The photomixer LO satisfies these requirements.

When the LO frequency is fixed, the spectral coverage of the receiver system is limited by the electrical bandwidth of the heterodyne mixer. Recently, heterodyne mixer technology in the THz region has been greatly progressed, and the electrical bandwidth of the superconductor devices such as the superconductor-insulator-superconductor (SIS) mixer and the superconducting hot-electron bolometer (HEB) mixer have reached to several GHz [79, 80]. However, the GHz bandwidth covers only a tiny fraction of the whole THz range. If highly tunable LO is available, the observable range can be dramatically pushed up to "optical" bandwidth (sensible frequency range) of the mixer, usually over 100 GHz. It is known that the most sensitive detectors

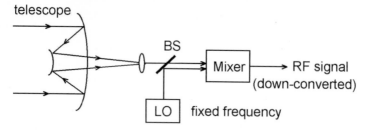

Fig. 33. Schematic view of heterodyne receiver system

in the THz range at present are semiconductor photoconductors, e.g., Ga-doped Ge. The semiconductor photoconductor mixers with bulk structures (not antenna-coupled) have wide optical bandwidth over 1 THz and have been used for astronomical observations, but they have suffered from narrow electrical bandwidth or slow response of $< 100\,\text{MHz}$ [81]. If the LO frequency can be tuned with an accuracy $\approx 10\,\text{MHz}$, the photoconductor mixer would be revived as a sensitive heterodyne receiver. The tunable LO would allow us to develop a novel type of mixer with little thought to the electrical bandwidth issue.

In the heterodyne receiver, the incoming THz signal is mixed with the LO signal and downconverted to the intermediate frequency (IF) signal, as shown in Fig. 33. Since the IF signal power, i.e., the receiver response, is linearly dependent on the LO power, the LO power is usually set to be much higher than the THz signal. The sensitivity (detection limit) of the heterodyne receiver is determined by the ratio between the IF signal power to the system noise power. For an ideal heterodyne system, the receiver sensitivity approaches the quantum-noise limit, i.e., the fundamental physical limit by the photon-number fluctuation, with increasing LO power. In most cases, the mixer is saturated by the LO power before reaching the quantum-noise limit, and the sensitivity is limited by the mixer/amplifier noise. It is noteworthy that the saturation behavior is, in some cases, important for the receiver to avoid the sensitivity change introduced by the LO-power fluctuation.

Present-day heterodyne mixers used for astronomy operated at $\approx 1\,\text{THz}$, e.g., SIS and HEB mixers, show sensitivity only a few times worse than the quantum-noise limit [78, 79]. To attain such high sensitivity, an LO power of $0.1\,\mu\text{W}$ to $1\,\mu\text{W}$ is required. A Gunn diode followed by a multiplier chain can provide such power levels at $\approx 1\,\text{THz}$, and it is commonly used for astronomical instruments. The LT-GaAs photomixer LO satisfies marginally the power requirement for the superconducting mixers. The semiconductor mixers require modest-level LO power, $\approx 10\,\mu\text{W}$, which is not achievable with the photomixer LO at present [81]. The metal-insulator-metal whisker-contact diodes and the GaAs Schottky diodes are available as the heterodyne mixer in the THz regime, but these mixers are quite power hungry, an LO power of

$\approx 1\,\mathrm{mW}$ is required [82]. This would be far beyond the scope of the LT-GaAs photomixer LO.

The THz output of the LT-GaAs photomixer was for the first time coupled to an SIS mixer at $630\,\mathrm{GHz}$ as described in [30]. In this experiment, one half of the THz output power, approximately $0.2\,\mu\mathrm{W}$, was coupled to the SIS mixer through a free-space diplexer. The noise temperature of the mixer pumped by the photomixer LO was measured by using the Y-factor method with a hot/cold load. The double-sideband noise temperature was $331\,\mathrm{K}$, in good agreement with the value obtained by using a Gunn diode followed by a varactor multiplier chain in place of the photomixer LO. Therefore, the photomixer LO is comparable to the Gunn oscillator and shows no apparent source noise, as discussed in Sect. 7. This result indicates that, even for high power laser input, the g–r noise in the THz output power is negligible compared with the total system noise including the mixer noise, because of the low photoconductive gain, g, of the LT-GaAs photomixer. The noise of photomixer in the ideal case is roughly estimated from (8) to be $P_\mathrm{n}/B \approx 10^{-21}\,\mathrm{W/Hz}$, which corresponds to the noise temperature of $T_\mathrm{n}(= P_\mathrm{n}/kB) \approx 70\,\mathrm{K}$. This value is comparable to the quantum-noise limit at $1\,\mathrm{THz}$, $T_\mathrm{limit}(= h\nu/k) = 48\,\mathrm{K}$. This would be the case for the SIS measurement that shows no excess noise from the photomixer.

The photomixer LO is expected to be applied to a large internationale mm and sub-mm telescope array for astronomy to be built in the Atacama desert, Chile, South America, named ALMA (Atacama Large Millimeter Array) [83]. For this project, about 100 array elements of $10\,\mathrm{m}$ size telescope will be constructed as an interferometer with the maximum baseline of $\approx 10\,\mathrm{km}$. The receivers for each telescope are designed to cover the frequency range from $\approx 100\,\mathrm{GHz}$ to $\approx 1\,\mathrm{THz}$. The interferometer will provide an angular resolution of $\approx 0.1''$ at $1\,\mathrm{THz}$ and a total telescope area of $\approx 10^4\,\mathrm{m}^2$, which are sufficient to unveil structure and evolution of distant galaxies, and physical phenomena around compact objects such as Black Holes. In order to operate such a large number of telescopes synchronously as an interferometer, the fiber-optic communication technology for linking all telescopes together will play the key role. The LO signal generated by a single oscillator located in a central station is distributed to separate telescopes via the optical fiber, the receiver signal from each telescope is sent back to the signal-processing station, and the image synthesis is done by using the phase difference between the receiver signals. This type of LO-fiber link system can be constructed only with "photonic" LO as the photomixer [84]. For the ALMA project, new development of the photonic LO with uni-traveling-carrier photodiodes (UTC-PD), described in Sect. 5, has already started.

Currently available UTC-PD have the electrical bandwidth of $\approx 300\,\mathrm{GHz}$, and are much more efficient and powerful than the LT-GaAs photomixers at lower frequencies below $1\,\mathrm{THz}$. The optical–THz conversion efficiency of UTC-PD is roughly 0.1 in the $100\,\mathrm{GHz}$ range, and the maximum output power reaches to the order of $100\,\mu W$ in the $300\,GHz$ range [66, 67] and a few microwatt in the $1\,THz$ range [68]. Ground-based telescopes do not require LO sources available at frequencies higher than $1\,THz$, because the observation in the THz range from the ground is very difficult due to the atmospheric absorption. UTC-PD is therefore suitable for this purpose. Recently, an astronomical observation of interstellar molecular lines with an SIS receiver pumped by the UTC-PD LO source was for the first time performed at $98\,GHz$ [85]. The results show that the performance of the UTC-PD, in terms of the linewidth and the noise temperature, is comparable to that of conventional Gunn LOs.

Although the maximum output power of UTC-PD is larger than that of the LT-GaAs photomixer at frequencies below $1\,THz$, the wider electrical bandwidth of the LT-GaAs photomixer over $1\,THz$ could be an advantage for the output power in the high-frequency regime. The LT-GaAs would be promising especially for space-based observations, in which there is no limitation on the observable frequency range [86]. Further development of the LT-GaAs photomixer to improve the efficiency is demanded for actual application to space astronomy.

9 Summary and Future Trends

In this Chapter, we have reviewed recent development of the photomixer as a CW coherent THz source. The principle of the difference-frequency generation, the output properties including the noise behavior, and examples of spectroscopic application have been described mainly on the diode-laser-pumped LT-GaAs photomixers.

Since the first demonstration of THz photomixing in LT-GaAs, many efforts to optimize the growth condition of the wafer and the device design, such as antenna shape and electrode structure, have been made. We have briefly surveyed a variety of device designs and their current best performance.

Spectroscopy of molecules and solid-state materials is a good test bench for the photomixer as a spectroscopic source. As expected, the frequency-stabilized photomixer system provided excellent tunability and stability that have never been attained with the other currently available THz sources. This system is suitable for applications in molecular spectroscopy and astronomy.

It is well known that the THz output power of the photomixer obeys the square-law dependence on the photocurrent. At very high photocurrent, a super-Manley–Rowe limit of optical–THz conversion efficiency is predicted, relying on this property. However, in reality the maximum output power of the LT-GaAs photomixer is limited by the thermal failure of the device and is currently insufficient for some important applications including the heterodyne LO application. To overcome the thermal limit practically, the AlAs-layered device and the traveling-wave design have been proposed. These device designs have successfully been demonstrated, but further improvement of the output power is still required.

The best solution for very high THz output power is a drastic improvement of the optical–THz conversion efficiency, which is proportional to the gain–bandwidth product. Recently developed UTC-PD provide both high responsivity (efficiency) and wide bandwidth consistently, while an ultrawide bandwidth of the LT-GaAs photomixer results from the sacrifice of the gain by shortening the carrier lifetime. As for the maximum output power, UTC-PD is already superior to the LT-GaAs photomixer at frequencies up to at least 1 THz. Of course, UTC-PD is not irrelevant to the thermal problem either. The traveling-wave design of UTC-PD to avoid the thermal problem would potentially provide the highest output power ever achieved. It is also noteworthy that the traveling-wave UTC-PD may yield "THz gain" due to the nonlinear current–voltage characteristics of the diode. Study of this type of interaction between THz waves and photocarriers is an interesting research direction in terms of the amplification of THz waves and new development of a THz modulator.

A solid-state THz laser as a fundamental oscillator, e.g. multiple-quantum-well laser, may be much more powerful than the photomixer and highly tunable as well. However, the photomixer has an advantage of the capability for synchronization of multiple devices located at distant places via the fiber-optical link, and this property is especially pronounced for the LO application in the long-baseline interferometer of multiple radio telescopes, e.g., ALMA. Communication with THz waves as the carrier waves would be inefficient, because low-loss transmission technology in the THz range is "light-years" behind the present optical communication technology. It seems to be better to use the THz laser as an amplifier that is injection-seeded by the photomixer LO. Thus, we conclude that the photomixer will remain as a useful THz source even after the THz laser is available.

References

[1] E. Brown, J. Söderström, C. Parker, L. Mahoney, K. Molvar, T. McGill: Oscillations up to 712 GHz in InAs/AlSb resonant-tunneling diodes, Appl. Phys. Lett. **58**, 2291–2293 (1991)
[2] J. Carlstrom, J. Zmuidzinas: *Review of Radio Science 1993–1996* (Oxford Univ. Press, New York 1996)
[3] K. Kawase, H. Minamide, K. Imai, J. Shikata, H. Ito: Injection-seeded terahertz-wave parametric generator with wide tenability, Appl. Phys. Lett. **80**, 195–197 (2002)
[4] R. Köhler, A. Tredicucci, F. Beltram, H. Beere, E. Linfield, A. Davies, D. Ritchie, R. Iotti, F. Rossi: Terahertz semiconductor-heterostructure laser, Nature **417**, 156–159 (2002)

[5] G. Scalari, L. Ajili, J. Faist, H. Beere, E. Linfield, D. Ritchie, G. Davies: Far-infrared ($\lambda \cong 87\,\mu m$) bound-to-continuum quantum-cascade lasers operating up to 90 k, Appl. Phys. Lett. **82**, 3165–3167 (2003)

[6] B. Williams, S. Kumar, H. Callebaut, Q. Hu, J. Remo: Terahertz quantum-cascade laser at 100 μm using metal waveguide for mode confinement, Appl. Phys. Lett. **83**, 2124–2126 (2003)

[7] S. Kumar, B. Williams, S. Kohen, Q. Hu, J. Reno: Continuous-wave operation of terahertz quantum-cascade lasers above liquid-nitrogen temperature, Appl. Phys. Lett. **84**, 2494–2496 (2004)

[8] E. Gornik, A. Andronov: Far-infrared semiconductor lasers, in E. Gornik, A. Andronov (Eds.): *Special Issue on Far-infrared Semiconductor Lasers*, Opt. Quantum Electron **23** (Chapman, Hall, New York 1991)

[9] G. Blake, K. Laughlin, R. Cohen, K. Busarow, D.-H. Gwo, C. Schmuttenmaer, D. Steyert, R. Saykally: The berkeley tunable far infrared laser spectrometers, Rev. Sci. Instrum. **62**, 1701–1716 (1991)

[10] H. Odashima, L. Zink, K. Evenson: Tunable far-infrared spectroscopy extended to 9.1 THz, Opt. Lett. **24**, 406–407 (1999)

[11] R. Kingston: *Detection of Optical and Infrared Radiation* (Springer, Berlin, Heidelberg 1978)

[12] E. Brown, F. Smith, K. McIntosh: Coherent millimeter-wave generation by heterodyne conversion in low-temperature-grown GaAs photoconductors, J. Appl. Phys. **73**, 1480–1484 (1993)

[13] E. Brown, K. McIntosh, F. Smith, M. Manfa, C. Dennis: Measurements of optical-heterodyne conversion in low-temperature-grown GaAs, Appl. Phys. Lett. **62**, 1206–1208 (1993)

[14] E. Brown, K. McIntosh, K. Nichols, C. Dennis: Photomixing up to 3.8 THz in low-temperature-grown GaAs, Appl. Phys. Lett. **66**, 285–287 (1995)

[15] K. McIntosh, E. Brown, K. Nichols, O. McMahon, W. DiNatole, T. Lyszczars: Terahertz photomixing with diode lasers in low-temperature-grown GaAs, Appl. Phys. Lett. **67**, 3844–3846 (1995)

[16] S. Verghese, K. McIntosh, E. Brown: Optical and terahertz power limits in low-temperature-grown GaAs photomixers, Appl. Phys. Lett. **71**, 2743–2745 (1997)

[17] S. Matsuura, M. Tani, K. Sakai: Generation of coherent terahertz radiation by photomixing in dipole photoconductive antennas, Appl. Phys. Lett. **70**, 559–561 (1997)

[18] S. Matsuura, M. Tani, H. Abe, K. Sakai, H. Ozeki, S. Saito: High-resolution terahertz spectroscopy by a compact radiation source based on photomixing with diode lasers in a photoconductive antenna, J. Mol. Spectrosc. **187**, 97–101 (1998)

[19] E. Peytavit, G. Mouret, J. Lampin, S. Arscott, P. Masselin, L. Desplanque, O. Vanbésien, R. Bocquet, F. Mollot, D. Lippens: Terahertz electromagnetic generation via optical frequency difference, IEE Proc.-Optoelectron. **149**, 82–87 (2002)

[20] F. Smith, A. Calawa, C.-L. Chen, M. Manfra, L. Mahoney: New mbe buffer used to eliminate backgating in GaAs MESFETs, IEEE Electron. Dev. Lett. **9**, 77–80 (1988)

[21] S. Gupta, M. Frankel, J. Valdmanis, J. Whitaker, G. Mourow, F. Smith, A. Calawa: Subpicosecond carrier lifetime in GaAs grown by molecular beam epitaxy at low temperatures, Appl. Phys. Lett. **59**, 3276–3278 (1991)

[22] N. Sekine, K. Hirakawa, F. Sogawa, Y. Arakawa, N. Usami, Y. Shiraki, T. Katoda: Ultrashort lifetime photocarriers in ge thin films, Appl. Phys. Lett. **68**, 3419–3421 (1996)

[23] C. Kadow, A. Jackson, A. Gossard, J. Bowers, S. Matsuura, G. Blake: Self-assembled ErAs islands in GaAs for THz applications, Physica E **7**, 97–100 (2000)

[24] C. Kadow, A. Jackson, A. Gossard, S. Matsuura, G. Blake: Self-assembled ErAs islands in GaAs for optical-heterodyne THz generation, Appl. Phys. Lett. **76**, 3510–3512 (2000)

[25] S. Verghese, K. McIntosh, E. Brown: Highly tunable fiber-coupled photomixers with coherent terahertz output power, IEEE Trans. Microwave Theory and Tech. **45**, 1301–1309 (1997)

[26] S. Duffy, S. Verghese, K. McIntosh: *Sensing with Terahertz Radiation* (Springer, Berlin, Heidelberg 2003) pp. 193–236

[27] S. Duffy, S. Verghese, K. McIntosh, A. Jackson, A. Gossard, S. Matsuura: Accurate modeling of dual dipole and slot elements used with photomixers for coherent terahertz output power, IEEE Trans. Microwave Theory and Tech. **49**, 1032–1038 (2001)

[28] A. Jackson: *Low-Temperature-Grown GaAs Photomixers Designed for Increased Terahertz Output Power*, Ph. D. thesis, Univ. California at Santa Barbara (1999)

[29] C. Brewitt-Taylor, D. Gunton, H. Rees: Planar antennas on a dielectric surface, Electron. Lett. **17**, 729–731 (1981)

[30] S. Verghese, E. Duerr, K. McIntosh, S. Duffy, S. Calawa, C.-Y. Tong, R. Kimberk, R. Blundell: A photomixer local oscillator for a 630-GHz heterodyne receiver, IEEE Microwave and Guided Wave Lett. **9**, 245–247 (1999)

[31] S. Matsuura, P. Chen, G. Blake, J. Pearson, H. Pickett: Simultaneous amplification of terahertz difference frequencies by an injection-seeded semiconductor laser amplifier at 850 nm, Int. J. Infrared Millim. Waves **19**, 849–858 (1998)

[32] M. Ohtsu: *Highly Coherent Semiconductor Lasers* (Artech House, Boston 1992)

[33] P. Zorabedian: Tunable external-cavity semiconductor lasers, in Duarte (Ed.): *Tunable Lasers Handbook* (Academic, San Diego 1995) pp. 349–442

[34] S. Matsuura, P. Chen, G. Blake, J. Pearson, H. Pickett: A tunable cavity-locked diode laser source for terahertz photomixing, IEEE Trans. Microwave Theory and Tech. **48**, 380–387 (2000)

[35] E. Brown: A photoconductive model for superior GaAs THz photomixers, Appl. Phys. Lett. **75**, 769–771 (1999)

[36] E. Peytavit, S. Arscott, D. Lippens, G. Mouret, S. Matteon, P. Masselin, R. Bocquet, J. Lampin, L. Desplanque, F. Mollot: Terahertz frequency difference from vertically integrated low-temperature-grown GaAs photodetector, Appl. Phys. Lett. **81**, 1174–1176 (2002)

[37] J. Darrow, X.-C. Zhang, D. Auston: Power scaling model of large-aperture photoconducting antennas, Appl. Phys. Lett. **58**, 25–27 (1991)

[38] E. Duerr, K. McIntosh, S. Verghese: in *Proc. 10th Int. Symp. on Space Terahertz Tech.* (Charlottesville 1999) pp. 29–37

[39] Y.-J. Chiu, S. Fleischer, J. Bowers: High-speed low-temperature-grown GaAs p-i-n traveling-wave photodetector, IEEE Photonics Tech. Lett. **10**, 1012–1014 (1998)

[40] J.-W. Shi, S.-W. Chu, M.-C. Tien, C.-K. Sun, Y.-J. Chiu, J. Bowers: Edge-coupled membrane terahertz photonic transmitters based on metal-semiconductor-metal traveling-wave photodetectors, Appl. Phys. Lett. **81**, 5108–5110 (2002)

[41] S. Matsuura, G. Blake, R. Wyss, J. Pearson, C. Kadow, A. Jackson, A. Gossard: A traveling-wave THz photomixer based on angle-tuned phase matching, Appl. Phys. Lett. **74**, 2872–2874 (1999)

[42] S. Matsuura, G. Blake, R. Wyss, J. Pearson, C. Kadow, A. Jackson, A. Gossard: Free-space traveling-wave THz photomixers, in *Proc. IEEE 7th Int. Terahertz Electron. Conf.* (Nara 1999) pp. 24–27

[43] D. Rutledge, D. Neikirk, D. Kasilingam: Integrated circuit antennas, in K. Button (Ed.): *Infrared and Millimeter Waves*, vol. 10 (Academic, New York 1983) pp. 1–90

[44] S. Matsuura: unpublished work

[45] S. Matsuura, G. Blake, R. Wyss, J. Pearson, C. Kadow, A. Jackson, A. Gossard: Design and characterization of optical-THz phase-matched traveling-wave photomixers, in (Proc. SPIE 1999) pp. 484–492

[46] R. Wyss, T. Lee, J. Pearson, S. Matsuura, G. Blake, C. Kadow, A. Gossard: Embedded coplanar strips traveling-wave photomixers, in (Proc. 12th Int. Symp. on Space Terahertz Tech., San Diego 2001) p. 91

[47] R. Wyss, S. Martin, B. Nakamura, A. Neto, D. Pasqualini, P. Siegel, C. Kadow, A. Gossard: Traveling-wave membrane photomixers, in (Proc. 12th Int. Symp. on Space Terahertz Tech., San Diego 2001) p. 233

[48] A. Stöhr, A. Malcoci, F. Siebe, K. Lill, P. van der Waal, R. Güsten, D. Jäger: Integrated photonic THz transmitter employing ultra-broadband traveling-wave 1.55 µm photodetectors, in (Tech. Dig. Int. Topical Meeting on Microwave Photonics, Awaji 2002) pp. 69–72

[49] A. Malcoci, A. Stöhr, A. Schulz, D. Jäger: Optical terahertz generation, in (Tech. Dig. Int. Topical Meeting on Microwave Photonics, Budapest 2003) pp. 179–182

[50] T. Ishibashi, N. Shimizu, S. Kodama, H. Ito, T. Nagatsuma, T. Furuta: Uni-traveling-carrier photodiodes, in (Tech. Dig. Ultrafast Electronics and Opto-electronics, Lake Tahoe 1997) pp. 83–87

[51] T. Ishibashi, T. Furuta, H. Fushimi, S. Kodama, H. Ito, T. Nagatsuma, N. Shimizu, Y. Miyamoto: InP/InGaAs uni-traveling-carrier photodiodes, IEICE Trans. Electron. **E83-C**, 938–949 (2000)

[52] H. Ito, T. Furuta, T. Ishibashi: High-speed and high-output uni-traveling-carrier photodiodes, IEICE Trans. Electron. **E84-C**, 1448–1454 (2001)

[53] T. Ishibashi, T. Furuta, H. Fushimi, H. Ito: Photoresponse characteristics of uni-traveling-carrier photodiodes, in (Proc. SPIE 2001) pp. 469–479

[54] T. Furuta, H. Ito, T. Ishibashi: Photocurrent dynamics of uni-traveling-carrier and conventional pin-photodiodes, Inst. Phys. Conf. Ser. **166**, 419–422 (2000)

[55] T. Ishibashi: GaAs-based and InP-based heterostructure bipolar transistors, in R. Kiehl, T. Sollner (Eds.): *High Speed Heterostructure Devices*, Semiconductors and Semimetals **41** (Academic, San Diego 1994) pp. 291–358

[56] K. Kato, S. Hata, K. Kawano, A. Kozen: Design of ultrawide-band, high-sensitivity p-i-n photodetectors, IEICE Trans. Electron. **E76-C**, 214–221 (1993)

[57] H. Ito, T. Furuta, S. Kodama, T. Ishibashi: InP/InGaAs uni-traveling-carrier photodiode with 310 GHz bandwidth, Electron. Lett. **36**, 1809–1810 (2000)

[58] T. Nagatsuma, M. Yaita, M. Shinagawa, K. Kato, A. Kozen, K. Iwatsuki, K. Suzuki: Electro-optic characterization of ultrafast photodetectors using adiabatically compressed soliton pulses, Electron. Lett. **30**, 814–816 (1994)

[59] H. Ito, Y. Hirota, A. Hirata, T. Nagatsuma, T. Ishibashi: 11 dBm photonic millimeter-wave generation at 100 GHz using uni-traveling-carrier photodiodes, Electron. Lett. **37**, 1225–1226 (2001)

[60] H. Ito, T. Nagatsuma, A. Hirata, T. Minotani, A. Sasaki, Y. Hirota, T. Ishibashi: High-power photonic millimeter-wave generation at 100 GHz using matching-circuit-integrated uni-traveling-carrier photodiodes, IEE Proc. Optoelectron. **150**, 138–142 (2002)

[61] T. Nagatsuma, T. Ishibashi, A. Hirata, Y. Hirota, T. Minotani, A. Sasaki, H. Ito: Characterization of uni-traveling-carrier photodiode monolithically integrated with matching circuit, Electron. Lett. **37**, 1246–1247 (2001)

[62] P. Huggard, B. Ellison, P. Shen, N. Gomes, P. Davis, W. Shillue, A. Vaccari, J. Payne: Efficient generation of guided millimeter-wave power by photomixing, IEEE Photonics Technol. Lett. **14**, 197–199 (2002)

[63] T. Noguchi, A. Ueda, H. Iwashita, S. Takano, Y. Sekimoto, M. Ishiguro, T. Ishibashi, H. Ito, T. Nagatsuma: Millimeter wave generation using a uni-traveling-carrier photodiode, in (Proc. 12th Int. Symp. on Space Terahertz Technol., San Diego 2001) pp. 73–80

[64] H. Ito, T. Ito, Y. Muramoto, T. Furuta, T. Ishibashi: Rectangular waveguide output uni-traveling-carrier photodiode module for high-power photonic millimeter-wave generation in the F-band, IEEE J. Lightwave Technol. **LT-21**, 3456–3462 (2003)

[65] A. Stöhr, R. Heinzelmann, K. Hagedorn, R. Güsten, F. Schäfer, H. Stüer, F. Siebe, P. van der Wal, V. Krozer, M. Feiginov, D. Jäger: Integrated 460 GHz photonic transmitter module, Electron. Lett. **37**, 1347–1348 (2001)

[66] H. Ito, T. Furuta, Y. Hirota, T. Ishibashi, A. Hirata, T. Nagatsuma, H. Matsuo, T. Noguchi, M. Ishiguro: Photonic millimeter-wave emission at 300 GHz using an antenna-integrated uni-traveling-carrier photodiode, Electron. Lett. **38**, 989–990 (2002)

[67] H. Ito, S. Kodama, Y. Muramoto, T. Furuta, T. Nagatsuma, T. Ishibashi: High-speed and high-output uni-traveling-carrier photodiodes, IEEE J. Selected Topics in Quantum Electron. **10**, 709–727 (2004)

[68] H. Ito, F. Nakajima, T. Furuta, K. Yoshino, T. Ishibashi: Photonic terahertz-wave generation using an antenna-integrated uni-traveling-carrier photodiode, Electron. Lett. **39**, 1828–1829 (2003)

[69] S. Verghese, K. McIntosh, S. Duffy, E. Duerr: Continuous-wave terahertz generation using photomixers, in *Terahertz Sources and Systems* (Kluwer, Dordrecht 2001)

[70] S. Sze: *Physics of Semiconductor Devices*, 2nd ed. (Wiley, New ork 1981)

[71] G. Rieke: *Detection of Light* (Cambridge, New ork 1994)

[72] A. Rose: *Concepts in Photoconductivity and Allied Problems* (Wiley, New York 1963)

[73] L. Coldren, S. Corzine: *Diode Lasers and Photonic Integrated Circuits* (Wiley, New York 1995)

[74] A. Pine, R. Suenram, E. Brown, K. McIntosh: A terahertz photomixing spectrometer – Application to SO_2 self-broadening, J. Mol. Spectrosc. **175**, 37–47 (1996)

[75] P. Chen, G. Blake, M. Gaidis, E. Brown, K. McIntosh, S. Chou, M. Nathan, F. Williamson: Spectroscopic applications and frequency locking of THz photomixing with distributed-bragg-reflector diode lasers in low-temperature-grown GaAs, Appl. Phys. Lett. **71**, 1601–1603 (1997)

[76] H. Pickett, P. Chen, J. Pearson, S. Matsuura, G. Blake: Construction of a three-diode-laser terahertz difference-frequency synthesizer, in (Proc. IEEE 7th Int. Terahertz Electron. Conf., Nara 1999) pp. 95–101

[77] P. Chen, J. Pearson, H. Pickett, S. Matsuura, G. Blake: Submillimeter-wave measurements and analysis of the ground and $\nu_2 = 1$ states of water, Astrophys. J. Suppl. Ser. **128**, 371–385 (2000)

[78] I. Kleiner, G. Tarrago, C. Cottaz, L. Sagui, L. Brwon, R. Poynter, H. Pickett, P. Chen, J. Pearson, R. Sams, G. Blake, S. Matsuura, V. Nemtchinov, P. Varanasi, L. Fusina, G. D. Lonardo: $^{14}NH_3$ and PH_3 line parameters: 2000 hitran update and new results, in (Proc. JQSRT 2003) to be published

[79] A. Karpov, J. Blondel, M. Voss, K. Gundlach: A three photon noise SIS heterodyne receiver at submillimeter wavelength, IEEE Trans. Appl. Supercond. **9**, 4456–4459 (1999)

[80] B. Karasik, M. Gaidis, W. McGrath, B. Bumble, H. LeDuc: A low-noise 2.5 THz superconductive nb hot-electron mixer, IEEE Trans. Appl. Supercond. **7**, 3580–3583 (1997)

[81] T. Philips, K. Jefferts: A low temperature bolometer heterodyne receiver for millimeter wave astronomy, Rev. Sci. Instrum. **44**, 1009–1014 (1973)

[82] S. Marazita, K. Hui, J. Hesler, W. Bishop, T. Crowe: Progress in submillimeter wavelength integrated mixer technology, in (Proc. 10th Int. Symp. on Space THz Tech., Charlottesville 1999) pp. 74–85

[83] M. Ishiguro, et al.: Japanese large millimeter and submillimeter array, in T. Philips (Ed.): *Advance Technology MMW, Radio, and Terahertz Telescopes*, vol. 3357 (Proc. SPIE 1998) p. 244

[84] J. Payne, L. D'Addario, D. Emerson, A. Kerr, B. Shillue: in *Proc. SPIE*, vol. 3357 (1998) p. 143

[85] S. Takano, A. Ueda, T. Yamamoto, S. Asayama, Y. Sekimoto, T. Noguchi, M. Ishiguro, H. Takara, S. Kawanishi, H. Ito, A. Hirata, T. Nagatsuma: The first radioastronomical observation with photonic local oscillator, Publ. Astron. Soc. Japan **55**, L53–L56 (2003)

[86] S. Matsuura, G. Blake, P. Chen, R. Wyss, J. Pearson, H. Pickett, A. Jackson, C. Kadow, A. Gossard: A Photonic Local Oscillator Source for Far-IR and Sub-mm Heterodyne Receivers, The Institute of Space Science and Astronautical Sciences Report SP No.14, 337–344 (2000)

Terahertz Time-Domain Spectroscopy

Seizi Nishizawa[1,2], Kiyomi Sakai[3],
Masanoi Hangyo[4], Takeshi Nagashima[4], Mitsuo Wada Takeda[5],
Keisuke Tominaga[6,7,8,9], Asako Oka[6,7,8,9], Koichiro Tanaka[10],
and Osamu Morikawa[11]

[1] Office of Technology Transfer, Japan Science Technology Agency: JST
   Advanced Infrared Spectroscopy Corporation: Aispec
   11-17 Hiyoshi, Hachioji, Tokyo 193-0836, Japan
   `nzw914@aispec.com`
[2] Research Center for Development of Far-Infrared Region, Fukui University,
   3-9-1, Bunkyo, Fukui 910-8507, Japan
[3] National Institute of Information and Communications Technology
   558-2 Iwaoka, Nishi-ku, Kobe, Hyogo 651-2392, Japan
   `sakai@nict.go.jp`
[4] Institute of Laser Engineering, Osaka University, 2-6 Yamadaoka, Suita, Osaka
   565-0871, Japan
[5] Department of Physics, Faculty of Science, Shinshu University, Matsumoto
   390-8621, Japan
[6] Molecular Photoscience Research Center, Kobe University, Nada, Kobe,
   657-8501, Japan
[7] Graduate School of Science and Technology, Kobe University, Nada, Kobe,
   657-8501, Japan
[8] CREST/JST, Nada, Kobe, 657-8501, Japan
[9] Department of Chemistry, Faculty of Science, Kobe University, 1-1, Rokkoudai,
   Nada, Kobe, Hyogo, 657-8501, Japan
[10] Department of Physics, Faculty of Science, Kyoto University, Kitashirakawa,
   Sakyo, Kyoto 606-8502, Japan
[11] Department of Science, Coast Guard Academy, 5-1 Wakabacho, Kure,
   Hiroshima 737-8512, Japan

Abstract. Terahertz time-domain spectroscopy (THz-TDS) is one of the most successful fields of THz optoelectronics. Measurements of semiconductors, ferroelectric crystals of current industrial interest, photonic crystals and biomolecules by means of standard THz-TDS are shown first, then modified versions such as attenuated total reflection spectroscopy and ellipsometry are described. In addition to the THz-TDS, generation and detection of THz radiation with the use of multimode laser diodes and applications in spectroscopy are described, although the radiation is emitted in CW mode. This Chapter is composed of topical sections written by specialists of the respective topics.

1 Introduction

The terahertz (THz) region of the electromagnetic (EM) spectrum is of great importance due to the rich physical and chemical processes in this range. The

K. Sakai (Ed.): Terahertz Optoelectronics, Topics Appl. Phys. **97**, 203–269 (2005)
© Springer-Verlag Berlin Heidelberg 2005

recently progressed THz time-domain spectroscopy (TDS) affords a powerful technique for the research of these physical and chemical processes [1, 2, 3].

The principle of the THz-TDS is described in the Chapter by *Sakai* and *Tani* (Sect. 3) showing two representative systems. The THz-TDS is, in principle, a coherent emission and detection system that emits single-cycle THz pulses and detects them at a repetition rate close to 100 MHz. The signal is detected in the form of an electric field and the Fourier transformation of the pulse signal gives rise to both amplitude and phase spectra over a wide spectral range.

The THz-TDS is reasonably compared with the well-established Fourier transform spectroscopy (FTS) that detects THz beams in the form of the power. It has been recognized that TDS is advantageous over FTS since it gives phase as well as amplitude information, which can avoid the uncertainty caused by the Kramers–Kronig analysis, but it has been restricted to a region less than several THz until recently. However, the progress of short-pulse laser technology has made it possible to extend the spectral range to over 60 THz, as is described in the Chapter by *Kono* et al. The second important advantage is that the signal is observed with higher signal-to-noise ratio by TDS than by FTS in spite of the fact that the former uses a detector operating at room temperature. There are other advantages: the signal is observed in the form of a time trace with subpicosecond time resolution, and the gated and coherent nature of THz detector dramatically reduces its minimum detectable power due to the thermal background.

The THz-TDS is becoming increasingly important and it is used not only for traditional physical or chemical materials but biological substances or novel materials.

In succession to the standard THz-TDS, cross-correlation THz-TDS is added, which in essence deals with CW THz radiation.

2 Principles of THz-TDS

2.1 Transmission Spectroscopy

Typical optical setups for transmission spectroscopy are shown in Fig. 15 in the Chapter by *Sakai* and *Tani*. The waveform of the THz pulse is obtained by changing the time delay (see Fig. 10 in the same chapter). Two waveforms with and without the sample, $E_{\mathrm{sam}}(t)$ and $E_{\mathrm{ref}}(t)$, are measured in the time domain and they are Fourier-transformed into the complex amplitudes $\tilde{E}_{\mathrm{sam}}(\omega)$ and $\tilde{E}_{\mathrm{ref}}(\omega)$ in the frequency domain, respectively. The ratio of $\tilde{E}_{\mathrm{sam}}(\omega)$ to $\tilde{E}_{\mathrm{ref}}(\omega)$ is given by

$$\frac{\tilde{E}_{\text{sam}}(\omega)}{\tilde{E}_{\text{ref}}(\omega)} = \left|\sqrt{T(\omega)}\right| \exp\left\{-i\left[\Delta\phi(\omega) - \frac{\omega}{c}d\right]\right\}$$

$$= \frac{4\tilde{n}(\omega)}{[\tilde{n}(\omega) + 1]^2} \frac{\exp\left\{-i[\tilde{n}(\omega) - 1]\frac{\omega}{c}d\right\}}{1 - \dfrac{[\tilde{n}(\omega) - 1]^2}{[\tilde{n}(\omega) + 1]^2} \exp\left[-i2\tilde{n}(\omega)\frac{\omega}{c}d\right]}, \tag{1}$$

where $\tilde{n}(\omega) = n(\omega) - ik(\omega)$ is the complex refractive index, $T(\omega)$ the measured power transmittance, $\Delta\phi(\omega)$ the intrinsic phase shift, d the thickness of the sample and c the speed of light in a vacuum.

From the experimentally obtained $\sqrt{T(\omega)}$ and $\Delta\phi(\omega)$, we can determine $n(\omega)$ and $k(\omega)$, starting with a roughly estimated value of the complex refractive index and iteratively modifying the value so that the difference between the measured value and the calculated value is minimized.

The complex refractive index can be easily translated into the complex relative dielectric constant (hereafter, complex dielectric constant) $\tilde{\varepsilon}(\omega) = \varepsilon_1(\omega) - i\varepsilon_2(\omega)$ or into complex electrical conductivity $\tilde{\sigma}(\omega) = \sigma_1(\omega) - i\sigma_2(\omega)$. The relations are $\tilde{\varepsilon}(\omega) = \tilde{n}^2(\omega)$ and $\sigma_1(\omega) = \varepsilon_0\omega\varepsilon_2(\omega)$, $\sigma_2(\omega) = -\varepsilon_0\omega[\varepsilon_1(\omega) - \varepsilon_\infty]$ with ε_∞ the dielectric constant of the material at high enough frequency and ε_0 the permittivity of free space.

The THz radiation can also be measured by means of interferometry, but, in this case, only amplitude information is obtained and phase information is lost as is stated in the first chapter, which makes it difficult to obtain complex refractive index.

2.2 Reflection Spectroscopy

It is necessary to make reflection measurements for optically thick samples (e.g., heavily carrier-doped semiconductors). The waveform measured on the sample and that on the reference mirror, $E_{\text{sam}}(t)$ and $E_{\text{ref}}(t)$, are Fourier-transformed into the complex amplitudes $\tilde{E}_{\text{sam}}(\omega)$ and $\tilde{E}_{\text{ref}}(\omega)$, respectively. For normal incidence the ratio of $\tilde{E}_{\text{sam}}(\omega)$ and $\tilde{E}_{\text{ref}}(\omega)$ is given by

$$\frac{\tilde{E}_{\text{sam}}(\omega)}{\tilde{E}_{\text{ref}}(\omega)} = \frac{|\sqrt{R(\omega)}| \exp\left[-i\Delta\phi(\omega)\right]}{|\sqrt{R_{\text{ref}}(\omega)}| \exp\left[-i\Delta\phi_{\text{ref}}(\omega)\right]} = \frac{[1 - \tilde{n}(\omega)][1 + \tilde{n}_{\text{ref}}(\omega)]}{[1 + \tilde{n}(\omega)][1 - \tilde{n}_{\text{ref}}(\omega)]}, \tag{2}$$

where $\tilde{n}_{\text{ref}}(\omega)$ is the refractive index of reference mirror. The surface of the reference mirror must be positioned at the same place as that of the sample. Any misalignment causes a severe error in the phase and it should be suppressed to less than $1\,\mu\text{m}$ [4]. In the case of contributions of internal multireflection, (2) is rewritten as

$$\frac{\tilde{E}_{\text{sam}}^{(R)}(\omega)}{\tilde{E}_{\text{ref}}^{(R)}(\omega)} = \frac{[\tilde{n}(\omega)^2 - 1]\left\{1 - \exp\left[-2i\frac{\omega}{c}\tilde{n}(\omega)d\right]\right\}}{[\tilde{n}(\omega) + 1]^2 - [\tilde{n}(\omega) - 1]^2 \exp\left[-2i\frac{\omega}{c}\tilde{n}(\omega)d\right]}$$

$$\times \frac{[\tilde{n}_{\text{ref}}(\omega) + 1]^2 - [\tilde{n}_{\text{ref}}(\omega) - 1]^2 \exp\left[-2i\frac{\omega}{c}\tilde{n}_{\text{ref}}(\omega)d\right]}{[\tilde{n}_{\text{ref}}(\omega)^2 - 1]\left\{1 - \exp\left[-2i\frac{\omega}{c}\tilde{n}_{\text{ref}}(\omega)d\right]\right\}}. \tag{3}$$

The misalignment problem in reflection measurements is resolved by combining the THz-TDS with the attenuated-total-reflection (ATR) and ellipsometry methods described in the following section.

3 Application of THz-TDS

3.1 Semiconductors*

There are absorptions associated with various excitations, for example, phonons and free carriers, in the THz region in semiconductors. The optical constants in the THz region are strongly affected by the carrier density and scattering mechanisms of doped free carriers. Therefore, the measurement of optical constants in the THz region has technological importance in addition to the interest from the viewpoint of the semiconductor physics. Before the invention of the THz-TDS, measurements of the complex constants or the absorption constant associated with doped carriers in semiconductors in the millimeter wave and far-infrared (FIR) regions were performed using somewhat specialized apparatus and they are time consuming. For example, both real and imaginary parts of the dielectric constant were measured by transmission and reflection measurements at some particular frequency (107.3 GHz radiation from a klystron) and the complex dielectric constant was deduced [5]. The absorption constant of n-type Si from $20\,\text{cm}^{-1}$ to $200\,\text{cm}^{-1}$ ($600\,\text{GHz}$ to $6\,\text{THz}$) was measured by Martin–Puplett (M–P)-type interferometer using synchrotron radiation [6]. In this measurement, the transmission measurements for the samples with two different thicknesses are necessary to deduce the absorption coefficient by removing the effect of the reflection and transmission at the surfaces. By using the THz-TDS, as described in Sect. 2.1, the complex refractive index $\tilde{n} = n - i\kappa$ and complex dielectric constant $\tilde{\varepsilon} = \varepsilon_1 - i\varepsilon_2$ are directly deduced from the measurement of the waveforms with and without samples and their Fourier transformation. With the THz-TDS, these quantities from $0.03\,\text{THz}$ to $3\,\text{THz}$ are obtained in a typical measurement time of $10\,\text{min}$. The effectiveness of the THz-TDS for measurements of the carrier density and mobility has been demonstrated by *Grischkowsky* and coworkers [7, 8, 9] and even a small deviation from the simple Drude model was found for doped Si [10, 11]. In this subsection, the

---

* by Masanori Hangyo, and Takeshi Nagashima

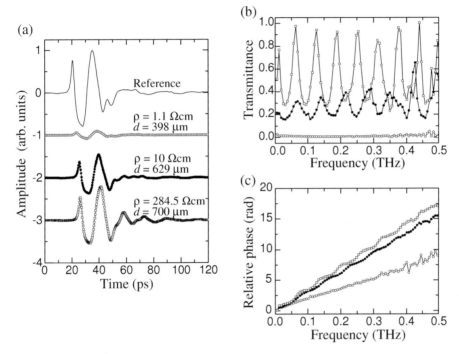

Fig. 1. (a) Waveforms of the THz pulse transmitted through Si wafer samples with various resistivities. **(b)** transmittance and **(c)** intrinsic phase shift obtained from the Fourier transformation of the waveforms in **(a)**

measurements for doped Si wafers with a thickness of $\approx 500\,\mu\mathrm{m}$ are described as a typical case.

Figure 1a shows the waveform of the THz pulses transmitted through n-type Si samples with various resistivities. In this measurement, the THz pulses are generated by exciting a LT-GaAs photoconductive antenna with a compact Er-doped fiber laser. The waveforms transmitted through the wafers are delayed and their amplitudes are reduced compared with the reference waveform. The amplitude is reduced with decreasing resistivity (increasing the carrier density) due to the free-carrier absorption. The transmittance and intrinsic phase shift calculated by (1) in Sect. 2.1 are shown in Figs. 1b and c, respectively. By applying the iterative procedure to (1), the complex refractive indices are deduced as shown in Figs. 2a, c and e. It is seen that the increase in the refractive index in the low-frequency region becomes prominent with decreasing resistivity. The complex electrical conductivity $\tilde{\sigma}(\omega) = \sigma_1(\omega) - i\sigma_2(\omega)$ due to free carriers in doped Si is given by the relation

$$\tilde{n}^2(\omega) = \tilde{\varepsilon}(\omega) = \varepsilon_{\mathrm{Si}} - i\tilde{\sigma}(\omega)/\omega\varepsilon_0 \,, \tag{4}$$

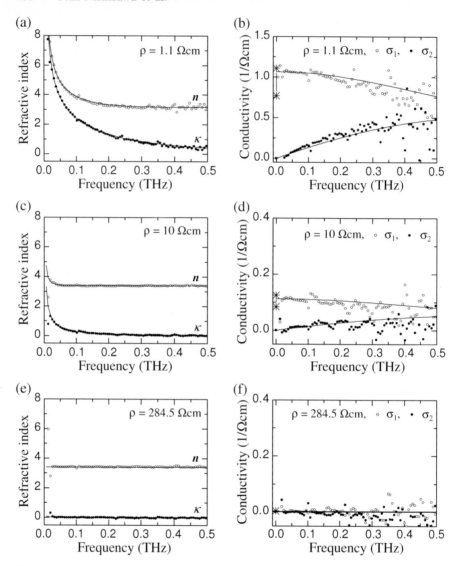

Fig. 2. Complex refractive indices (**a, c, e**) and complex conductivities (**b, d, f**) of Si with various resistivities obtained by the THz-TDS. The *asterisks* in (**b**), (**d**) and (**f**) are the DC conductivities measured by the four-point contact method

where $\tilde{\varepsilon}(\omega) = \varepsilon_1(\omega) - i\varepsilon_2(\omega)$ is the complex dielectric constant, $\varepsilon_{\mathrm{Si}} = 11.6$ is the relative dielectric constant of undoped Si. The complex conductivity obtained from the complex refractive indices in Fig. 2 are shown in Figs. 2b, d and f. The asterisks at zero frequency in Fig. 2 are the DC conductivities obtained by the DC four-point contact method, which agree well with the DC conductivities obtained by extending the dispersion of the conductivity

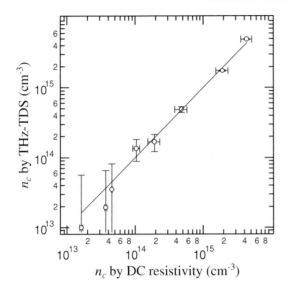

Fig. 3. Comparison of the carrier densities obtained by the THz-TDS and those by the four-point contact method

measured by the THz-TDS to zero frequency. Usually, the optical constants of the semiconductor in the FIR region are analyzed by the Drude model. According to the Drude model, the complex conductivity is expressed as

$$\tilde{\sigma}(\omega) = \varepsilon_0 \frac{\omega_{\mathrm{p}}^2}{i\omega + \frac{1}{\tau}}, \tag{5}$$

where τ is the momentum relaxation time of free carriers, ω_{p} the plasma angular frequency defined by $\omega_{\mathrm{p}} = \sqrt{n_c e^2 / \varepsilon_0 m^*}$; n_c the carrier density; e the elementary charge; m^* the effective carrier mass taken as $0.26 m_0$ for Si, where m_0 is the rest mass of free electrons. On fixing $\tau = 0.2\,\mathrm{ps}$, the dispersion curves of the conductivity are fitted to (5) taking n_c as a fitting parameter. The obtained carrier density is plotted in Fig. 3 as a function of the carrier density measured by the DC four-point contact method. The two carrier densities agree well over nearly three orders of magnitude. The results in Fig. 3 show that noncontact characterization of electrical properties of doped semiconductor wafers is possible if the carrier density is in the proper range. For the relatively highly doped Si wafers with a thickness of $\approx$ 0.5 mm, the sample becomes opaque in the THz region for carrier densities larger than $10^{16}\,\mathrm{cm}^{-3}$. In such a case, the carrier density can be deduced from the reflection measurement [4, 12]. On the other hand, for the sample with carrier density less than $10^{13}\,\mathrm{cm}^{-3}$, it becomes very difficult to estimate the carrier density because the difference of the transmittance from undoped Si becomes quite small.

Next, the temperature dependence of the optical constants of a doped Si wafer is shown [13]. The sample is a $400\,\mu\mathrm{m}$ thick n-type Si wafer with a resistivity of $1.1\,\Omega\cdot\mathrm{cm}$ at room temperature. The sample is mounted on a

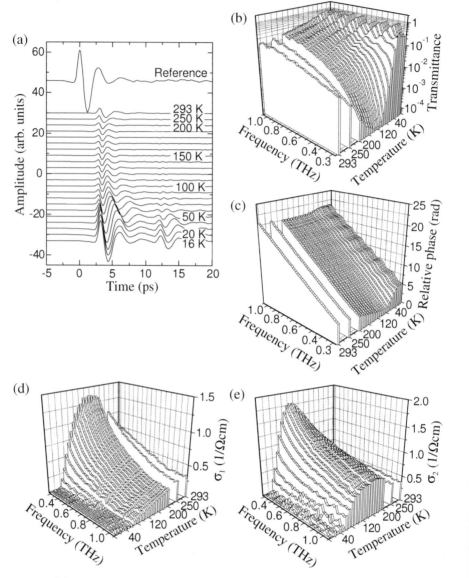

Fig. 4. (a) Temperature dependence of the waveforms transmitted through a doped Si wafer. (b), (c) The temperature dependence of the transmittance and intrinsic phase shift. (d), (e) The temperature dependence of the real and imaginary parts of the complex conductivity

cold finger of a cryostat with optical windows made of fused quartz (although opaque in the visible region, optical windows made of high-resistivity Si are better for the THz region). Figure 4 shows the temperature dependence of the

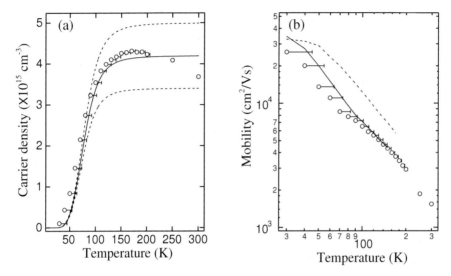

Fig. 5. Temperature dependence of (**a**) the carrier density and (**b**) the mobility obtained from the Drude fit to the complex conductivity data. The *solid line* in (**a**) shows the temperature dependence of the carrier density calculated based on the Fermi–Dirac distribution assuming a donor density of $4.2 \times 10^{15}\,\mathrm{cm}^{-3}$ and the *dashed lines* show that assuming donor densities of $3.4 \times 10^{15}\,\mathrm{cm}^{-3}$ and $4.2 \times 10^{15}\,\mathrm{cm}^{-3}$. The *solid* and *dashed lines* in (**b**) show the results from Hall measurements for Si with resistivities of $0.66\,\Omega \cdot \mathrm{cm}$ and $4.76\,\Omega \cdot \mathrm{cm}$, respectively

waveform transmitted through the sample. The amplitude of the waveform at first slightly decreases and then increases with decreasing temperature. The pulse reflected at the backward and forward surfaces is clearly observed at $\approx 13\,\mathrm{ps}$ after the main pulse at $\approx 4\,\mathrm{ps}$ at low temperatures, indicating that the sample becomes transparent at low temperatures. The transmittance and intrinsic phase shift obtained by the Fourier transformation are shown in Figs. 4b and c, respectively. The dispersion of the transmittance becomes apparent in the temperature range from $80\,\mathrm{K}$ to $200\,\mathrm{K}$. The complex conductivity deduced using (4) is shown in Figs. 4d and e. Both real and imaginary parts change drastically with temperature. From the Drude model analysis, by taking the carrier density and scattering time as fitting parameters, the temperature dependence of the carrier density and mobility $\mu(= e\tau/m^*)$ is obtained, as shown in Figs. 5a and b, respectively. The solid line in Fig. 5a is the temperature dependence of the carrier density calculated based on charge neutrality condition with the Fermi–Dirac distribution assuming a donor density of $4.2 \times 10^{15}\,\mathrm{cm}^{-3}$, and the dashed lines are calculations assuming densities of $3.4 \times 10^{15}\,\mathrm{cm}^{-3}$ and $4.2 \times 10^{15}\,\mathrm{cm}^{-3}$. A donor ionization energy of $45\,\mathrm{meV}$ is used for all calculations. The decrease of carrier density by freezing out is well reproduced by the calculation. The solid and dashed lines in Fig. 5b show the mobility from the Hall measurement for

Si with resistivities of $0.66\,\Omega\cdot\text{cm}$ and $4.76\,\Omega\cdot\text{cm}$, respectively. The temperature dependence obtained by the THz-TDS agrees with the Hall mobilities. The above measurements demonstrate the effectiveness of the THz-TDS for measuring the electrical properties of semiconductors in a noncontact way.

The THz-TDS has been applied not only to bulk semiconductors [14, 15], but also to thin films of semiconductors [16, 17]. In order to apply the transmission-type THz-TDS to semiconductor films (100 nm to 10 μm thick), it is necessary that the semiconductor films should have appropriate carrier densities. Too low a carrier density, for example below $10^{16}\,\text{cm}^{-3}$, results in too small a change of the transmittance compared with that of the substrate only, making the estimation of carrier density and mobility quite difficult. Although not touched upon in this book, optical-pump and THz-probe measurement of semiconductors using the THz-TDS is very effective to clarify the ultrafast carrier dynamics after photoexcitation [18, 19].

3.2 Ferroelectrics*

In ferroelectric crystals, the lowest optical modes, namely the soft optic modes, appears in the THz region below $150\,\text{cm}^{-1}$, in which the ferroelectricity originates dominantly from the instability of polar soft modes in a ferroelectric transition. The ferroelectric soft modes propagate as polaritons, which are the propagating EM waves coupled to a polar optical phonon. The polariton dispersion provides important information for both fundamental and technical problems in ferroelectrics. By the THz-TDS, both the exact measurements of power transmittance $T(\omega)$ and intrinsic phase shifts $\Delta\phi(\omega)$ make experimentally direct estimation of both the real and imaginary parts of the complex dielectric constants $\tilde{\varepsilon}(\omega) = \varepsilon_1(\omega) - i\varepsilon_2(\omega)$ and the complex refractive index $\tilde{n}(\omega) = n(\omega) - i\kappa(\omega)$ free from the uncertainty caused by the Kramers–Kronig analysis in the conventional FTS. Furthermore, the intrinsic phase shifts $\Delta\phi(\omega)$ on sample specimens enable us to make analytical estimation of the dispersion relations for various excitations coupled with the propagating THz radiations. These advantageous features of the THz-TDS are effectively adapted for the characterization of ferroelectric materials.

Recently, a compact instrument of THz-TDS [20, 21] has been developed with adaptation for the quantitative routine measurements. Complex transmission spectra were measured by using the newly developed THz-TDS spectrometer in a range between $3\,\text{cm}^{-1}$ and $100\,\text{cm}^{-1}$ on some ferroelectric crystals, which include bismuth titanate (BIT) [21, 22], lithium heptagermanate (LGO) [21, 23], and lithium niobate (LN) [21, 24], as the ferroelectric crystals of current industrial interest. From the power transmittance $T(\omega)$ and the intrinsic phase shift $\Delta\phi(\omega)$ of the transmission spectra, the magnitudes of the complex refractive index $\tilde{n}(\omega)$, the complex dielectric con-

---

* by Seizi Nishizawa

stant $\tilde{\varepsilon}(\omega)$ and the wavevector $\tilde{k}(\omega)$ of the propagating waves in the sample specimen were directly estimated.

3.2.1 Transmission Spectra of Ferroelectric Crystals

The waveform $E(t)$ measured in the time domain is Fourier-transformed into the complex electric field $\tilde{E}(\omega)$ in the frequency domain, which has the amplitude component $E(\omega)$ and the phase component $\phi(\omega)$, as shown by $E(\omega)\exp\left[-i\phi(\omega)\right]$. From the electric fields measured with and without the sample specimens, $\tilde{E}_{\mathrm{sam}}(\omega)$ and $\tilde{E}_{\mathrm{ref}}(\omega)$, the power transmittance $T(\omega)$ and the intrinsic phase shift $\Delta\phi(\omega)$ are given as $T(\omega) = \left[E_{\mathrm{sam}}(\omega)/E_{\mathrm{ref}}(\omega)\right]^2$ and $\Delta\phi(\omega) = \phi_{\mathrm{sam}}(\omega) - \left[\phi_{\mathrm{ref}}(\omega) - \omega d/c\right]$, respectively. When the additional factors produced by the internal multireflection and the surface reflection in the sample specimen are eliminated, the power transmittance $T(\omega)$ and the intrinsic phase shift $\Delta\phi(\omega)$ are expressed by the imaginary part $\kappa(\omega)$ and the real part $n(\omega)$ of the complex refractive index $\tilde{n}(\omega)$ in $T(\omega) = \exp\left[-\omega\kappa(\omega)d/c\right]$ and $\Delta\phi(\omega) = \omega n(\omega)d/c$, respectively, where the real part $k(\omega)$ of the wavevector $\tilde{k}(\omega)$ is directly proportional to the intrinsic phase shift $\Delta\phi(\omega)$ as expressed by the relation

$$k(\omega) = \frac{\Delta\phi(\omega)}{d}. \tag{6}$$

The internal multireflection and the surface reflection make the analytical expressions more complicated, in which expressions the ratio of E_{sam} and E_{ref} is written by (1). On the basis of (1), the magnitudes of complex refractive index $\tilde{n}(\omega)$ are experimentally estimated from the measured values of power transmittance $T(\omega)$, intrinsic phase shift $\Delta\phi(\omega)$, and sample thickness d. The estimated values of $\tilde{n}(\omega)$ are reduced to the magnitudes of the dielectric constant $\tilde{\varepsilon}(\omega)$ through $\tilde{\varepsilon}(\omega) = \left[\tilde{n}(\omega)\right]^2$, and to those of the wavevector $\tilde{k}(\omega)$ through $\tilde{k}(\omega) = \omega\tilde{n}(\omega)/c$. In the high-frequency region of $\omega n(\omega)d/c \gg \pi$, the additional phase shifts caused by the internal multireflection and the surface reflection become ineffectual to be negligible enough, where the magnitude of $k(\omega)$ is simply estimated through (6) for the measured values of $\Delta\phi(\omega)$. In the ferroelectric crystals, the frequency dependence of $k(\omega)$ is responsible for the propagation of phonon–polariton excitation.

Bismuth titanate $\mathrm{Bi_4Ti_3O_{12}}$ *(BIT) [21, 22]*

BIT is one of the most important key materials in FeRAM that has recently attracted much attention. The sample specimen used for the THz-TDS measurements is a thin c-plate of a flux-grown single crystal with an area of $15\,\mathrm{mm}^2 \times 15\,\mathrm{mm}^2$ and a thickness of $d = 0.225\,\mathrm{mm}$. Since the cleavage perpendicular to the c-axis is very strong, very flat cleaved surfaces are used for the measurements. Complex transmission spectra of the c-plate BIT were measured at room temperature by using the newly developed THz-TDS spectrometer. BIT has monoclinic-layered perovskite structure with the point group m

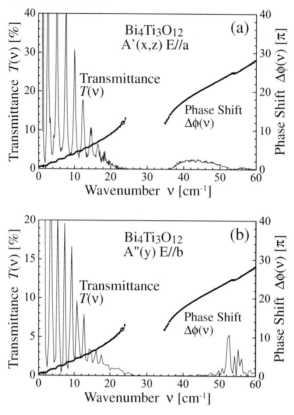

Fig. 6. Transmittance and intrinsic phase shift measured on the c-plate of $Bi_4Ti_3O_{12}$ (BIT) crystal for the light polarization parallel to (**a**) the a-axis and (**b**) the b-axis

at room temperature. It undergoes a ferroelectric phase transition at 675 °C, and a high-temperature paraelectric phase is tetragonal with the point group $4/mmm$ [25]. The direction of spontaneous polarization is inclined at a small angle of about 4.5° toward the monoclinic a-axis in the ac-plane. The a- and c-axis components of the spontaneous polarization are 50 $\mu C/cm^2$ and 4 $\mu C/cm^2$, respectively. In the ferroelectric phase, the displacive nature was observed with the underdamped soft optic mode at 28 cm^{-1} and at room temperature [26].

As to optical phonon modes, $A'(x, z)$ and $A''(y)$ modes are both infrared (IR)-active where the mirror plane is perpendicular to the y-axis. The soft optic mode has an $A'(x, z)$ symmetry where the coordinate z is parallel to the a-axis. The $A'(x, z)$ and $A''(y)$ modes are active when the light polarization is parallel to the a-axis ($E \parallel a$) and b-axis ($E \parallel b$), respectively. For the polarizations $E \parallel a$ and $E \parallel b$, the power transmittance $T(\nu)$ and the intrinsic phase shift $\Delta\phi(\nu)$ of the transmission spectra measured in the

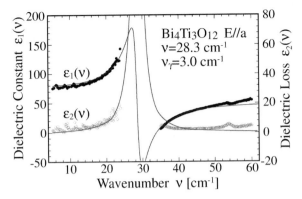

Fig. 7. Complex dielectric constants measured on a *c*-plate of $Bi_4Ti_3O_{12}$ (BIT) crystal for the light polarization parallel to the *a*-axis

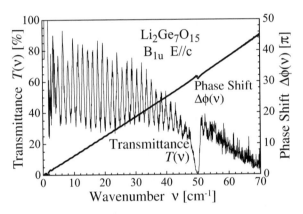

Fig. 8. Transmittance and intrinsic phase shift measured on $Li_2Ge_7O_{15}$ (LGO) crystal for the light polarization parallel to *c*-axis

unit of wave number ν [cm^{-1}] are shown in Figs. 6a and b, respectively. In Fig. 7, the complex dielectric constants $\tilde{\varepsilon}(\nu)$ estimated from the power transmittance $T(\nu)$ and the intrinsic phase shift $\Delta\phi(\nu)$ of transmission spectra are shown for the polarization $E \parallel a$. It is found that the wavenumber dependence of $\Delta\phi(\nu)$ and $\tilde{\varepsilon}(\nu)$ shows nonlinear behavior near the gap edges at TO- and LO-mode frequencies. The increase of the gradient in $\Delta\phi(\nu)$ at the edges indicates the slowing down of group velocity reasonable for the propagation of phonon–polariton (as described analytically in "polariton dispersion of the ferroelectric crystals").

Lithium heptagermanate $Li_2Ge_7O_{15}$ *(LGO) [21, 23]*

LGO is one of the weak ferroelectric crystals for surface acoustic wave filter applications of current industrial interest. LGO undergoes a ferroelectric

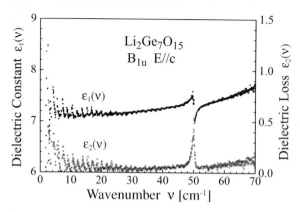

Fig. 9. Complex dielectric constants measured on $Li_2Ge_7O_{15}$ (LGO) crystal for the light polarization parallel to c-axis

phase transition at $T_{mpc} = 283.5$ K. At the transition, the lattice structure of space group *Pbcn* in the paraelectric phase transits to that of *Pbc2$_1$* with the spontaneous polarization along the c-axis in the ferroelectric phase where the unit cell size is maintained ($z = 4$). The soft phonon modes of the B_{1u} irreducible representation in the paraelectric and A_1 representation in the ferroelectric phases were observed by IR-absorption and Raman-scattering measurements, and a dielectric critical slowing down was also observed above 1 GHz from dielectric dispersion measurements [27].

The sample specimen used for the THz-TDS measurements is a plate of LGO single crystal with a thickness of 1.18 mm. The transmission spectra of the LGO plate were measured in units of wave number ν at the temperature of 295 K. The power transmittance $T(\nu)$ and intrinsic phase shift $\Delta\phi(\nu)$ of the transmission spectra measured on the plate specimen of LGO are shown in Fig. 8. The electric field of incident THz radiation was polarized parallel to the polar c-axis of LGO so that the response of the polar $B_{1u}(z)$ phonon mode was measured. The complex dielectric constant $\tilde{\varepsilon}(\nu)$ estimated from the transmission spectra is given in Fig. 9. The sharp response is at 50 cm^{-1}, and an anomalous increase is seen above 60 cm^{-1} and below 10 cm^{-1}. It is found that an absorption peak appears near 50 cm^{-1} and an anomalous increase of gradient indicates the propagation of phonon–polaritons.

Lithium niobate LiNbO$_3$ *(LN)[21, 24]*

LN is called a multifunctional crystal because of its many applications. LN has a rhombohedral structure with the point group $3m$ at room temperature. It undergoes a ferroelectric phase transition at 1402 K, and a high-temperature paraelectric phase has the point group $\bar{3}m$ [28]. The direction of spontaneous polarization is along the three-fold z-axis. At the Γ point of reciprocal lattices, the symmetry of the optical modes is given by $4A_1(z) + 9E(x, y) + 5A_2$,

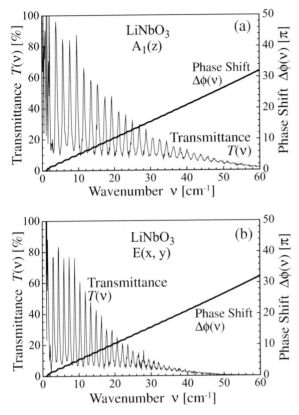

Fig. 10. Transmittance and intrinsic phase shift measured on an x-plate of a poled LiNbO$_3$ (LN) crystal for the light polarization parallel to the (**a**) z-axis and (**b**) y-axis

where $A_1(z)$ and $E(x, y)$ modes are both IR- and Raman-active polar modes, while A_2 modes are Raman- and IR-inactive nonpolar modes. For uniaxial crystals such as LN with a unique z-axis, $\tilde{\varepsilon}(\omega)$ becomes anisotropic, in which $\tilde{\varepsilon}_x(\omega) = \tilde{\varepsilon}_y(\omega) = \tilde{\varepsilon}_\perp(\omega), \tilde{\varepsilon}_z(\omega) = \tilde{\varepsilon}_\parallel(\omega)$.

The sample specimen used was a thin x-plate of a congruent LN single crystal with an area of $10 \, \text{mm}^2 \times 10 \, \text{mm}^2$ and a thickness d of 0.508 mm. Transmission spectra of an x-plate of a poled LN crystal were observed at room temperature by using the THz-TDS. $A_1(z)$ and $E(x, y)$ modes are active for light polarization parallel to the z-axis ($E \parallel z$) and the y-axis ($E \parallel y$), respectively. For two polarization directions, two different transmission spectra with $A_1(z)$ and $E(y)$ symmetries were separately observed down to $3 \, \text{cm}^{-1}$, as shown in Figs. 10a and b, respectively. It is found that the frequency dependence of $\phi(\nu)$ shows a deviation from linearity as the frequency increases. From the power transmittance $T(\nu)$ and the intrinsic phase shift $\Delta\phi(\nu)$ from $3 \, \text{cm}^{-1}$ to $55 \, \text{cm}^{-1}$, the real and imaginary parts of complex

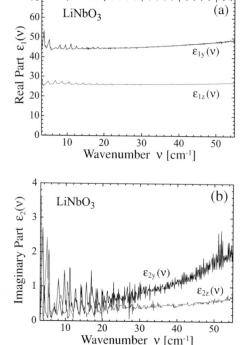

Fig. 11. (a) Real and (b) imaginary parts of complex dielectric constants, $\tilde{\varepsilon}_y(\nu)$ and $\tilde{\varepsilon}_z(\nu)$, of a poled LiNbO$_3$ (LN) crystal x-plate

dielectric constants $\tilde{\varepsilon}_y(\nu)$ and $\tilde{\varepsilon}_z(\nu)$ were estimated as shown in Figs. 11a and b, respectively. Such high-frequency complex dielectric constants of LN are very important in the current research of optical communications. However, they have not yet been reported, to our knowledge. Although the lowest TO-modes are far above this frequency window, the imaginary parts increase as frequency increases, and $\varepsilon_{2y}(\nu)$ increases faster than $\varepsilon_{2z}(\nu)$. This finding reflects the difference in the lowest TO-mode frequency between $A_1(z)$ and $E(x,y)$ symmetries, because the lowest TO-mode with $E(x,y)$ symmetry is at $151\,\text{cm}^{-1}$, while that of $A_1(z)$ symmetry is $256\,\text{cm}^{-1}$. These increases are caused mainly by phonon damping originating from the defect-induced scattering [29]. When the order–disorder nature is valid, the response of a relaxation mode can be observed in the high-frequency dielectric constants. In contrast, the observed dielectric behavior does not show any evidence of such a relaxation mode.

3.2.2 Phonon–Polariton Dispersion

Propagating radiations in crystals are coupled with polar optical phonons, and those coupled together to excite phonon–polaritons. The nonlinear increases found in the intrinsic phase shifts $\Delta\phi(\nu)$ (shown in Figs. 6, 8 and 10) indicate the slowing of the propagating group velocity, which is reasonable

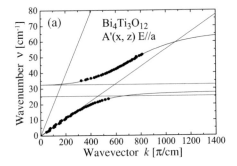

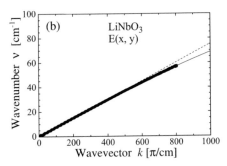

Fig. 12. Dispersion relation measured on a c-plate of $Bi_4Ti_3O_{12}$ (BIT) crystal for the light polarization parallel to (**a**) the a-axis and (**b**) the b-axis. *Solid curves* denote the calculated curves on the basis of the Kurosawa formula

for the propagation of phonon–polaritons. The special dielectric feature of ferroelectric materials in the THz region is controlled by phonon–polaritons [21, 22, 23, 24]. From the intrinsic phase shift $\Delta\phi(\nu)$ shown in Figs. 6a and b on BIT, the polariton dispersion relations estimated through (6) are shown in Figs. 12a and b [21, 22], respectively. In the same procedure, Fig. 13 shows the polariton dispersion relation estimated for LGO [21, 23], and also Figs. 14a and b show those estimated for LN [21, 24]. The observed dispersion curves using solid circles are well reproduced by the phonon–polariton dispersion curves calculated through the following equation based on the Kurosawa formula [30]. When the damping of phonons is negligible, the wavevector $k(\omega)$ is expressed as

$$k(\omega) = \frac{\sqrt{\varepsilon(1)}}{c}\omega \prod_{i=1}^{2}\left(\left\{\frac{\omega_{LO_i}^2 - \omega^2}{\omega_{TO_i}^2 - \omega^2}\right\}^{\frac{1}{2}}\right), \qquad (7)$$

where ω_{LO_i} and ω_{TO_i} are the frequencies of LO- and TO-phonons, respectively. The contribution of modes $i \geq 3$ is renormalized by the fitting parameter $\varepsilon(1)$ as given by

$$\varepsilon(1) = \varepsilon(\infty) \prod_{i=3}^{N} \frac{\omega_{LO_i}^2}{\omega_{TO_i}^2}. \qquad (8)$$

In BIT, $\varepsilon(\nu)$ is anisotropic and depends on the polarization direction. For the two directions of light polarization parallel to the a-axis ($E \parallel a$) and the b-axis

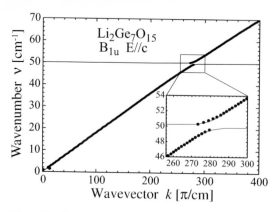

Fig. 13. Dispersion relation measured on $Li_2Ge_7O_{15}$ (LGO) crystal for the light polarization parallel to the c-axis. *Solid curves* denote the calculated curves on the basis of the Kurosawa formula

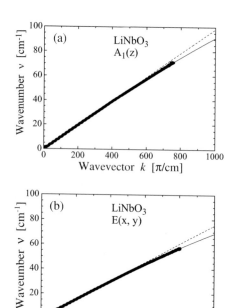

Fig. 14. Dispersion relation measured on an x-plate of a poled $LiNbO_3$ (LN) crystal for the light polarization parallel to the **(a)** z-axis and **(b)** y-axis. *Solid curves* denote the calculated curves on the basis of the Kurosawa formula

$(E \parallel b)$, two different low-frequency polariton branches were separately observed down to $3\,cm^{-1}$. The solid curves in Figs. 12a and b denote the curves calculated by (7) and (8) using mode frequencies of $A'(x, z)$ and $A''(y)$ symmetries, respectively. A good agreement can be seen between the observed and calculated polariton dispersions. Therefore, it is concluded that this nonlinear relation reflects the dispersion relation of bulk phonon–polaritons. Table 1 shows the values of fitting parameters for observed polariton dispersions.

Table 1. Values of the fitting parameters. Phonon–polariton dispersion relation of bismuth titanate $Bi_4Ti_3O_{12}$ (BIT)

	$A'(x,y)$ polariton $(E \parallel a)$	$A''(y)$ polariton $(E \parallel b)$
ν_{TO1}	28.3 cm^{-1}	35.9 cm^{-1}
ν_{LO1}	34.42 cm^{-1}	42.0 cm^{-1}
ν_{TO2}	53.76 cm^{-1}	85.0 cm^{-1}
ν_{LO2}	53.78 cm^{-1}	98.4 cm^{-1}
ν_{TO3}	70.5 cm^{-1}	
ν_{LO3}	74.5 cm^{-1}	
$\varepsilon(\infty)$	6.76	6.76
$\varepsilon(1)$	45.97	74.74
$\varepsilon(0)$	75.99	146.41

It is possible to estimate the static dielectric constant $\varepsilon(0)$ at the low-frequency limit through the asymptote of the polariton lowest branch, in which

$$\omega = ck/\sqrt{\varepsilon(0)}. \qquad (9)$$

The solid curves in Fig. 13 denote the calculated curves for $B_{1u}(z)$ symmetry polaritons in LGO, where the values of phonon parameters are $\nu_{TO1} = 49.5$ cm^{-1}, $\nu_{LO1} = 50.1$ cm^{-1}, $\nu_{TO2} = 87.5$ cm^{-1}, and the values of dielectric parameters of $\varepsilon_c(1) = 7.0$ and $\varepsilon_c(0) = 7.8$, respectively. A good agreement can be found between the observed and calculated polariton dispersions. The obtained value of $\varepsilon_c(0) = 7.8$ is slightly smaller than that of $\varepsilon_c(0) = 9.8$ measured at 10 kHz by an LCR meter, where it is suggested that soft phonons exist below 10 cm^{-1}.

Table 2. Values of the fitting parameters. Phonon–polariton dispersion relation of lithium niobate $LiNbO_3$ (LN)

	$A_1(z)$ polariton $(E \parallel z)$	$E(y)$ polariton $(E \parallel y)$
$\varepsilon(1)$	12.1	19.9
$\varepsilon(0)$	25.9	44.0

In Figs. 14a and b, the calculated dispersion curves for $A_1(z)$ and $E(x,y)$ symmetry polaritons in LN crystal are shown, respectively. The upper dotted lines are those calculated by (9) and the lower solid curves are obtained using (7) and (8). A good agreement is found between the observed and calculated polariton dispersions. Table 2 shows the values of dielectric parameters of $\varepsilon(1)$ and $\varepsilon(0)$. The obtained value of $\varepsilon_z(0)$ is slightly smaller than the dielectric constant of $\varepsilon_z(0) = 29$ at $0.5\,\text{kHz}$ reported previously. Although the order–disorder nature was predicted by the first principles calculation, the obtained dispersion relation indicates no evidence of a relaxation mode, at least above $3\,\text{cm}^{-1}$.

Finally, we summarize our main results. The THz-TDS measurements on some important key materials in the ferroelectric crystals of current industrial interest and the observation of the phonon–polariton propagations from which the ferroelectricity originated were presented. With a newly developed compact THz-TDS instrument adapted for accurate quantitative measurements, the spectral power transmittances $T(\omega)$ and the intrinsic phase shifts $\Delta\phi(\omega)$ were carefully measured on high-quality ferroelectric crystals of bismuth titanate (BIT), lithum heptagermanate (LGO) and lithum niobate (LN) of current industrial interest. The nonlinear dispersion relations observed in the intrinsic phase shifts $\Delta\phi(\omega)$ and in the complex dielectric constants $\tilde{\varepsilon}(\omega)$ were well reproduced by the phonon–polariton dispersion relations analytically calculated on the basis of Kurosawa's formula. The anisotropy of polariton dispersion was also clearly observed by rotating the polarization plane of the incident beam. The present measurement indicates the possibility of accurate determination of the damping factor of phonon–polaritons in the FIR region by making analytical calculations for the wavevector $\tilde{k}(\omega)$ dependence on spectral transmission. The lowest branch of phonon–polariton dispersion was accurately determined experimentally down to low wavenumbers less than $3\,\text{cm}^{-1}$, which has not been attained by other previous spectrometries, especially for extraordinary polaritons. The present approach of the THz-TDS has achieved the first observation of the nonlinear dispersion caused by the propagation of phonon–polaritons in the THz region. This suggests that the THz-TDS has the technological advantage of noncontact measurements on FIR dielectric properties, whose potential utilization applies for both fundamental studies of various FIR elementary excitations and utilitarian developments of electronics devices, and further for quality control of industrial products in the near future.

3.3 Photonic Crystals*

The optical properties of periodic dielectric structures, called photonic crystals, have been attracting a great deal of attention [31, 32]. The main reason for these recent intensive investigations is that a frequency range can be achieved where there are no propagating EM modes by choosing the proper lattice structure and the dielectric constants. This frequency region is called a photonic bandgap (PBG). Since there are no EM modes with frequencies in the absolute gaps, spontaneous emission cannot occur in situations where the bandgap overlaps the electronic band edge. The inhabitation of spontaneous

---

* by Mitsuo Wada Takeda

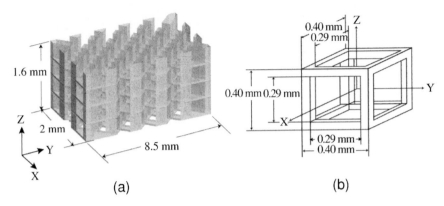

Fig. 15. (a) Schematic illustration of simple cubic air-rod lattice photonic crystal, and (b) its unit cell

emission can improve the performance of many optoelectronic devices, such as high-efficiency semiconductor lasers. It is therefore necessary to develop a lattice with a gap frequency in the visible and near infrared (NIR) regions.

The first studies on photonic crystals started in the microwave range because of relatively easy preparation [33]. However, the focus of studies is now directed at the NIR and visible regions because of the importance of controlling spontaneous emissions in the field of pure physics and the importance of developing applications for laser oscillation or nonlinear optical effects. However, developments in the THz region are not just a step toward progressing from longer to shorter wavelengths. Rather, considering the development of practical application techniques, great opportunities lie in the THz region, including the millimeter-wave region. The dispersion relation of THz waves in photonic crystals has been revealed by THz-TDS [34, 35].

The purpose of photonic-crystal studies in the THz region is to directly acquire knowledge concerning the propagation characteristics of EM waves, such as the dispersion relation of EM waves in a photonic crystal, which is difficult to measure directly by using any techniques in the IR or visible regions with devices developed for this wavelength region. Attention has now turned to the effectiveness whereby the knowledge obtained in the THz region can be applied to other wavelength regions, such as the IR or visible region, as the scaling law is well established between the lattice constant and the wavelength of an EM wave in the photonic crystal.

3.3.1 Transmission Spectra

A photonic crystal with a pseudosimple cubic lattice composed of square air-rods was fabricated with silicon (Fig. 15). The lattice constant is 0.40 mm, the dielectric constant is 11.4, and the filling factor is about 0.82 [34].

Figure 16 shows typical THz transmission spectra of a pseudosimple cubic photonic crystal with eight unit cells along the z-axis shown. The solid line indicates the transmission intensity spectrum for the Γ–Z direction. It should be noted that there are opaque regions from $6.5\,\mathrm{cm}^{-1}$ to $10.0\,\mathrm{cm}^{-1}$, which correspond to the photonic bandgap between the first and second bands predicted by a plane wave. The transmission region below $6.5\,\mathrm{cm}^{-1}$ corresponds to the first band and that from $10.0\,\mathrm{cm}^{-1}$ to $12.3\,\mathrm{cm}^{-1}$ corresponds to the second band. The periodic or modulated transmittance features in both regions are due to the Fabry–Perot interference effects as discussed below.

The open circles in Fig. 16 indicate the intrinsic phase shift spectrum for the Γ–Z direction. The row data for the intrinsic phase shift include an ambiguity of $n\pi$ radians. The intrinsic phase shift in the second-band region above $10.0\,\mathrm{cm}^{-1}$ is added of π radian to those in the first-band region, because the intrinsic phase shifts π between the bandgaps, as will be discussed in the following section.

We could not determine the change in phase between the first and second branches by directly measuring the intrinsic phase shift with THz-TDS. However, we could infer this from its similarity to the electronic band structure in semiconductors, where the conduction band and the valence band surround the fundamental gap. The gap between the first and second bands occurs at the Brillouin zone boundary at $k = \pi/a$. The modes are standing waves with a wavelength of $2a$, twice the lattice constant for $k = \pi/a$. There are two types of standing waves. The first is where the nodes are located at the center of the high-ε (dielectric constant) layer, and the second is where the nodes are located at the center of the low-ε layer. Due to the EM variational theorem, the low-frequency modes concentrate their energy in the high-ε region (dielectric band), whereas the high-frequency modes concentrate their energy in the low-ε regions (air band). Therefore, the difference between the phase of the mode at the top of the first branch (dielectric band) and that at the bottom of the second branch (air band) must be π. Therefore we can shift the phase π during the first bandgap as shown in Fig. 16.

The transmittance is maximum at frequencies corresponding to standing waves in the specimen (Fabry–Perot effect). Similar to the standing wave of the monocode, the difference in phase between the neighboring mode should be π. This feature is seen in both the observed spectra in Fig. 16. Here, the wave-like oscillations in the transmittance spectrum are due to the interference between the THz wave reflected by the front and end surfaces of the specimen, the so-called Fabry–Perot mode. The influence of interference can also be observed in the intrinsic phase-shift spectrum. That is, the intrinsic phase shift remains almost constant in the region with the low transmission coefficient.

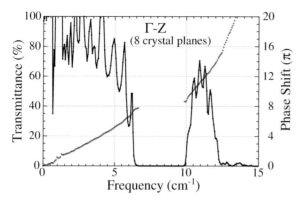

Fig. 16. Terahertz transmission spectra of the pseudosimple cubic air-rod lattice with 8 crystal planes along the z-axis. *Solid line* denotes transmission intensity and *open circles* indicate intrinsic phase shift

3.3.2 Dispersion Relations of Photonic Bands

The intrinsic phase shift, $\Delta\phi(\omega)$, is proportional to the magnitude of wavevector, $k(\omega)$, as written by (6). The open circles in Fig. 16 represent the intrinsic phase-shift spectra. The intrinsic phase shift increases rapidly at the frequencies where the transmission intensities show a maximum and relatively slowly at the frequencies around the minimum, as is demonstrated in the intrinsic phase-shift spectra. This indicates that the optical density of states (ODOS) is very large around the frequency of each standing wave [36]. These characteristics are also observed for the branches around the Brillouin zone center and boundaries. It should be noted that in the vicinity of bandgap edges the (ODOS) must be very large.

The dispersion relations of THz waves in the photonic crystal can be estimated from the intrinsic phase-shift spectrum. Figure 17 plots the dispersion relation of (a) E- and H-polarized EM waves in a square-lattice photonic crystal made of methylpentene polymer (TPX) with circular air-rods [35]. The solid line indicates the calculated results obtained from the plane-wave expansion method using 289 plane waves and the open circles indicate the measured results estimated from the intrinsic phase-shift spectra. Both show very good coincidence to each other not only qualitatively but also quantitatively. The bands are labeled by the assignment of the symmetry of the band's eigenfunctions. The band labeled A can be coupled to the external plane-wave radiation and be observed by the time-domain measurement. The B labeled bands have a symmetry that is incompatible with the incident plane wave and cannot be observed in our experiments. It should be noted that the gradient of the dispersion curves becomes small around the zone center and zone boundaries. This shows the group velocities of the EM waves become very small around these points.

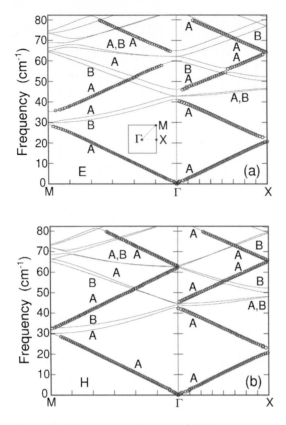

Fig. 17. Dispersion relations of THz waves in a square-lattice photonic crystal made of TPX polymer with circular air-rods for (**a**) *E*-polarization and (**b**) *H*-polarizations. The *solid line* shows results calculated by the plane-wave expansion method using 289 plane waves and the *open circles* show those estimated from the measurements

The other anomalies are anticipated around the regions where level repulsion causes anticrossing between the modes belonging to the same symmetry. The anticrossing characteristic is clearly seen between the fifth and seventh bands along the Γ–X direction for the *E*-polarization of the square lattice (Fig. 17a). Consequently, THz-TDS is one of the most powerful methods to investigate the propagation characteristics of the EM wave in photonic crystals.

3.4 Biological Molecules*

There has been considerable interest in both the experimental and theoretical investigation of the low-frequency motion of biomolecules. Vibrational

---

* by Keisuke Tominaga, and Asako Oka

motions with resonance frequencies in the THz frequency range are characterized by weaker potential forces and/or larger reduced masses, which are in sharp contrast to strongly localized vibrations having resonance frequencies in the mid-IR (MIR) region [37]. In this section, we overview recent developments on application of pulsed THz radiation (hereafter, called simply THz radiation) spectroscopy to biologically important molecules.

By the MIR spectroscopic methods, we investigate well-resolved bands, which correspond to coherent oscillation of a limited number of atoms within a molecule. From analysis of the spectrum, such as normal mode analysis or *ab initio* calculation, we can assign each band as specific vibrational motions of atoms. On the other hand, microwave spectroscopy is a powerful tool to investigate molecular motions in condensed phases in lower frequency region, where a large number of molecules and atoms are involved, and motions of constituent species are almost random. Contrary to MIR spectroscopy where coherent motions of individual atoms can be described, in microwave spectroscopy collective natures such as density of states or dielectric relaxation are discussed from the experimentally obtained absorption and dispersion spectra. Since the THz region is between the IR and microwave regions, THz-radiation spectroscopy has both of the unique features. As described in the Chapter by *Sakai* et al., the THz region almost overlaps with the traditional definition of FIR, the word 'FIR' is sometimes used in substitution of THz respecting for the referred authors. One approach of this spectroscopy is to investigate sharp, distinct bands in the THz region and assign what kind of nuclear motions are associated with these bands. Molecules investigated in this approach are inevitably somewhat small. The other approach is to investigate broad structureless bands of macromolecules in the THz region to discuss the collective nature of dynamics. The obtained THz spectra can be compared with other low-frequency measurements such as neutron scattering, Raman scattering, molecular dynamics calculation, and so on. In these studies it is almost impossible to identify which atoms move in which directions. We first overview the steady-state measurement of refractive index and absorption coefficient or extinction coefficient, mainly obtained by THz-TDS on biomoleucles. We briefly describe other spectroscopic studies using the THz radiation on biomolecules.

Since THz-TDS is a technique to detect waveforms of the pulsed THz radiation, the delay and the decrease of the amplitude of the waveform by the existence of the sample correspond to changes of the refractive index and extinction coefficient, respectively. We can obtain frequency-dependent refractive index $n(\omega)$ and absorption coefficient $\alpha(\omega)$ or extinction coefficient $\kappa(\omega)$ by THz-TDS; $\alpha(\omega)$ and $\kappa(\omega)$ are related by $\alpha(\omega) \propto \omega\kappa(\omega)$.

The FIR spectrum corresponds to the Fourier transform of the correlation function of the total dipole moment of the system, $\mathbf{M}(t)$, expressed as [38],

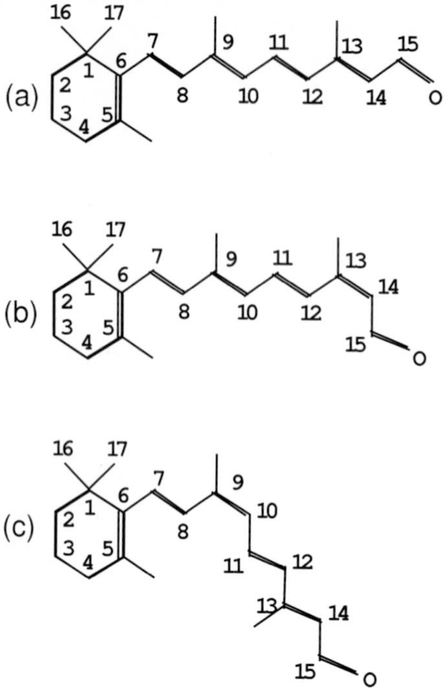

Fig. 18. Molecular structures of three different isomers of retinal: (a) all-trans; (b) 13-cis; (c) 9-cis configuration [39]

$$\alpha(\omega) = \frac{2\pi\omega\left[1 - \exp\left(-\hbar\omega/k_B T_{mp}\right)\right]}{3\hbar c n(\omega)}$$

$$\times \int_{-\infty}^{+\infty} dt \exp(-i\omega t) \langle \mathbf{M}(t) \cdot \mathbf{M}(0) \rangle, \quad (10)$$

where c is the speed of light in vacuum, and k_B is the Boltzmann constant, and T_{mp} is the temperature. Two types of molecular motions in condensed phases contribute to the FIR spectrum; one is the reorientational relaxation of a permanent and/or induced dipole of the molecule, and the other is inter- or intramolecular vibrational motion. If the molecule or the complex in solution does not possess a permanent dipole moment, the reorientational relaxation contribution should be minor, unless an induced dipole moment is significantly large.

3.4.1 Small Biological Molecules

In this section, we describe studies on THz radiation spectroscopy of biologically important small molecules such as amino acids, DNA components, and so on. Some of these molecules show sharp, well-resolved bands in the THz frequency region. Such a low-frequency region is sometimes very important for these molecules, because the intermolecular interactions such as hydrogen-bonding interaction are crucial for structural stabilization and their function, and the resonance frequency for these modes often exists in the THz region. Furthermore, for some biological molecules, large-amplitude motions are important for their function. These motions are sometimes associated with isomerization reactions by photoexcitation in proteins, which, eventually, induces structural change of the whole protein. These well-resolved bands are compared with prediction by the molecular orbital calculation such as *ab initio* calculation to assign the vibrational bands.

We first describe studies on large-amplitude motions of biological molecules in the THz region. Low-frequency torsional modes of retinal, a conjugated polyene chain, are important for understanding its photoisomerization dynamics. *Walther* et al. [39] measured the FIR absorption and dispersion of three retinal isomers (all-*trans*, 9-*cis*, and 13-*cis* retinal, Fig. 18) in the region between $10\,\mathrm{cm}^{-1}$ and $100\,\mathrm{cm}^{-1}$. In rhodopsin the conformation change 11-*cis* $\rightarrow$ all-*trans* triggers the primary step of the vision process, and the all-*trans* $\rightarrow$ 13-*cis* isomerization is the initial step of the photosynthesis cycle in bacteriorhodopsin. The reaction coordinate of this process is a well-defined torsion along one of the C=C double bonds of the polyene chain. The time-resolved experiments revealed that the 11-*cis* $\rightarrow$ all-*trans* isomerization in rhodopsin takes place within $200\,\mathrm{fs}$ [40]. Damped oscillatory features in the visible absorption with a period of $550\,\mathrm{fs}$, corresponding to a frequency of $60\,\mathrm{cm}^{-1}$, have been observed in transient absorption measurements after photoexcitation [41, 42].

In Fig. 19, the frequency dependences of the molar absorptivity α_M ($\alpha_\mathrm{M} = \alpha/K$, where α is the absorption coefficient and K is the sample concentration) and refractive index n of all-*trans* retinals are shown at $298\,\mathrm{K}$ and $10\,\mathrm{K}$. Samples were prepared by mixing retinal with polyethylene powder and pressing the mixtures to disks of thickness of sub-mm. All samples show broad absorption features at room temperature. At low temperatures, the absorption peaks shift to slightly higher frequencies and the absorption bands show narrow peaks correlating to torsional modes of the molecule. To analyze the data they used the frequency-dependent complex dielectric function [37, 42],

$$\varepsilon(\nu) = \varepsilon_\infty + \sum_j \frac{S_j \nu_j^2}{\nu_j^2 - \nu^2 + \mathrm{i}\nu\Gamma_j} = (n - \mathrm{i}\kappa)^2\,, \tag{11}$$

where ν is the wavenumber, and the sum is taken over the different oscillators. Here, ε_∞ is the high-frequency contribution to the dielectric function, ν_j the center frequency, S_j the oscillator strength, and Γ_j the linewidth of the j-th oscillator. Comparison of the absorption spectra of different retinal isomers enables the approximate localization of the modes within the molecule to be found. Their results show that the bands at $54\,\mathrm{cm}^{-1}$ exist in all three isomers, independent of the configuration of the polyene chain C_7–C_{15}. This implies a localization of this mode at the ring C_1–C_6. In contrast, the mode at $47\,\mathrm{cm}^{-1}$, which is very distinct in the all-*trans* isomer, is shifted by $4\,\mathrm{cm}^{-1}$ to

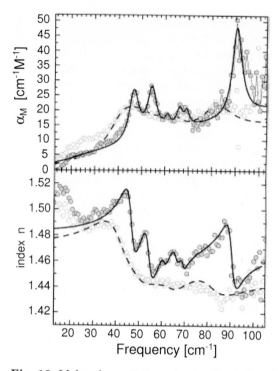

Fig. 19. Molar absorptivity and refractive index of all-*trans* retinal in polyethylene powder at 298 K (*open circles*) and 10 K (*closed circles*). The *solid line* is a fit according to a sum of six Lorentzian oscillators fitted simultaneously to refractive index and absorption data of the cooled sample. The *dashed line* is a fit to the room-temperature data [39]

$5\,\mathrm{cm}^{-1}$ to lower frequencies and is only weakly active in the 9-*cis* and 13-*cis* configurations. This suggests a localization of the mode at the terminal part of the chain C_9–C_{14}. Since the modes at $66\,\mathrm{cm}^{-1}$ and $69\,\mathrm{cm}^{-1}$ have equal strength only in the all-*trans* and 13-*cis* retinal, they concluded that the modes are localized at the middle part of the chain C_6–C_9.

Another example of studies on large-amplitude motions of biological molecules is on tryptophan. *Yu* et al. [43] measured the refractive index and absorption coefficient of a tryptophan film and pressed powders from $0.2\,\mathrm{THz}$ to $2.0\,\mathrm{THz}$ ($6.6\,\mathrm{cm}^{-1}$ to $66\,\mathrm{cm}^{-1}$) by THz-TDS at room temperature. They found two distinct modes at $1.435\,\mathrm{THz}$ ($47.4\,\mathrm{cm}^{-1}$) and $1.842\,\mathrm{THz}$ ($60.8\,\mathrm{cm}^{-1}$). The origin of the observed torsional vibrations was assigned to the chain and ring of the tryptophan molecule by a density-functional calculation. The mode at $1.435\,\mathrm{THz}$ was assigned to torsional motion of the terminal part, and the mode at $1.842\,\mathrm{THz}$ was considered to be localized at the ring. The broadband THz absorption is due to a large density of low-frequency torsional modes with picosecond to subpicosecond relaxation lifetimes. The refractive index of tryptophan is determined to be $n \approx 1.18$ for film and 1.27 for powders in the 0.2 THz to 2.0 THz region.

Korter and *Plusquellic* [44] employed a high-resolution CW THz spectrometer to investigate the absorption spectrum of biotin in a polyethylene matrix in the spectral region from $6\,cm^{-1}$ to $115\,cm^{-1}$. Biotin was chosen to investigate the conformational flexibility, which is responsible for its bioactivity. At both 4.2 K and room temperature the spectrum of biotin displays several resolved absorption features, and at 4 K these are at $18.12\,cm^{-1}$, $34.34\,cm^{-1}$, $41.85\,cm^{-1}$, $44.44\,cm^{-1}$, and so on. Vibrational anharmonicity adequately accounts for the observed line shapes over this temperature range, suggesting other sources of line broadening are small in comparison. While the majority of the temperature-dependent features are explained by this model, the room-temperature fit is further improved if small center-frequency shifts are permitted that presumably accounts for size changes in the crystalline-cage environment with temperature.

Another important piece of information obtained by studies on the low-frequency mode in biomolecules is intermolecular interactions such as hydrogen bonding and other low-energy interactions including dipole–dipole and dispersion interactions [45, 46, 47, 48, 49, 50, 51, 52, 53, 54, 55, 56, 57, 58]. Many biological macromolecules are allowed to be stable, but still highly flexible, in their natural environments due to the interplay between hydrogen bonding and dispersion forces. We next describe studies of THz radiation spectroscopy on intermolecular hydrogen bonding in biomolecules.

Intermolecular hydrogen bonding is of great importance in structural stabilization and flexibility of the DNA double helix. Several groups conducted THz-TDS on DNA component molecules to obtain information regarding the intermolecular hydrogen bond [45, 46, 47]. *Fischer* et al. [45] studied the FIR dielectric function of the four nucleobases [adenine (A), guanine (G), cytosine (C), and thymine (T)] and corresponding nucleosides (dA, dG, dC, and dT (d = deoxyribose)) forming the building blocks of DNA at temperatures of 10 K (solid curves) and 300 K (dashed curves) in the range 0.5 THz to 4.0 THz. They observed numerous distinct spectral features with large differences between the molecules in both frequency-dependent absorption coefficient and index of refraction. At a sample temperature of 10 K they observed that the broad room-temperature resonances split up into several narrow bands. Due to decreasing bond lengths in the system at low temperatures the position of the bands generally move to higher frequencies when cooled. They have performed a vibrational analysis based on the density-functional theory of thymine, as shown in Fig. 20.

The spatial arrangement of the molecules in the crystal structure was taken into account. The full unit cell contains four thymine molecules with the hydrogen-bonded sheets stacked, and there are two types of hydrogen bonds termed η_a and η_b in the unit cell. Figure 20 compares the calculated and observed absorption and refractive index of thymine. The calculated

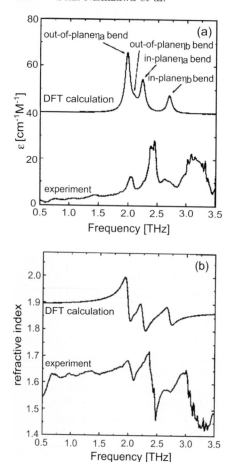

Fig. 20. Comparison between a DFT calculation and experimental determination of (**a**) absorption spectrum and (**b**) refractive index of thymine [45]

spectrum was produced by assuming that each resonance can be described as a damped oscillator, resulting in a characteristic absorption and index profile described by (11). A vibrational analysis shows that the four lowest-frequency, IR-active modes arising from intermolecular motion in the form of out-of-plane and inplane vibrations of the hydrogen-bond systems denoted η_a and η_b. Associated with the bending motion of one of the hydrogen-bond systems is a torsional/stretch motion of the other hydrogen-bond system, bringing their interpretation into agreement with assignments of some of the low-frequency vibrational modes in thymine. The results show that the extended hydrogen bonding in the crystalline structure of the nucleobases and nucleosides is responsible for the observed vibrational features.

Shen et al. [46, 47] studied the temperature dependence of the absorption spectra in the THz region (0.2 THz to 3.0 THz) for several molecules associated with DNA. The samples investigated are polycrystalline nucleosides

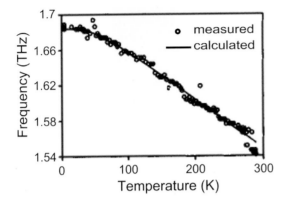

Fig. 21. Temperature dependence of the resonance frequency of the purine vibrational mode centered at 1.68 K (at 4 K). *Open circles* are experimental data and the *solid line* is calculated using the empirical expression given in (12) [47]

(adenosine, thymidine, cytidine, and guanosine) [46], polycrystalline nucleic acid bases (purine and adenine) [47]. A number of well-resolved absorption peaks are observed for polycrystalline nucleosides. They found that all absorption bands become broader and less intense as the sample temperature increases. Furthermore, most absorption bands shift to lower frequencies as the temperature is increased, and they found that the frequency shift is quantitatively described by the following expression,

$$\Delta\omega(T_{\mathrm{mp}}) = \omega_0 - \frac{AT_{\mathrm{mpc}}}{\exp(T_{\mathrm{mpc}}/T_{\mathrm{mp}}) - 1}, \tag{12}$$

where A is a constant and ω_0 is the center frequency of the vibration mode at $0\,\mathrm{K}$. T_{mpc} is the characteristic temperature related to the energy of the effective mode. Figure 21 shows the temperature dependence of the resonance frequency of purine centered at $1.68\,\mathrm{THz}$ (at $4\,\mathrm{K}$) [47]. The strong temperature dependence of these absorption features was thought to be related to the temperature dependence of the sample density and the average hydrogen-bond strength, as well as the anharmonic distribution of vibrational states.

Another class of biologically important molecules in which intermolecular hydrogen bond plays an essential role in its biological activity is saccharides [48, 49, 50, 51]. *Upadhya* et al. [48, 49] reported THz absorption spectra of D- and L-glucose, sucrose, uric acid, its derivative, allantoin from $0.1\,\mathrm{THz}$ to $3.0\,\mathrm{THz}$ at room temperature. They proposed that the stereoisomers of glucose showed spectral features originating from intermolecular vibrational modes, as did uric acid and allantoin [48]. They also presented FIR absorption spectra of a range of monosaccharide molecules, with hexopyranose structures, together with different disaccharides with glycosidic linkage. Figure 22 shows absorption spectra of several saccharides both at room temperature ($295\,\mathrm{K}$) and $4\,\mathrm{K}$ in the spectral range $0.1\,\mathrm{THz}$ to $3.0\,\mathrm{THz}$. Many distinct vibrational features were observed. The samples are expected to show an intermolecular vibrational mode as well as intramolecular modes in this frequency range. Both amorphous and a saturated aqueous solution of D-glucose showed

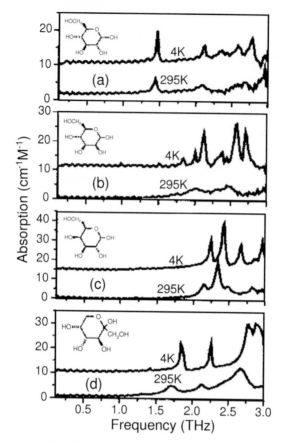

Fig. 22. The FIR spectra of (**a**) D-glucose, (**b**) D-mannose, (**c**) D-galactose, and (**d**) D-fructose measured at 4 K and 295 K. The spectra at 4 K are vertically offset for clarity [48]

no sharp spectral absorption features in the same spectral range [49]. Their observed peaks at 4 K centered at 1.46 THz and 2.11 THz for D-glucose, and those (at 4 K) centered at 2.23 THz and 2.65 THz for D-galactose agree well with previous vibrational assignments of the low-frequency translational and rotational lattice vibrational modes in crystalline structures. These modes, therefore, appear to originate from the collective motion of molecules linked by hydrogen bonds.

The low-frequency spectra of amino acids have been studied by THz-TDS, because they provide information on intermolecular interactions and conformational fluctuations, which are fundamental quantities to understand the formation mechanisms of three-dimensional structures and biological functions of proteins and polypeptides. *Taday* and coworkers [53] reported the THz absorption spectrum of L-glutamic acid, and there are a number of well-resolved transitions in the 1.75 THz to 2.5 THz ($58\,\mathrm{cm}^{-1}$ to $83\,\mathrm{cm}^{-1}$) region. These were compared with published theoretical data on intra- and intermolecular transitions. *Kutteruf* et al. [54] reported absorption spectra in the THz region of 1 THz to 15 THz ($33\,\mathrm{cm}^{-1}$ to $500\,\mathrm{cm}^{-1}$) for a variety of dipeptides and tripeptides in the solid state measured by a FTIR spectrometer and DTGS detector at 77 K and 298 K. They concluded that the degree of complexity and structure of the THz spectra indicated that for small peptides, molecular solid-state structure, and sequence-dependent information might be contained in this spectral region. *Yamamoto* and *Tominaga* [55] also studied low-frequency modes of sulfur-containing amino acids, L-cystine at room temperature and reported that there were well-resolved, sharp bands at $23.8\,\mathrm{cm}^{-1}$, $49.8\,\mathrm{cm}^{-1}$, and $68.9\,\mathrm{cm}^{-1}$.

THz radiation spectroscopy has been applied to pharmaceutical compounds in polymorphic, liquid crystalline, and amorphous forms [56, 57, 58]. The molecules investigated are carbamazepine, enalapril maleate, indomethacin, fenoprofen calcium, and so on. The absence of distinct modes in the amorphous and liquid-crystalline samples suggests that the absorbances in more ordered samples are due to crystalline phonon vibrations. THz spectroscopy easily differentiates different crystalline polymorphs, even when the crystalline structures of polymorphs are very similar. It also differentiates crystalline forms from liquid crystalline and amorphous forms of drugs. The large spectral differences between different forms of the compounds studied are evidence that THz spectroscopy is well suited to distinguish crystallinity differences in pharmaceutical compounds.

3.4.2 Biological Macromolecules

In this subsection we overview application of THz radiation spectroscopy to biologically important macromolecules. The significance of THz-frequency motions in biopolymers such as proteins has been theoretically emphasized through studies of relations between low-frequency modes and protein functions [59, 60, 61, 62, 63]. The results from the normal-mode analysis [64] and the principal-component analysis [60] on several proteins indicate the presence of various vibrations in a picosecond time scale with frequencies down to several wavenumbers. The vibrations have large amplitudes and are globally delocalized. Motions at these frequencies were shown to contribute to a major part of protein fluctuations, which were characterized by mean-square displacements of protein atoms from their vibrational equilibria [60, 61].

THz-frequency bands around $20\,\mathrm{cm}^{-1}$ to $30\,\mathrm{cm}^{-1}$ of proteins were previously observed by Raman scattering [65, 66] and inelastic neutron scattering [67, 68, 69, 70, 71]. *Brown* et al. [65] observed a broad band at $29\,\mathrm{cm}^{-1}$ in Raman spectra of α-hymotrypsin and considered that this band originated in intramolecular vibrations that involved large portions of atoms. This band was relatively independent of the sample preparation of α-chymotrypsin but was dependent on the structures of this protein; the band at $29\,\mathrm{cm}^{-1}$ vanished

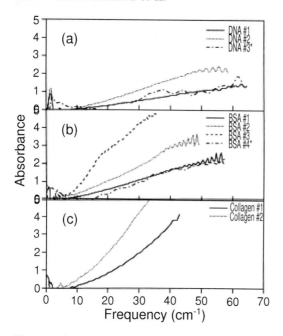

Fig. 23. Absorbance of biomolecular pressed pellet samples at r.h. < 5%: **(a)** DNA #1 and #2 DNA lyophilized powder mixed with PET (polyethylene), whereas #3 is the absorbance for a pure DNA lyophilized powder pellet renormalized to the equivalent molecular path length to DNA #1. **(b)** BSA #1, #2 and #3 are BSA lyophilized powder mixed with PET, whereas #4 is pure BSA lyophilized powder pellet renormalized to the equivalent molecular path length to BSA #1. **(c)** Collagen #1 and #2 are collagen lyophilized powder mixed with PET. The data are truncated to remove values lying outside the dynamic range of the pulsed THz spectroscopy system, defined by the ratio of the noise-floor intensity to the reference transmitted intensity [78]

upon SDS denaturation (SDS; sodium dodecyl sulfate). *Genzel* et al. [66] also reported a Raman band at 25 cm^{-1} of lysozyme. They assigned this band to an intermolecular vibration between proteins because the band was observed in crystalline states but not in solution. In inelastic neutron scattering of lyophilized proteins, a single, broad band centered at about 25 cm^{-1} was frequently observed [67, 68, 69, 71]. From the experimental observation of this band, *Smith* [67] concluded that there existed some underdamped vibrations of low-frequency modes in proteins.

FIR absorption spectra of polypeptides [72, 73] and proteins [74, 75, 76] down to the THz-frequency region were also measured in the 1970s. *Schotts* and *Sievers* [75] reported detailed studies on FIR absorption spectra of polyalanines from room temperature to 4 K. They found that the spectral dependence on temperature was related to the number of residues in alanine-oligomers. FIR transmission spectra of lysozyme at different hydration conditions were measured by using synchrotron radiation, and a band at $19 \, \text{cm}^{-1}$ was observed [77]. This band was considered to originate in low-frequency vibrations of the protein and/or water molecules perturbed in hydration. Previous studies [72, 73, 74, 75, 76, 77] on FIR absorption spectroscopy of biopolymers have mainly dealt with the existence of bands and some qualitative features of these bands.

Markels et al. [78] reported the first use of THz-radiation spectroscopy to examine low-frequency modes of biomolecules. The samples were lyophilized powder samples of calf thymus DNA, bovine serum albumin (BSA), and collagen in the $0.06 \, \text{THz}$ to $2.00 \, \text{THz}$ ($2 \, \text{cm}^{-1}$ to $67 \, \text{cm}^{-1}$) frequency range. Figure 23 shows the measured absorbance for several biomolecular samples. For all samples examined they observed good agreement with Beer's law behavior. The absorbance at higher frequencies is strong and broadband for all samples. For both DNA and BSA the absorbance increases nearly linearly with frequency. For collagen, however, the absorbance increases more rapidly with increasing frequency. All these samples show broadband absorption increasing with frequency, suggesting that a large number of the low-frequency collective modes for these systems are IR-active. This gives rise to the possibility of preferentially exciting a particular mode via initiating a conformational change associated with that mode.

The hydration dependence of the DNA spectra was also investigated. The spectra are different from those in Fig. 23a for $< 5\%$ relative humidity (r.h.). The low-frequency absorption below $15 \, \text{cm}^{-1}$ diminishes from that in Fig. 23. Similarly for BSA, the entire transmission spectrum shifts down in frequency as the native BSA is hydrated, as was seen in the DNA samples. As DNA is hydrated the absorption features in the FIR transmission spectra appear to shift to lower frequencies.

Spectra of DNA macromolecules and related biological materials in the region of $\sim 10 \, \text{cm}^{-1}$ to $500 \, \text{cm}^{-1}$ were also reported by *Globus* et al. [79, 80]. Several different techniques were used for sample preparation. Film samples made from commercial DNA fibers, polyadenylic acid potassium salt, and cellular agents such as the spore form of Bacillus subtillis were prepared. They observed fine features in the spectra, and these spectral results confirm earlier theoretical predictions for the existence of phonon modes in DNA macromolecules. Their theoretical study of two double-helical DNA fragments predicted that most normal modes were at frequencies below $220 \, \text{cm}^{-1}$, with the density higher than one mode per cm^{-1}, which was consistent with what was observed experimentally. They concluded that the long-wavelength vibrational spectra could potentially be used for identification of DNA signatures.

Brucherseifer et al. [81, 82, 83] demonstrated a strong dependence of the complex refractive index in the THz range on the binding state of DNA molecules. They compared the THz spectra of hybridized (double-stranded) DNA with that of denatured (single-stranded) DNA as shown in Fig. 24. In order to ensure identical experimental parameters when comparing hybridized and

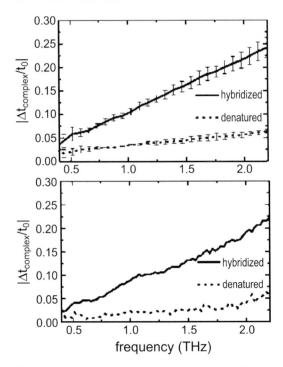

Fig. 24. Normalized transmission change induced by hybridized and denatured DNA films. The *upper* and *lower* plots illustrate measurements on samples made from distinct DNA solutions with different laser systems, to verify the reproducibility of the results. The *error bars* in the *upper* plot are the standard deviation of three consecutive measurements at different positions within the same sample. The higher noise in the *lower* plot is due to the lower stability of the Ar^+-ion laser pumped setups [81]

denatured DNA samples, they used exactly the same material for both types of samples. A much larger change of the THz transmission induced by hybridized DNA samples in comparison to denatured samples is observed. This can plausibly be due to the predicted phonon or plasmon modes for DNA in the THz frequency range, which are absent or strongly modified in denatured material, or by a change in the eventually present hydration chain along the polynucleotide molecules. They concluded that THz-TDS had the capability for detecting the binding state of genetic material directly, potentially giving rise to label-free methods for genetic analysis.

Yamamoto et al. [84, 85, 86] reported FIR absorption spectra of polypeptides (polyglycine and poly-L-alanine) and cytochrome c. They used THz radiation to measure the wavenumber region from $7\,cm^{-1}$ to $55\,cm^{-1}$ and blackbody radiation for the region from $15\,cm^{-1}$ to $160\,cm^{-1}$. FIR absorption, Raman scattering, and inelastic neutron scattering can be applied to

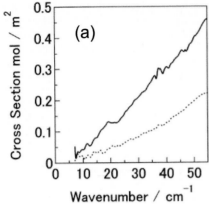

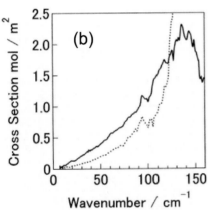

Fig. 25. FIR absorption spectra of polyglycine I and glycine; (**a**) is an enlarged figure of (**b**) below $55\,\mathrm{cm}^{-1}$. *Solid lines* represent average absorption cross section of polyglycine per one mole of glycine residues and *dotted lines* represent absorption cross section of one mole of glycine. Both samples are pressed pellets without any dispersing powders [84]

investigate low-frequency motions of biopolymers. However, different nuclear motions contribute to the spectra of the three methods; FIR absorption, Raman scattering, and inelastic neutron scattering originate in dipole-, polarizability-, and density-fluctuations, respectively. Comparison of spectra obtained by these methods is, therefore, important for studies on the low-frequency motions. At the same time, it is essential to investigate relations between structures of biopolymers and spectral features in the low-frequency region. Biopolymers and proteins, whose structures have been studied extensively, should be chosen as samples for this purpose. Polyglycine, poly-L-alanine, and cytochrome c are the examples whose structures are well known. In the region from $7\,\mathrm{cm}^{-1}$ to $55\,\mathrm{cm}^{-1}$, FIR absorption cross sections of polyglycine (Fig. 25) and poly-L-alanine (Fig. 26) in powder are greater than those of glycine and L-alanine in powder. On the other hand, FIR absorption spectra of cytochrome c in lyophilized powder show little dependence on protein structures, as shown in Fig. 27. The structures of biopolymers are investigated by MIR absorption (polypeptides and cytochrome c) and by resonance

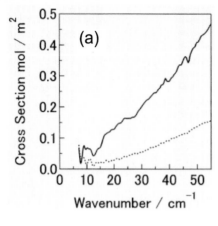

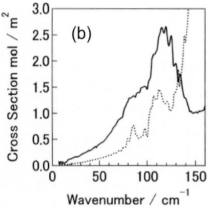

Fig. 26. FIR absorption spectra of α-poly-L-alanine and L-alanine; (**a**) is an enlarged figure of (**b**) below $55\,\mathrm{cm}^{-1}$. *Solid lines* represent average absorption cross section of α-poly-L-alanine per mole of L-alanine residues and *dotted lines* represent absorption cross section of one mole of Lalanine. Both samples are pressed pellets without any dispersing powders [84]

Raman scattering (cytochrome *c*). FIR spectral features of biopolymers in the THz frequency region are qualitatively discussed in terms of density of states and homogeneous/inhomogeneous broadening.

3.4.3 Related Studies and Future Perspectives

To conclude this section we briefly introduce other promising techniques using THz radiation and related experimental methods to study dynamics, structure, and functions of biological molecules. One of the most exciting applications of the THz radiation to biomolecules is to obtain time-resolved THz absorption spectra after photoinitiation of a biopolymer conformational change. These measurements will tell us what kind of low-frequency motions are important for biological functions initiated by photoexcitation. The time-resolved measurements, in which a visible or UV short pulse excites the molecule into the excited state and THz radiation probes changes of absorbance in the FIR range induced by the photoexcitation [87, 88], have

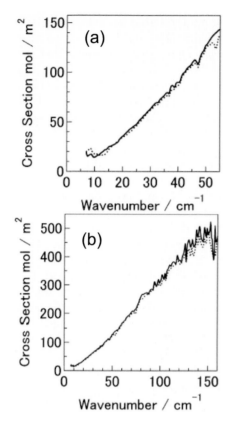

Fig. 27. FIR absorption spectra of cytochrome c in lyophilized powder; (**a**) is an enlarged figure of (**b**) below 55 cm^{-1}. *Solid lines* and *dotted lines* represent absorption cross sections of cytochrome c in the native and pH-denatured states, respectively [84]

been already demonstrated for photodynamics of semiconductors [89, 90] and solute–solvent systems [91, 92, 93, 94].

For example, since *Walther* et al. observed low-frequency vibrational modes of retinal, which may be important for photoisomerization, it is definitely interesting to observe the molecular dynamics directly by studying the low-frequency modes during the isomerization process of retinal in a time-resolved experiment. *Markelz* et al. [95] measured absorption spectra in the THz region of bacteriorhodopsin in a steady-state condition with and without illumination in order to investigate the dependence on protein conformation and mutation. The sample is a thin film and hydrated to ensure bioactivity. They further developed a low time resolution visible pump/THz probe measurement system for studying the photocycle of bacteriorhodopsin and demonstrated that the technique holds promise.

A free-electron laser is another source delivering pulsed radiation in the THz region. The pulse energy of the radiation from a free-electron laser is extremely high, which enables unique experiments to be designed in the THz region. *Xie* et al. [96] measured the lifetime of vibrational excitations of the collective modes at $115\,\mathrm{cm}^{-1}$ in bacteriorhodopsin by a time-resolved picosecond FIR pump-probe technique. The free-electron laser they utilized is capable of delivering $10\,\mu\mathrm{J}$ pulses in $10\,\mathrm{ps}$ in the FIR. They found that the modes had extremely long lifetime of vibrational excitations, over $500\,\mathrm{ps}$, and suggested that the slow rate is related to the similar rate of conformational transitions in protein. *Xu* et al. [97,98] proposed THz circular dichroism (CD) spectroscopy as a tool of a biospecific sensor, since biological materials absorb strongly across the THz part of the spectrum and are enantiomerically pure. They developed a broadband THz CD spectrometer based on a free-electron laser and a polarization-sensitive interferometer and succeeded in measuring CD in a biomolecule, lysozyme, in aqueous solution.

While it would be of the highest priority to obtain low-frequency spectra for biomolecules in aqueous solutions in which these molecules often have their biological function, this is not immediately feasible with IR spectroscopy because water is a highly absorbing EM radiation field in the FIR region. There have been a few attempts to measure THz absorption spectra of biomolecules in aqueous solutions [99,100]. Using hydrated films of proteins is one possible method to measure absorption spectra in the THz region to reduce the contributions of water absorption from the spectra [95]. It should also be noted that Raman-active low-frequency modes of biological molecules in aqueous solutions have been recently measured by optical heterodyne-detected Raman-induced Kerr-effect spectroscopy [101, 102]. *Mickan* et al. [103] conducted THz absorption experiments on protein suspension in organic solvents. In certain organic solvent suspensions, biomolecules have been shown to retain their activity and it will be possible to monitor chemical interactions in real time with THz-TDS. They used the protein subtilisin *Carlsberg* (SC) in a mixture of 1,4-dioxane and water. They found that the addition of SC to the mixture lowers the dielectric constant of the mixture. The water molecules bound to the protein are hindered in their motion and have a lower dielectric constant than that of free water. The measurements are valuable in understanding the THz response of biological systems and in studying the interaction between bound water and proteins.

The spectroscopic application of the THz radiation to biologically important molecules has started quite recently. Fruitful information obtained for biological molecules by THz radiation spectroscopy clearly indicates that the techniques demonstrated so far are quite promising for exploring structures, dynamics, and interactions relevant to biological function.

3.5 Attenuated Total Reflection Spectroscopy*

One finds difficulty in applying THz-TDS in optically thick materials, such as heavily carrier-doped semiconductors. One also encounters difficulty applying it to polar liquids or powder samples. Important biomolecules, such

---

* by Koichiro Tanaka

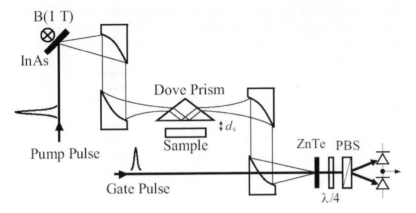

Fig. 28. Schematic diagram of the experimental setup of THz-ATR [104] for THz-ATR. The system is basically the same as that in conventional THz time-domain transmission spectroscopy. Femtosecond (fs) pulses from Ti:sapphire laser are used to generate a THz wave (pump pulse) and to detect the temporal response (gate pulse). In this case, the THz wave is generated on the InAs surface under a magnetic field and detected by a standard EO sampling method. ATR spectroscopy is achieved by inserting a so-called Dove prism into the focus position of the THz wave. The sample is set at a distance of d_s from the total reflection surface

as DNA and proteins, can be obtained only as liquid solutions or powders since it is difficult to fabricate durable free-standing samples. The attenuated total reflection (ATR) spectroscopy affords a powerful alternative to measure the spectrum almost equivalent to the transmission spectrum, regardless of the shape of the sample. In ATR, the evanescent wave travels in parallel with the total reflection plane (Goos–Hanchen effect). One can expect a finite interaction length between the evanescent THz wave and the sample. The reference wave can be obtained without arranging the optical pathways, leading to an accurate evaluation of the complex reflection coefficients. In addition, ATR enables us to observe a longitudinal mode that is difficult to explore by conventional spectroscopy.

An example of the experimental setup is shown in Fig. 28. ATR spectroscopy is achieved just by inserting a special 'magic prism', a kind of Dove prism, into the focus position of the THz wave in Fig. 28.

Teflon, Si, and MgO prisms were examined and confirmed to work satisfactorily in the 0.2 THz to 2.5 THz region. The total reflection condition is satisfied at the bottom plane of the prisms. The sample is set at a distance of d_s from the total-reflection surface. The THz wave attenuates exponentially into the air as an evanescent wave. The evanescent wave is transverse magnetic (TM) or transverse electric (TE) according to the polarization of the incident THz pulse (p- or s-polarized). TM modes may excite longitudinal modes, such as the surface plasmon-polaritons.

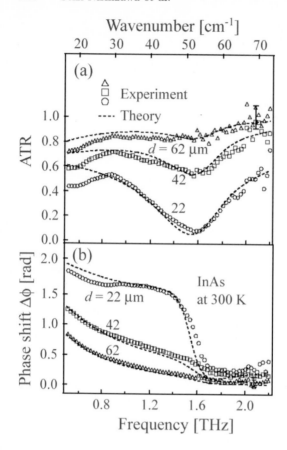

Fig. 29. (a) The attenuated total reflection spectra ATR and (b) the intrinsic phase shift in an *n*-type (100)-oriented InAs crystal. *Open circles* represent ATR data for several distances between the prism and the sample. *Dashed curves* are theoretical ones based on the Drude model [104]

To obtain ATR spectra, a temporal profile of the total-reflected THz wave is measured as a function of the distance d_s between the prism and the sample. Amplitude and phase spectra, $E(d_s, \omega)$ and $\phi(d_s, \omega)$, are obtained by Fourier transformation of the temporal profile. As a reference, the data for $d_s \to \infty$ is adopted. The ATR spectrum and intrinsic phase shift can be determined by the following formula,

$$\mathrm{ATR}(d_s, \omega) = \left| \frac{E(d_s, \omega)}{E(d_s = \infty, \omega)} \right|^2 \tag{13}$$

$$\Delta\phi^{\mathrm{Me}}(d_s, \omega) = \mathrm{Arg}(R(d_s, \omega)) = \mathrm{Arg}\left(\frac{E(d_s, \omega)}{E(d_s = \infty, \omega)} \right). \tag{14}$$

These quantities are theoretically estimated by a simple two-interface model including a prism, an air gap, and a sample. In the plain-wave approximation, the total-reflection coefficient r_{123} is as follows,

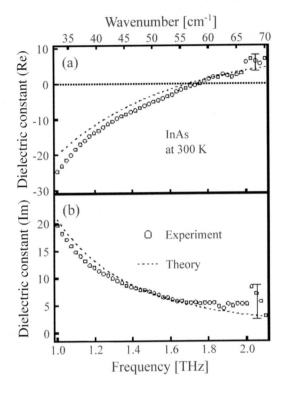

Fig. 30. (a) The real part and (b) the imaginary part of the complex dielectric constant in InAs. The *open circles* are derived directly from the THz-ATR technique. *Broken lines* are calculated curves based on the Drude model [104]

$$r_{123}(d_s, \omega) = \frac{r_{12} + r_{23} \exp(-2ik \cos \theta_2 d_s)}{1 + r_{12}r_{23} \exp(-2ik \cos \theta_2 d_s)}, \qquad (15)$$

where r_{12} and r_{23} are reflection coefficients for prism/air and air/sample interfaces: k is the magnitude of the wavevector in the air:

$$\cos \theta_2 = \sqrt{1 - \varepsilon_1^{(1)}/\varepsilon_1^{(2)} \sin^2 \theta_1}$$

is the cosine of the refractive angle in the air gap, expressing $\varepsilon_1^{(1)}$ the real part of the complex dielectric constant of the prism, $\varepsilon_1^{(2)}$ that of the air gap, and θ_1 the incidence angle at the prism and air-gap boundary. The ATR and intrinsic phase shift are theoretically given by

$$\text{ATR}^{\text{Theo}}(d_s, \omega) = \left| \frac{r_{123}(d_s, \omega)}{r_{12}(\omega)} \right|^2, \qquad (16)$$

$$\Delta \phi^{\text{Theo}}(d_s, \omega) = \text{Arg}\left(\frac{r_{123}(d_s, \omega) SE_{\text{IN}}(\omega)}{r_{123}(d_s = \infty, \omega) SE_{\text{IN}}(\omega)} \right) = \text{Arg}\left(\frac{r_{123}(d_s, \omega)}{r_{12}(\omega)} \right),$$

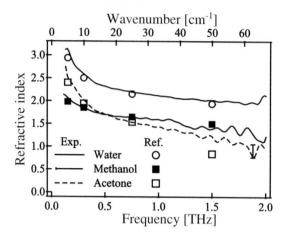

Fig. 31. Frequency dependence of the refraction index measured by ATR at room temperature (water: *solid line*, methanol: *dotted line*, acetone: *dashed line*) [104]. Data reported in [105] are also shown as *open circles* (water), *filled squares* (methanol), and (*open triangles*) acetone

$$\tag{17}$$

where $E_{\mathrm{IN}}(\omega)$ and S are Fourier components of the incident THz wave and an instrumental function of the system, respectively. By equalizing experimental and theoretical quantities, one can deduce r_{123} from the experimental quantities

$$r_{123}(d_{\mathrm{s}},\omega) = \frac{E(d_{\mathrm{s}},\omega)}{E(d_{\mathrm{s}} = \infty,\omega)} r_{12}(\omega)\,. \tag{18}$$

The complex dielectric constant of the sample is obtained from r_{23} using (15). In this way, one can estimate a complex dielectric constant by the time-domain ATR technique.

Time-domain THz-ATR spectroscopy is applied to determine the complex dielectric constants in the undoped (100)-oriented InAs crystal at room temperature [104]. Figure 29 shows ATR and intrinsic phase-shift spectra for several sample–prism distances d_{s}. These data are obtained by a Teflon prism with a p-polarized THz pulse. One can see that ATR has a characteristic minimum and the intrinsic phase shift jumps abruptly around 1.6 THz. These features are attributed to the surface plasma resonance. The broken curves are theoretical calculations based on the Drude model and reproduce experimental features. In Fig. 30 are shown complex dielectric constants in InAs directly obtained using (18) and (15). The broken curve is a calculated curve based on the Drude model with plasma frequency ($\omega_{\mathrm{p}}/2\pi = 1.9\,\mathrm{THz}$) and damping constant ($\gamma/2\pi = 0.62\,\mathrm{THz}$). The zero-crossing point in the real part of the dielectric constant is almost identical to that of the theory. This result justifies the Drude model for the electron plasma in InAs.

Liquids are one of the most interesting samples, since they are used as solvents that define the environment for the solute. However, since most liquids are strong absorbers, we must prepare an extremely thin liquid container (cell) for the transmission measurement. A particular advantage of the ATR method is that reflection spectra can be observed just by dropping the necessary amount of liquid on the prism surface. Figure 31 shows refractive indices determined by the ATR method in several liquids such as distilled liquid water, methanol, and acetone [104]. These data are in agreement with values in previous reports.

3.6 Ellipsometry*

In order to obtain complex optical constants, measurements of the two physical quantities are required at each frequency. The amplitude and phase spectra of the samples are these quantities, for example. For opaque samples, reflection measurements are needed rather than transmission measurements. In the time-domain measurements with the reflection geometry, the accurate phase measurements are difficult because it needs to locate the sample and the reference mirror surfaces at exactly the same position. In order to resolve the difficulties, a technique called "THz ellipsometry" has been proposed and demonstrated. In this technique, the THz-TDS is combined with the ellipsometry in which there is no need for a reference measurement for deducing the complex optical constants of the samples [106].

The ellipsometry itself has been developed in the ultraviolet, visible, and IR regions so far. In conventional ellipsometry [107, 108], the ellipsometric angles $\Psi(\omega) = \tan^{-1}|r_p(\omega)|/|r_s(\omega)|$ and $\theta_{el}(\omega) = \phi_p(\omega) - \phi_s(\omega)$ are experimentally obtained by rotating an analyzing polarizer and fitting the angle dependence with parameters, where r and ϕ denote the amplitude reflectivity and phase of the polarized EM wave, respectively. Then, the complex optical constants are derived from the ellipsometric angles [107]. In the THz-TDS, we can obtain not only the amplitude but also the phase at each frequency. Therefore, the ellipsometric angles can be obtained by simply measuring the waveforms reflected from a sample for s- and p-polarizations and by the complex Fourier transformation [106]. It should be noted that the ellipsometric angles are obtained without the reference measurement using the metallic mirror since they are expressed as the ratio of the amplitude reflectivity and the difference of the intrinsic phase shift for s- and p-polarizations.

Figure 32 shows a schematic diagram of the THz ellipsometry system [106, 109]. The THz wave emitted from a photoconductive (PC) antenna is p- or s-polarized by a wire-grid polarizer (WG1) in front of the sample. For the detection of the THz wave, the PC switch is used. Since the PC

---

* by Takeshi Nagashima and Masanori Hangyo

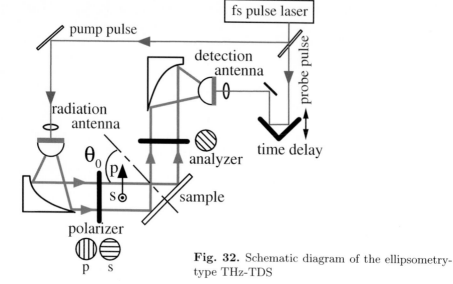

Fig. 32. Schematic diagram of the ellipsometry-type THz-TDS

switch for detection is sensitive to the polarization of the THz wave, a wire-grid analyzer (WG2) is inserted in order to detect the p- and s-waves with equal sensitivity. The incident angle θ_{in} is important for obtaining data with good signal-to-noise ratio (SNR). The incident angle near the Brewster (or principal) angle gives a good SNR because the difference of the reflectivity between p- and s-polarizations is large at this angle. The complex refractive index of the sample can be deduced by the following equations:

$$n^2 - \kappa^2 = \sin^2 \theta_{\text{in}} \left(1 + \frac{\tan^2 \theta_{\text{in}} \left(\cos^2 2\psi - \sin^2 2\psi \sin^2 \theta_{\text{el}} \right)}{\left(1 + \sin 2\psi \cos \theta_{\text{el}} \right)^2} \right), \tag{19}$$

$$2n\kappa = \sin^2 \theta_{\text{in}} \frac{\tan^2 \theta_{\text{in}} \sin 4\psi \sin \theta_{\text{el}}}{\left(1 + \sin 2\psi \cos \theta_{\text{el}} \right)^2}, \tag{20}$$

where $\tan \psi = 1 / \tan \Psi$. An example of the measurement is shown in Figs. 33 and 34. The sample is an n-type Si wafer with a resistivity of $0.136\,\Omega \cdot \text{cm}$. Since this sample is opaque in the THz region, it is not necessary to take into account the internal multireflection in the sample. The incident angle is 45°. Figures 33a and b show $|r_p| / |r_s|$ and $\phi_p - \phi_s$, respectively. The complex refractive index deduced using (19) and (20) is shown in Fig. 34 together with the Drude-model calculation denoted by full curves (see Sect. 3.1 for the Drude model). Although the SNR ratio is insufficient in the high-frequency region, this is improved by changing the incident angle to be near the Brewster or principal angle.

(a)

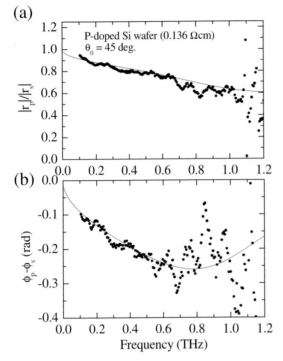

Fig. 33. The quantities (a) $|r_p| / |r_s|$ and (b) $\phi_p - \phi_s$ obtained for the doped Si sample by THz ellipsometry

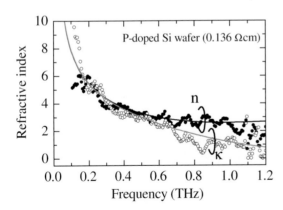

Fig. 34. The complex refractive index of a doped Si sample obtained by THz ellipsometry

3.7 Generation of Electromagnetic Radiation with Multimode Laser Diode*

In general, the intensity of a CW multimode laser diode (CW-MLD) shows fluctuations due to optical beats between many modes in the optical spectrum of MLD. The structure of the spectrum of the MLD can be regarded as a superposition of various spectral lines corresponding to combinations of longitudinal modes and transverse modes in the laser cavity, the latter of which determines the lateral spatial mode patterns of the output beam. Since the spectrum of an MLD spans typically about 2 nm, the highest frequency of the optical beats is around 1 THz. The intensity-fluctuated multimode laser is used as a pump source of a PC antenna to generate the THz EM radiation [110, 111]. It should be noted that this EM radiation is CW and not pulsed radiation.

The EM radiation emitted in this way is useful as a broadband radiation source in the sub-THz region. We can construct a Fourier transform spectrometer (FTS) using this radiation source with an interferometer and a sensitive detector. This system can measure the transmittance of samples with good signal-to-noise ratio in the sub-THz region including the range between 20 GHz and 100 GHz, where sufficient brightness is difficult to provide with thermal radiation sources, such as a high-pressure mercury arc lamp.

Furthermore, it was also demonstrated very recently that a cost-effective TDS system can be constructed by just replacing the Ti:sapphire laser in the THz-TDS with the CW-MLD (hereafter referenced to as MLD-TDS) as the pump source.

In this section, first the spectrum of CW-MLD is described and the relation to the spectrum of sub-THz EM radiation generated by a PC antenna is discussed. Then, the conversion efficiency of this method is compared to those by photomixing of two single-mode LDs and by fs-laser excitation. Spectroscopic characterization of Si wafers using a photoconductive antenna pumped with a CW-MLD is also described as an example of applications. Lastly, a THz-TDS system using a CW-MLD as the pump and probe source (MLD-TDS system) is described.

3.7.1 Spectrum of the Source MLD, Photomixing Efficiency and Output Spectrum

In the standard THz-TDS system described in the previous sections, the fs laser dominates the system cost and size: A Ti:sapphire laser system including the pump source (watt-class CW green laser) is expensive and large in size. A fs-pulse laser based on an Er-doped fiber is much smaller than the Ti:sapphire laser and does not require any adjustment in daily operation. However, it is still bulky and expensive compared to LD systems. In the

---

* by Osamu Morikawa

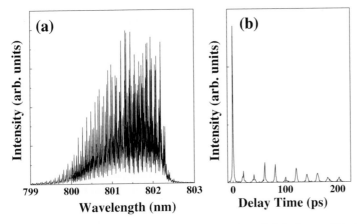

Fig. 35. (a) Spectrum of the multimode LD, and (b) the amplitude of the Fourier transformation of (a). Peaks stand at multiples of 20 ps, which indicates that the complicated spectrum in (a) is the superposition of combs with the spacing of 50 GHz [111]

standard THz-TDS systems, other components (optics, delay line, PC antennas, general measurement electronics, etc.) are much less expensive, whose cost approximately amounts to only one fifth that of the fs laser.

In the standard THz-TDS system, generation of THz radiation from a PC antenna is usually achieved by the ultrafast current modulation by excitation with an ultrashort laser pulse. This implies that a light source can be used to generate the EM radiation, if the light source is accompanied by fast intensity modulation. The MLD is such an optical source because its intensity randomly and rapidly fluctuates due to an optical beat between multiple modes. The spectral shape of emitted EM radiation is given by the product of the autocorrelation of the optical spectrum of the pump laser and the PC response of the emitter.

Figure 35a shows the emission spectrum of an MLD (SDL-2352). The spectrum has a width of about 2 nm, which indicates that the frequencies of the intensity fluctuation should distribute from near DC to about 1 THz. The structure of the spectrum is complicated because the MLD has a broad active layer of 50 μm width and thus there are a large number of laser modes with various longitudinal modes and various mode patterns (transverse modes). The wavelength of a laser mode is determined by the resonant condition, where the intrinsic phase shift of the light wave coincides with an integer multiple of 2π after the complete round trip in the optical resonator of the laser [112]. Since the integers are very large, the resonance frequencies $\omega_{m,n}$ distribute with a constant frequency spacing $\Delta\omega_n$ for an n-th mode pattern, where the subscripts m and n are integers labeling longitudinal modes and mode patterns, respectively. The spacings $\Delta\omega_n$ are almost the same for various mode patterns because the width of the optical resonator is much smaller than the length in the MLDs. Therefore, one mode pattern corresponds to a set of lines with a constant spacing $\Delta\omega$ in the optical spectrum, which appears like a comb. Since our MLD permits various mode patterns, its spectrum can be regarded as a superposition of many combs (spectral lines at $\omega_{m,n}$, with various m for the n-th mode pattern) shifted randomly from each other. Thus the resonance frequency $\omega_{m,n}$ can be written as

$$\frac{\omega_{m,n}}{\Delta\omega} = m + C_n \,, \tag{21}$$

where C_n is a quantity depending on the mode pattern.

To confirm this, we carried out Fourier transformation of the MLD spectrum to analyze its periodicity, and the amplitude of the Fourier-transformed reslut is shown in Fig. 35b. We found a large peak at 0 ps and small peaks at multiples of 20 ps. Note that a comb-shaped spectrum in the frequency domain is Fourier-transformed into a comb-shaped spectrum in the time domain with components at multiples of a certain time interval, including that at 0 ps. Therefore, the comb shape of Fig. 35b suggests that the spectrum in Fig. 35a is the superposition of many combs with a constant spacing.

The spectral structure of laser-intensity fluctuation due to the optical beat can be inferred from that of the laser light. Since the frequency of an optical beat corresponds to the frequency difference between components in the laser light, a comb-shaped spectrum causes intensity fluctuation with frequencies of the comb spacing and its multiples. In the case of the MLD with the spectrum of a superposition of combs as shown in Fig. 35a, the optical beat includes components at the frequency $\Delta\omega$ and its multiples by the mixing of spectral lines in common combs (intracomb mixing). It also contains a component by the mixing of spectral lines in different combs (intercomb mixing), which results in a broad spectrum since the quantities C_n randomly distribute. Therefore, the laser-intensity fluctuation is expected to have a spectrum consisting of discrete peaks and continuous components from near DC to about 1 THz.

Figure 36 shows the schematic configuration of the experimental setup for generation of the sub-THz radiation using the MLD. The PC antenna used here was a bow-tie type that has higher radiation efficiency than the short dipole type in the sub-THz region. The bow-tie structure with a narrow PC gap (5 µm) and a bow angle of 60° was located at the center of a 6 mm long coplanar transmission line on the LT-GaAs substrate. The transmission line separation or the bow-tie length was 2 mm. The MLD beam was collimated and focused onto the biased PC gap of the antenna, from which the EM radiation was emitted by the fluctuating photocurrent. Since the MLD has a large active region of 1×50 µm^2, it was difficult to focus the beam onto the small PC gap. As is shown in the figure, a focusing lens with shorter focal length than the collimating lens was used to make a small focusing spot on the PC antenna. The MLD pump intensity was kept below 300 mW to avoid damage to the PC antenna.

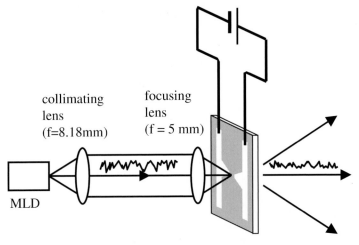

Fig. 36. Schematic configuration of the EM radiation source using the MLD and PC antenna (MLD-PC). A laser beam with intensity fluctuation is focused on the PC gap to generate EM radiation

The emitted EM radiation intensity from the MLD-pumped PC antenna (MLD-PC) was compared to that from the same PC antenna pumped with two single-mode LDs (see the Chapter by *Matsuura* et al.) and that from the PC antenna pumped with a fs laser. The PC antenna was irradiated by the three types of laser sources under the same condition with a bias voltage of 10 V and an average laser power of 20 mW. The EM radiation was detected with a cooled bolometer by directly focusing the THz beam to the bolometer with a pair of paraboloidal mirrors. Although the same pump power and bias were used in these measurements, the irradiating condition might be different with the three laser sources, since the irradiating beam can fail to be focused at the same spot and size on the PC gap because of different beam profiles. To correct this effect, the photocurrent in the PC antenna was measured and the EM-radiation intensity was scaled with the square of the photocurrent, since the radiation intensity is proportional to the square of the photocurrent in the PC antenna. In the case of the photomixing of two lasers, laser beams from two single-mode LDs were mixed, where their wavelengths of near 830 nm were adjusted to make an optical beat of 170 GHz. For the fs-laser excitation, a mode-locked Ti:sapphire laser with a repetition rate of 82 MHz was employed.

Table 3 shows the measured radiation intensity and the photocurrent. The generation efficiency of the EM wave by the MLD was about 1/13 of that by the photomixing of two single-mode LDs [113] and about $1/50\,000 \times (8/20)^2 = 1/300\,000$ of that by the Ti:sapphire laser pumping after the photocurrent correction.

Table 3. Irradiation condition and measured result of the EM radiation

Light source	Irradiating intensity mW	Photocurrent μA	Radiation intensity nW
MLD	20	20	0.02
Photomixing	10 + 10	20	0.25
Ti:sapphire	20	8	1 000

The ratio of about 1/300 000 can be explained by the large difference in the peak intensity of the MLD and fs laser. The fluctuating MLD output can be regarded as a continuous pulse train with a DC component (Fig. 37). In this model, the fluctuating beam from the MLD can be approximately regarded as a continuous train of optical pulses. The duration of the pulses is taken as about 1 ps because the MLD has a spectrum width of about 2 nm, which corresponds to the optical beat frequency of about 1 THz. On the other hand, the Ti:sapphire laser emits discrete optical pulses that irradiate the PC antenna about once every 10 ns since the repetition rate is about 100 MHz. In the case of the MLD, about 10 000 pulses irradiate the PC antenna every 10 ns. The intensity of one optical pulse in the MLD beam (Fig. 37b) is weaker by 1/10 000 than that of the Ti:sapphire laser (Fig. 37a) because the average intensity of the emission is the same. Since the radiation intensity is proportional to the square of the irradiating beam intensity, one pulse in the MLD beam generates radiation $(1/10\,000)^2$ times weaker than that by one pulse in the Ti:sapphire laser beam. In 10 ns, 10 000 pulses in the MLD beam generates radiation with an intensity of $10\,000 \times (1/10\,000)^2 = 1/10\,000$ times compared to that by one pulse of the Ti:sapphire laser. The remaining factor 1/30 compared to the Ti:sapphire laser excitation and 1/13 compared to the photomixing can be attributed to the DC component in the MLD beam that does not contribute to the EM radiation.

The EM-radiation spectrum of the MLD-PC was measured by using an M–P-type FTS, as shown in Fig. 38 [114]. The radiation was detected by a hot-electron InSb bolometer (QMC type QFI/2BI), whose output was mea-

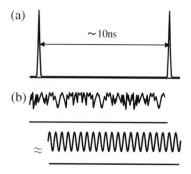

(a)

~10ns

(b)

≈

Fig. 37. Approximation of the pumping beam with intensity fluctuation for a rough estimation of the EM-radiation intensity. (a) The repetition rate of the fs-laser pulses is 82 MHz (≈ 100 MHz), hence the interval of the emitted laser pulses is about 10 ns. (b) The fluctuating laser beam from the MLD can be approximately regarded as a continuous train of optical pulses with an interval of 20 ps with a duration of about 1 ps

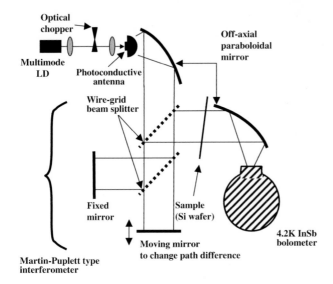

Fig. 38. Experimental configuration to measure the radiation spectrum by using the M–P-type FTS. This configuration was also used to measure transmittance of Si wafers [114]

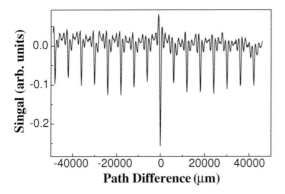

Fig. 39. The interferrogram of the MLD-PC obtained by using the M–P-type FTS shown in Fig. 38

sured with a lock-in amplifier (with a time constant of 300 ms) with signal modulation by mechanical chopping of the irradiating MLD beam (at 1 kHz). The moving mirror was scanned for a path difference of 96 mm with a step of 186 μm with recording the bolometer output at each step. The obtained interferogram is shown in Fig. 39, and the Fourier-transformed spectra using the Blackman window function is shown in Fig. 40. The instrumental resolu-

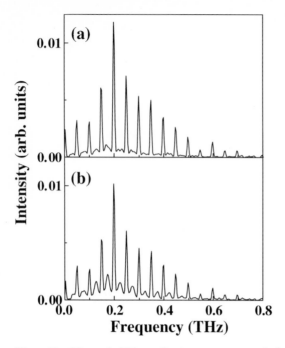

Fig. 40. The sub-THz radiation spectrum of the MLD-PC under irradiation-focusing condition (**a**) of maximum photocurrent and (**b**) of maximum radiation [114]

tion was about 3.2 GHz and the highest frequency limit was about 0.8 THz, which is well covered by the spectral range of the bolometer.

Figure 40a shows the spectrum of sub-THz radiation generated from the MLD-pumped PC antenna when the objective lens was adjusted to produce the largest photocurrent, and Fig. 40b shows the spectrum when the objective lens was adjusted to produce the most intense radiation. The output voltages from the bolometer and the photocurrents were (a) 156 μV and 180 μA, or (b) 179 μV and 166 μA, respectively. Under the maximum-photocurrent condition the laser focus on the antenna gap was smallest, while under the maximum-radiation condition the laser spot was defocused. Both the spectra contain discrete peaks with a constant spacing and continuous components. It is noteworthy that the intensity of the continuous components depends on the objective-lens position. The continuous components are larger under the maximum-radiation condition, while the peaks are higher under the maximum-photocurrent condition. The appearance of the continuous components improves the spectral coverage of the radiation, which is favorable for the broadband spectroscopic applications.

The spectral feature of the sub-THz radiation in Fig. 40 can be explained by considering the spectral structure of the MLD light and the spatial overlapping of the MLD laser modes on the PC antenna gap. The discrete peaks and the continuous components in the radiation spectrum can be attributed to the intercomb mixing and the intracomb mixing, respectively. According to this interpretation, larger continuous components suggest a larger contribution from the intercomb mixing. Under the maximum-photocurrent condition, the optical beam from the MLD is well collimated and tightly focused on the PC gap by lenses. Thus, the optical image on the antenna gap has an electric-field distribution that is a reduced copy of that on the emission surface of the MLD. Since different mode patterns in a laser cavity have patterns of optical electric field orthogonal to each other at the reflecting mirrors of the cavity (the emission surface of an LD) [115], the intercomb mixing does not cause photocurrent modulation because of the destructive interference between orthogonal modes. The photocurrent modulation is caused only from the intracomb mixing in the maximum-photocurrent condition, and the radiation spectrum mainly contains discrete peaks. On the other hand, in the defocused maximum-radiation condition, the orthogonality is weakened so that the intercomb mixing can cause photocurrent modulations in addition to the intracomb mixing and the radiation spectrum contains significant continuous components, which increases the total radiation intensity.

3.7.2 Spectroscopic Applications: Characterization of Si Wafers

The MLD-PC radiation can be used as the broadband light source in an interferometric spectrometer to measure optical properties of materials in the sub-THz region. For an experimental example, we describe a measurement for doped (n-type) Si wafers to estimate doping levels [114]. Since the free carriers in semiconductors have characteristic absorption and dispersion in the sub-THz region, which are well described by the Drude model, the doping level is deduced from transmittance in the sub-THz region. Comparing the transmittance obtained experimentally with that calculated by using the Drude model, the doping levels of the Si wafers were estimated quantitatively.

For the measurement of transmittance of the Si wafers, we used the M–P interferometer shown in Fig. 38. A sample wafer was inserted in the radiation path just after the interferometer, where the sub-THz beam is collimated. The sample wafer was tilted by $10°$ to avoid effects of standing waves between the sample and optical elements. Since the radiation was vertically polarized after the interferometer, it transmitted the sample with s-polarization. The transmittances were calculated from the spectra obtained with and without the sample.

Figures 41a and b show the spectra obtained without and with a Si wafer inserted in the radiation path, respectively, and Fig. 41c shows the measured transmittance. The Si wafer had a resistivity of $45\,\Omega \cdot cm$ and a thickness of $0.503\,mm$. The transmittance curve shows an interference pattern due to the internal multireflection within the Si wafer. The interference peak interval of $88\,GHz$ corresponds to a wavelength of $3.4\,mm$ and coincides with the optical

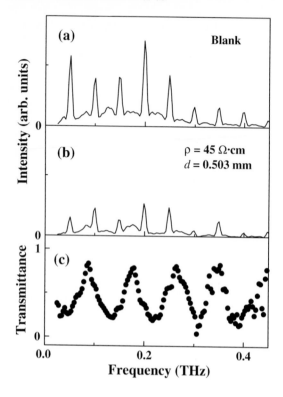

Fig. 41. Radiation spectra **(a)** without sample and **(b)** with sample inserted in the radiation path in Fig. 38. Transmittance obtained by taking ratio of **(b)** and **(a)** is shown in **(c)** [114].

path length for a round trip in the wafer, $2 \times 0.503 \times 3.4$ mm, where 3.4 is the approximated value of the refractive index of Si in the sub-THz frequency range.

The experimental transmittance was compared with that calculated by using the Drude model, which is known to well describe the optical response of doped Si in the sub-THz region [116]. The complex dielectric constant $\tilde{\varepsilon} = \varepsilon_1 - i\varepsilon_2$ of n-type Si can be written as [117]

$$\varepsilon_1 = \varepsilon_\infty - \frac{n_c e^2 \tau^2}{m^* \varepsilon_0 (1 + \omega^2 \tau^2)}, \tag{22}$$

$$\varepsilon_2 = \frac{n_c e^2 \tau}{m^* \varepsilon_0 \omega (1 + \omega^2 \tau^2)}, \tag{23}$$

where m^* is the effective carrier mass, n_c the carrier density, ω the EM radiation angular frequency, ε_∞ the contribution of the bound electrons to the relative dielectric constant, which is 11.6 for Si [118], τ the momentum relaxation time of free carriers, and ε_0 the permittivity of free space. We use the known value $m^* = 0.26 m_0$, where m_0 is the rest mass of the free electron [117]. The momentum relaxation time τ was assumed to be 0.20 ps that is calculated from the electron mobility of 1350 cm^2/V · s. This assumption is

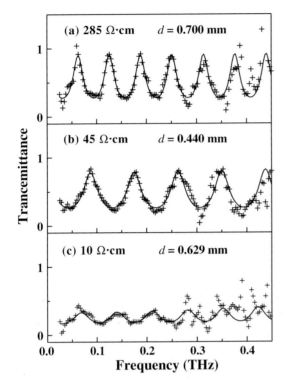

Fig. 42. Transmittance of three samples of Si wafers with different doping levels (+). *Solid lines* are fitting curves calculated by using the Drude model [114]

valid when the carrier density in the Si wafer is below 2×10^{15} cm^{-3} because the electron mobility is almost constant in this region of carrier density [119]. The transmittance T for s-polarization can be expressed as

$$T = \left| \frac{(1+\Gamma)(1-\gamma)}{e^{i\Theta} - \Gamma^2 e^{-i\Theta}} \right|^2, \tag{24}$$

where $\Theta = (2\pi d/\lambda_0)(\varepsilon - \sin^2\theta_{\text{in}})^{1/2}$, $\Gamma = (Z-1)/(Z+1)$, $Z = \cos\theta_{\text{in}}/(\varepsilon - \sin^2\theta_{\text{in}})^{1/2}$, and θ_{in} is the angle of incidence, d the thickness of the sample, and λ_0 the free-space wavelength of the incident radiation. We estimated the carrier density n_c by a fitting calculation in which n_c was taken as the fitting parameter in the frequency range from 60 GHz to 380 GHz. Figure 42 shows the fitting results for three samples with different doping levels.

The carrier densities determined by the sub-THz transmittance were compared to those estimated from the DC conductivity in Fig. 43. The estimation by the transmittance measurement was difficult for samples with carrier density less than 4×10^{13} cm^{-3}, since the free carriers cause little change in the transmittance spectrum of the Si wafers with a thickness of about 0.5 mm. The values agree well over two orders of magnitude of the carrier density, demonstrating the usefulness of the MLD-PC radiation source in spectroscopic measurements.

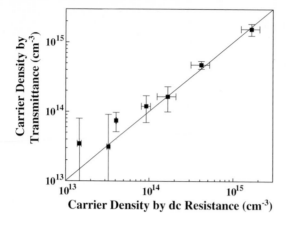

Fig. 43. Comparison between the carrier density obtained from this transmittance measurement and the DC resistivity measurement [114]

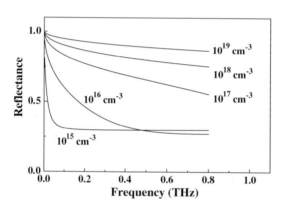

Fig. 44. Reflectance of n-type Si with various carrier densities calculated by the Drude model for normal incidence. The used values of scattering time are listed in Table 4

Characterization of samples with a resistivity less than $1\,\Omega\cdot$ cm is difficult for the transmittance measurement by this system, since the transmittance is very low in the frequency region of sub-THz. For such samples, *Jeon* et al. and *Nashima* et al. [120, 121] have reported that reflection measurements by THz-TDS is useful to characterize the samples. Similarly, the radiation source described here is able to characterize samples with doping levels up to about $10^{18}\,\mathrm{cm}^{-3}$ or $10^{19}\,\mathrm{cm}^{-3}$ by the reflection measurement, since the reflection has characteristic dispersions in the sub-THz region even for such high doping levels (Fig. 44).

3.7.3 TDS System with MLD-PC

In the experiments mentioned above the radiation spectra were measured by the M–P FTS and a cooled bolometer. However, the interferometric measurements lack the phase information, and it requires cooling of the detector with liquid He, which makes the measurements rather laborious. It is preferable to detect the signal with the PC antenna as in the THz-TDS system based on

Table 4. Scattering time τ used in the calculation of Fig. 44, which is calculated from the mobility [119]

carrier density (cm^{-1})	mobility $(cm^2/(V \cdot s))$	scattering time τ (ps)
10^{15}	1 350	0.20
10^{16}	1 270	0.19
10^{17}	800	0.12
10^{18}	260	0.039
10^{19}	100	0.015

fs lasers. In the standard THz-TDS system, the photocurrent induced in the detector PC antenna is proportional to the product of the intensity of the optical pulse gating the detector and the incident THz-radiation amplitude (in the low-intensity regime without a saturation effect). Since the radiation field emitted from the PC emitter is proportional to the time derivative of the photocurrent in the antenna, which is proportional to the optical intensity irradiating on the emitter, the photocurrent detected in the PC detector is given as the cross-correlation of the time derivative of the optical intensity on the emitter antenna and the optical intensity on the detector with convolution of the response function of the system. Therefore, we can detect time-domain signals with a PC-antenna detector gated with the MLD beam split from the one irradiating the emitter like the standard THz-TDS based on fs lasers (defined as MLD-TDS at the begining of this subsection) [122].

In the MLD-TDS and the standard THz-TDS, the light pulses are continuous and discrete, respectively, but optical constants of sample materials are calculated in the same manner (through comparison of the Fourier components of the signal waveforms before and after the sample transmission). This is because a single-frequency component in the optical-intensity fluctuation generates a single-frequency component in the EM radiation and a single Fourier component in the signal waveform as will be shown in the following. The fluctuating MLD intensities at the emitter $I(t)$ and at the detector $I'(t)$ can be written as a sum of various frequency components,

$$I(t) = \sum_j I_j(t) = \sum_j A_j \left[1 + \cos\left(\omega_j t + \phi_j\right)\right], \tag{25}$$

$$I'(t) = \sum_j I'_j(t) = \sum_j A'_j \left\{1 + \cos\left\{\omega_j t + \left(\phi'_j + \omega_j t_d\right)\right\}\right\}, \tag{26}$$

where ω_j is an angular frequency, A_j and A'_j amplitudes of a frequency component at the emitter and detector, respectively, ϕ_j and $\phi'_j + \omega_j t_d$ phase constants at the emitter and detector, respectively. The phase constant $\phi'_j + \omega_j t_d$ depends on the optical time delay t_d, which corresponds to the difference between the optical path length of the EM radiation and that of the MLD beam gating the detector. Since the PC-antenna photocurrent is linear to the irradiating optical intensity, a frequency component $I_j(t)$ generates a frequency

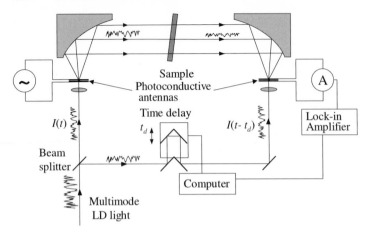

Fig. 45. Schematic configuration of the MLD-TDS [122]

component in the EM radiation, whose electric field at the detector $E_j(t)$ is written as

$$E_j(t) = -B_j\omega_j \sin(\omega_j t + \phi_j), \tag{27}$$

where B_j is the amplitude including the effect of the instrumental functions. When a sample is inserted, the amplitude B_j and phase constant ϕ_j include the effect of the sample. The photocurrent component caused by $I'_j(t)$ and $E_j(t)$ in the detector PC antenna $J_j(t_\mathrm{d}, t)$ is written as

$$\begin{aligned}
J_j(t_\mathrm{d}, t) &= aE(t)I'(t) \\
&= -\frac{aA'_j B_j\omega_j}{2} \Big[\sin(\omega_j t + \phi_j) + \sin\big(2\omega_j t + \phi'_j + \phi_j + \omega_j t_\mathrm{d}\big) \\
&\qquad\qquad\qquad + \sin\big(\phi'_j - \phi_j + \omega_j t_\mathrm{d}\big)\Big],
\end{aligned} \tag{28}$$

where a is a proportionality constant. While the first and second terms including $\omega_j t$ oscillates too rapidly to be detected with a current amplifier with a slow response time and the time-averaged signal $\bar{J}_j(t_\mathrm{d}, t) = J_j(t_\mathrm{d})$ or the third term gives us the signal waveform. Its Fourier transformation with respect to the time delay t_d gives us the amplitude B_j and phase ϕ_j of the frequency component ω_j. We can obtain the information of the transmission of a sample by comparing B_j's and ϕ_j's before and after the sample insertion.

Figure 45 shows a schematic configuration of the MLD-TDS system. The MLD beam irradiated a voltage-biased PC antenna, which emitted the sub-THz radiation. A pair of off-axis paraboloidal mirrors collimated and focused the sub-THz radiation onto a detector PC antenna. A part of the MLD beam was split by a beam splitter and was introduced to the detector PC antenna with a time delay t_d. The signal was modulated at 20 kHz through square waves of the bias voltages ($\pm$10 V) applied to the emitter PC antenna. The

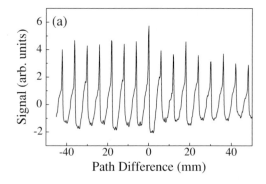

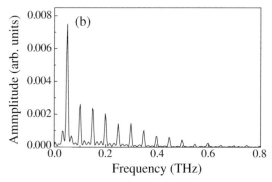

Fig. 46. (a) Time-domain waveform of the MLD-TDS. (b) Fourier-transformed spectrum of the time-domain signal in (a). The horizontal axis denotes the frequency of the corresponding radiation [122]

modulation through the emitter bias has advantages that (1) it does not cause a mechanical vibration in the system that may produce noise in the signal and that (2) it yields a factor-of-2 enhancement in the signal relative to applying a 10 V DC bias to the emitter and simply chopping the pump beam [123]. A lock-in amplifier detected the photocurrent induced in the detector PC antenna and the signal waveform was obtained by scanning the time delay t_d while monitoring the photocurrent. For a spectroscopic measurement, a sample was inserted into the sub-THz beam path between two paraboloidal mirrors. The PC antennas for the emitter and detector were the same as that used in the interferometric measurement (Fig. 38).

Figure 46a shows the signal waveform and Fig. 46b shows its Fourier-transformed spectrum. The noise level was about 1/100 of the maximum amplitude of the signal. The waveform has a periodic shape with a period of about 6 mm optical path difference and its spectrum has discrete peaks at 50 GHz and its multiples. Since periodic structures with the same period are found in Figs. 39 and 40, the periodicity in the waveform and discrete spectral peaks can be attributed to the optical beat among laser modes with common mode patterns (intracomb mixing). However, the continuous components are very small in Fig. 46b, while they are large in Fig. 40. For a demonstration of the MLD-TDS, transmission spectra of Si wafers were measured. The optical constants of the Si wafers were deduced and were in good agreement with the calculation by the Drude model. By the curve fittings based on the Drude model, the doping levels were also estimated [122].

However, the SNR in these transmittance measurements by the MLD-TDS was not very good and the available frequency range was limited below 1 THz. The poor SNR is due to the absence of the continuous spectral components, as shown in Fig. 46b. This is due to the variation of the mode patterns, which results in variation of optical phases among the laser modes with various mode patterns and causes destructive interference in the signal components from the intercomb mixing [124]. To remove the mode-pattern variation we used a single-mode optical fiber and observed that the spectral coverage and SNR were improved. The frequency range of the MLD-TDS is limited by (a) the low efficiencies of the bow-tie PC antenna at high frequencies [125] and (b) the narrow spectrum of the MLD ($\approx 2\,\mathrm{nm}$) compared to that of a standard fs laser ($\approx 10\,\mathrm{nm}$). The use of electro-optic crystals as the emitter and detector may improve the efficiency in the higher-frequency region. The bandwidth limitation in the light source can be improved by employing another laser as the second photomixing light source. Actually, higher-frequency components have been found to appear by using a narrow-band LD and a MLD [126].

THz emission–detection systems based on photomixing of two single-mode lasers on PC antennas have been reported by two groups [123,127]. Since THz radiation generated and detected in such a system has very narrow band, it is suitable for spectroscopy of materials with sharp absorption spectra, such as gaseous molecules. However, the photomixing system with two lasers is not suitable for broadband measurements, because the continuous tuning of the laser frequency over a wide range is usually difficult due to mode hopping. The MLD-TDS system is suitable for measurements of a broad absorption and dispersion spectrum in the sub-THz region.

4 Summary

In this Chapter, some of the pioneering and novel applications of the THz-TDS have been presented in the style of an omnibus by each contributor. The applications of the THz-TDS have included the utilization for noncontact characterization of the electric properties of doped Si wafers, and of the dielectric properties of ferroelectric crystals and photonic crystals. In the application to biological molecules, we have overviewed the results of measurements on the intermolecular interactions of small molecules and the low-frequency actions of macromolecules, and discussed future prospects. In addition to the standard spectrometry of the THz-TDS, attenuated total reflection (ATR) spectrometry and ellipsometry have been considered as advanced measurement techniques. Furthermore, a new spectrometry of cross-correlation THz-TDS with the use of CW multimode laser diode has been mentioned for expanding spectral frequency coverage to the sub-THz region.

References

[1] D. Grischkowsky, S. Keiding, M. van Exter, C. Fattinger: J. Opt. Soc. Am. B **7**, 2006 (1990)

[2] M. van Exter, D. R. Grischkowsky: IEEE Trans. **MMT38**, 1684 (1990)

[3] P. Y. Han, M. Tani, M. Usami, S. Kono, R. Kersting, X.-C. Zhang: J. Appl. Phys. **89**, 2357 (2001)

[4] S. Nashima, O. Morikawa, K. Takata, M. Hangyo: Appl. Phys. Lett. **79**, 3923 (2001)

[5] R. T. Kinasewitz, B. Senitzky: J. Appl. Phys. **54**, 3394 (1983)

[6] T. Ohba, S. Ikawa: J. Appl. Phys. **64**, 4141 (1988)

[7] M. van Exter, D. Grischkowsky: Phys. Rev. B **41**, 12140 (1990)

[8] M. van Exter, D. Grischkowsky: Appl. Phys. Lett. **56**, 1694 (1990)

[9] K. Katzenellenbogen, D. Grischkowsky: Appl. Phys. Lett. **61**, 840 (1992)

[10] T.-I. Jeon, D. Grischkowsky: Phys. Rev. Lett. **78**, 1106 (1997)

[11] T.-I. Jeon, D. Grischkowsky: Appl. Phys. Lett. **72**, 2259 (1998)

[12] S. C. Howells, L. A. Schlie: Appl. Phys. Lett. **69**, 550 (1996)

[13] S. Nashima, O. Morikawa, K. Takata, M. Hangyo: J. Appl. Phys. **90**, 837 (2001)

[14] G. Gallot, J. Zhang, R. W. McGowan, T.-I. Jeon, D. Grischkowsky: Appl. Phys. Lett. **74**, 3450 (1999)

[15] M. Schall, M. Walther, P. U. Jepsen: Phys. Rev. Lett. **64**, 094301 (2001)

[16] S. Labbé-Lavigne, S. Barret, F. Garet, L. Duvillaret, J.-L. Coutaz: J. Appl. Phys. **83**, 6007 (1998)

[17] W. Zhang, A. K. Azad, D. Grischkowsky: Appl. Phys. Lett. **82**, 2841 (2003)

[18] P. N. Saeta, J. F. Federici, B. I. Greene, D. R. Dykaar: Appl. Phys. Lett. **60**, 1477 (1992)

[19] R. Huber, F. Tauser, A. Brodschelm, M. Bichler, G. Abstreiter, A. Leitenstorfer: Nature **414**, 286 (2001)

[20] S. Nishizawa, T. Iwamoto, K. Shirawachi, M. W. Takeda, M. Tani, K. Sakai: in M. Hangyo, K. Sakai (Eds.): (Proc. 1999 IEEE 7th Int. Conf. Terahertz Electronics, Nara, Japan 1999) p. 308

[21] S. Nishizawa, N. Tsumura, H. Kitahara, M. W. Takeda, S. Kojima: in Phys. Med. Biol. **47** (Proc. 1st Int. Conf. Biomedical Imaging and Sensing Application of THz Technology 2002) p. 3771

[22] S. Kojima, N. Tsumura, M. W. Takeda, S. Nishizawa: Phys. Rev. B **67**, 035102 (2003)

[23] M. W. Takeda, S. Nishizawa, S. Kojima: J. Korean Phys. Soc. **42**, S1220 (2003)

[24] S. Kojima, N. Tsumura, M. W. Takeda, S. Nishizawa: Jpn. J. Appl. Phys. **41**, 7033 (2002)

[25] E. C. Subarao: J. Phys. Chem. Solids **23**, 665 (1962)

[26] S. Kojima, S. Shimada: Physica B **219 & 220**, 617 (1996)

[27] M. Horioka, M. Wada, A. Sawada: J. Phys. Soc. Jpn. **58**, 3793 (1989)

[28] R. L. Byer, J. F. Young, R. S. Feigelson: J. Appl. Phys. **41**, 2320 (1970)

[29] S. Kojima: J. Appl. Phys. **32**, 4373 (1993)

[30] T. Kurosawa: J. Phys. Soc. Jpn. **16**, 1298 (1961)

[31] K. Ohtaka: Phys. Rev. B **19**, 5057 (1979)

[32] E. Ozbay, A. Abeyta, G. Tuttle, M. Tringides, R. Biswas, C. T. Chan, C. M. Soukoulis, K. M. Ho: Phys. Rev. B **50**, 1945 (1994) ; E. Ozbay, E. Michel, G. Tuttle, R. Biswas, K. M. Ho, J. Bostak, and D. M. Bloom: Appl. Phys. Lett. **65**, 1617 (1994)

[33] E. Yablonovich: Phys Rev. Lett. **58**, 2059 (1987)

[34] T. Aoki, M. W. Takeda, J. W. Haus, Z. Yuan, M. Tani, K. Sakai, N. Kawai, K. Inoue: Phys. Rev. B **64**, 045106 (2001)

[35] H. Kitahara, N. Tsumura, H. Kondo, M. W. Takeda, J. W. Haus, Z. Yuan, N. Kawai, K. Sakoda, K. Inoue: Phys. Rev. B **64**, 045202 (2001)

[36] K. Ohtaka, Y. Suda, S. Nagano, T. Ueta, A. Imada, T. Toda, J. Bae, K. Mizuno, S. Yano, Y. Segawa: Phys. Rev. B **61**, 5267 (2000)

[37] K. D. Möller, W. G. Rothschild: *Far-infrared Spectroscopy* (Wiley, New York 1971)

[38] D. A. McQuarrie: *Statistical Mechanics* (University Science Books, California 2000)

[39] M. Walther, B. Fischer, M. Schall, H. Helm, P. U. Jepsen: Chem. Phys. Lett. **332**, 389 (2000)

[40] R. W. Schoenlein, L. A. Peteanu, R. A. Mathies, C. V. Shank: Science **254**, 412 (1991)

[41] Q. Wang, R. W. Schoenlein, L. A. Peteanu, R. A. Mathies, C. V. Shank: Science **266**, 422 (1994)

[42] G. Haran, E. A. Morlino, J. Matthes, R. H. Callender, R. M. Hochstrasser: J. Phys. Chem. A **103**, 2202 (1999)

[43] B. Yu, F. Zeng, Y. Yang, Q. Xing, A. Chechin, X. Xin, I. Zeylikovich, R. R. Alfano: Biophys. J. **86**, 1649 (2004)

[44] T. M. Korter, D. F. Plusquellic: Chem. Phys. Lett. **385**, 45 (2004)

[45] B. M. Fischer, M. Walther, P. U. Jepsen: Phys. Med. Biol. **47**, 3807 (2002)

[46] Y. C. Shen, P. C. Upadhya, E. H. Linfield, A. G. Davies: Vib. Spectrosc. **35**, 111 (2004)

[47] Y. C. Shen, P. C. Upadhya, E. H. Linfield, A. G. Davies: Appl. Phys. Lett. **82**, 2350 (2003)

[48] P. C. Upadhya, Y. C. Shen, A. G. Davies, E. H. Linfield: J. Biol. Phys. **29**, 117 (2003)

[49] P. C. Upadhya, Y. C. Shen, A. G. Davies, E. H. Linfield: Vib. Spectrosc. **35**, 139 (2004)

[50] M. Walther, B. M. Fischer, P. Uhd Jepsen: Chem. Phys. **288**, 261 (2003)

[51] J. Nishizawa, K. Suto, T. Sasaki, T. Tanabe, T. Kimura: J. Phys. D **36**, 2958 (2003)

[52] M. Walther, P. Plochocka, B. Fischer, H. Helm, P. U. Jepsen: Biopolymers **67**, 310 (2002)

[53] P. F. Taday, I. V. Bradley, D.D. Arnone: J. Biol. Phys. **29**, 109 (2003)

[54] M. R. Kutteruf, C. M. Brown, L. K. Iwaki, M. B. Campbell, T. M. Korter, E. J. Heilweil: Chem. Phys. Lett. **375**, 337 (2003)

[55] K. Yamamoto, K. Tominaga: in N. Hiromoto (Ed.): *Conf. Dig. 28th International Conference on Infrared and Millimeter Waves* (Ohtsu, 2003) p. 523

[56] C. J. Strachan, T. Rades, D. A. Newnham, K. C. Gordon, M. Pepper, P. F. Taday: Chem. Phys. Lett. **390**, 20 (2004)

[57] V. P. Wallace, P. F. Taday, A. J. Fitzgerald, R. M. Woodward, J. Cluff, R. J. Pye, D. D. Arnone: Faraday Discussions **126**, 255 (2004)

[58] P. F. Taday, I. V. Bradley, D. D. Arnone, M. Pepper: J. Pharm. Sci. **92**, 831 (2003)

[59] Y. Seno, N. Go: J. Mol. Biol. **216**, 111 (1990)

[60] S. Hayward, A. Kitao, H. Hirata, N. Go: J. Mol. Biol. **234**, 1207 (1993)

[61] S. Hayward, N. Go: Annu. Rev. Phys. Chem. **46**, 223 (1995)

[62] A. Kitao, N. Go: Curr. Opin. Struct. Biol. **9**, 164 (1999)

[63] H. J. Berendsen, S. Hayward: Curr. Opin. Struct. Biol. **10**, 165 (2000)

[64] N. Go, T. Noguchi, T. Nishikawa: P. Natl. Acad. Sci. USA **80**, 3696 (1983)

[65] K. G. Brown, S. C. Erfurth, E. W. Small, W. L. Peticolas: P. Nat. Acad. Sci. USA **69**, 1467 (1972)

[66] L. Genzel, F. Keilmann, T. P. Martin, G. Winterling, Y. Yacoby, H. Frohlich, M. W. Makinen: Biopolymers **15**, 219 (1976)

[67] J. C. Smith: Quart. Rev. Biophys. **24**, 227 (1991)

[68] W. Doster, S. Cusack, W. Petry: Phys. Rev. Lett. **65**, 1080 (1990)

[69] M. Diehl, W. Doster, W. Petry, H. Schober: Biophys. J. **73**, 2726 (1997)

[70] M. Ferrand, A. J. Dianoux, W. Petry, G. Zaccai: P. Natl. Acad. Sci. USA **90**, 9668 (1993)

[71] A. Paciaroni, A. R. Bizzarri, S. Cannistraro: J. Mol. Liq. **84**, 3 (2000)

[72] B. Fanconi: Biopolymers **12**, 2759 (1973)

[73] W. J. Shotts, A. J. Sievers: Chem. Phys. Lett. **21**, 586 (1973)

[74] U. Buontempo, G. Careri, P. Fasella, A. Ferraro: Biopolymers **10**, 2377 (1971)

[75] W. J. Shotts, A. J. Sievers: Biopolymers **13**, 2593 (1974)

[76] M. Ataka, S. Tanaka: Biopolymers **18**, 507 (1979)

[77] K. D. Moeller, G. P. Williams, S. Steinhauser, C. Hirschmugl, J. C. Smith: Biophys. J. **61**, 276 (1992)

[78] A. G. Markelz, A. Roitberg, E. J. Heilweil: Chem. Phys. Lett. **320**, 42 (2000)

[79] T. R. Globus, D. L. Woolard, A. C. Samuels, B. L. Gelmont, J. Hesler, T. W. Crowe, M. Bykhovskaia: J. Appl. Phys. **91**, 6105 (2002)

[80] T. R. Globus, D. L. Woolard, T. Khromova, T. W. Crowe, M. Bykhovskaia, B. L. Gelmont, J. Hesler, A. C. Samuels: J. Biol. Phys. **29**, 89 (2003)

[81] M. Brucherseifer, M. Nagel, P. Haring Bolivar, H. Kurz, A. Bosserhoff, R. Büttner: Appl. Phys. Lett. **77**, 4049 (2000)

[82] M. Nagel, P. Haring Bolivar, M. Brucherseifer, H. Kurz, A. Bosserhoff, R. Büttner: Appl. Opt. **41**, 2074 (2002)

[83] A. Menikh, R. MacColll, C. A. Mannella, X.-C. Zhang: Chem. Phys. Chem **3**, 655 (2002)

[84] K. Yamamoto, K. Tominaga, H. Sasakawa, A. Tamura, H. Murakami, H. Ohtake, N. Sarukura: Bull. Chem. Soc. Jpn. **75**, 1083 (2002)

[85] K. Yamamoto, K. Tominaga, H. Sasakawa, A. Tamura: in (CLEO, Chiba 2001) p. 200

[86] K. Yamamoto, K. Tominaga, H. Sasakawa, A. Tamura, H. Murakami, H. Ohtake, N. Sarukura: in (CLEO, Baltimore 2001) p. 312

[87] M. C. Beard, G. M. Turner, C. A. Schmuttenmaer: J. Phys. Chem. B **106**, 7146 (2002)

[88] C. A. Schmuttenmaer: Chem. Rev. **104**, 1759 (2004)

[89] M. Schael, P. U. Jepsen: Opt. Lett. **25**, 13 (2000)

[90] J. Zielbauer, M. Wegener: Appl. Phys. Lett. **68**, 1223 (1996)

[91] G. Haran, W. D. Sun, K. Wynne, R. M. Hochstrasser: Chem. Phys. Lett. **274**, 365 (1997)

[92] R. McElroy, K. Wynne: Phys. Rev. Lett. **79**, 3078 (1997)

[93] E. Knoesel, M. Bonn, J. Shan, T. F. Heinz: Phys. Rev. Lett. **86**, 340 (2001)

[94] Y. C. Shen, P. C. Upadhya, A. G. Davies, E. H. Linfield: J. Biol. Phys. **29**, 135 (2003)

[95] A. G. Markelz, J. Chen, J. R. Hillebrecht, R. R. Birge: in (Proc. SPIE 2003) p. 146

[96] A. Xie, A. F. G. v. d. Meer, R. H. Austin: Phys. Rev. Lett. **88**, 018102 (2002)

[97] J. Xu, J. Galan, G. Ramian, P. Savvidis, A. Scopatz, R. R. Birge, S. J. Allen, K. Plaxco: in (Proc. SPIE 2004) p. 19

[98] J. Xu, G. J. Ramian, J. F. Galan, P. G. Savvidis, A. M. Scopatz, R. Birge, S. J. Allen, K. W. Plaxco: Astrobiology **3**, 489 (2003)

[99] A. Matei, M. Dressel: J. Biol. Phys. **29**, 101 (2003)

[100] K. Tominaga, H. Ohtake, N. Sarukura, K. Saitow, H. Sasakawa, A. Tamura, I. V. Rubtsov, K. Yoshihara: *Advances in Multi-Photon Processes and Spectroscopy*, vol. 14 (World Scientific, Singapore 2001) pp. 317–338

[101] G. Giraud, K. Wynne: J. Am. Chem. Soc. **124**, 12110 (2002)

[102] G. Giraud, J. Karolin, K. Wynne: Biophys. J. **85**, 1903 (2003)

[103] S. P. Mickan, J. Dordick, J. Munch, D. Abbott, X.-C. Zahng: in (Proc. SPIE 2002) p. 49

[104] H. Hiroki, K. Yamashita, M. Nagai, K. Tanaka: Jpn. J. Appl. Phys. **43**, 1287

[105] L. Thrane, R. H. Jacobsen, P. U. Jepsen, S. R. Keiding: Chem. Phys. Lett. **240**, 330 (1995)

[106] T. Nagashima, M. Hangyo: Appl. Phys. Lett. **79**, 3917 (2001)

[107] K. L. Barth, F. Keilmann: Rev. Sci. Instrum. **64**, 870 (1993)

[108] C. Bernhard, T. Holden, A. Golnik, C. T. Lin, M. Cardona: Phys. Rev. B **62**, 9138 (2000)

[109] M. Hangyo, T. Nagashima, S. Nashima: Meas. Sci. Technol. **13**, 1727 (2002)

[110] M. Tani, S. Matsuura, K. Sakai, M. Hangyo: IEEE Microwave Guided Wave Lett. **7**, 282 (1997)

[111] O. Morikawa, M. Tonouchi, M. Tani, K. Sakai, M. Hangyo: Jpn. J. Appl. Phys. **38**, 1388 (1999) part 1

[112] A. Yariv: *Quantum Electronics*, 3rd ed. (Wiley, New York 1989)

[113] E. R. Brown, K. A. McIntosh, K. B. Nichols, C. L. Dennis: Appl. Phys. Lett. **66**, 285 (1995)

[114] O. Morikawa, M. Tonouchi, M. Hangyo: Appl. Phys Lett. **75**, 3772 (1999)

[115] A. G. Fox, T. Li: in (Proc. IEEE 1963) p. 80

[116] M. van Exter, D. Grischkowsky: Phys. Rev. B **41**, 12140 (1990)

[117] R. T. Kinasewitz, B. Senitzky: J. Appl. Phys. **54**, 3394 (1983)

[118] E. D. Palik: *Handbook of Optical Constants of Solids* (Academic, Orlando 1985)

[119] S. M. Sze: *Physics of Semiconductor Devices*, 2nd ed. (Wiley, New York 1981)

[120] T.-I. Jeon, D. Grischkowsky: Appl. Phys. Lett. **72**, 3032 (1998)

[121] S. Nashima, O. Morikawa, K. Takata, M. Hangyo: Appl. Phys. Lett. **79**, 3923 (2001)

[122] O. Morikawa, M. Tonouchi, M. Hangyo: Appl. Phys. Lett. **76**, 1519 (2000)

[123] A. Nahata, J. T. Yardley, T. F. Heinz: Appl. Phys. Lett. **75**, 2524 (1999)

[124] O. Morikawa, M. Fujita, M. Hangyo: Appl. Phys. Lett. **85**, 881 (2004)

[125] S. Matsuura, M. Tani, K. Sakai: Appl. Phys. Lett. **70**, 559 (1997)

[126] O. Morikawa: unpublished

[127] S. Verghese, K. A. McIntosh, S. Calawa, W. F. Dinatale, E. K. Duerr, K. A. Molvar: Appl. Phys. Lett. **73**, 3824 (1998)

[128] M. C. Beard, G. M. Turner, C. A. Schmuttenmaer: J. Appl. Phys. **90**, 5915 (2001)

[129] P. G. Huggard, J. A. Cluff, G. P. Moore, C. J. Shaw, S. R. Andrews, S. R. Keiding, E. H. Linfield, D. A. Ritchie: J. Appl. Phys. **87**, 2382 (2000)

[130] T. Iwamoto, K. Shirawachi, S. Nishizawa, M. W. Takeda, M. Tani, K. Sakai: in K. Itoh, M. Tasumi (Eds.): (Proc. 12th Int. Conf. Fourier Transform Spectroscopy, Tokyo, Japan 1999) p. 211

[131] T.-I. Jeon, D. Grischkowsky: Appl. Phys. Lett. **72**, 3032 (1998)

[132] P. U. Jepsen, W. Schairer, I. H. Libon, U. Lammer, N. E. Hecker, M. Birkholz, K. Lips, M. Scall: Appl. Phys. Lett. **79**, 1291 (2001)

[133] J. Kircher, R. Henn, M. Cardona, P. L. Richards, G. P. Williams: J. Opt. Soc. Am. B **14**, 705 (1997)

[134] S. S. Prabhu, S. E. Ralph, M. R. Melloch, E. S. Harmon: Appl. Phys. Lett. **70**, 2419 (1997)

[135] K. Sakoda: *Optical Properties of Photonic Crystals* (Springer, bh 2001)

[136] M. Schall, P. U. Jepsen: Opt. Lett. **25**, 13 (2000)

[137] Terahertz electromagnetic pulse generation, physics, and applications, J. Opt. Soc. Am. B **11**, 2454 (1994) special review issues

Terahertz Optics in Strongly Correlated Electron Systems

Noriaki Kida, Hironaru Murakami, and Masayoshi Tonouchi

Institute of Laser Engineering, Osaka University
2-6 Yamadaoka, Suita, Osaka 565-0871, Japan
tonouchi@ile.osaka-u.ac.jp

Abstract. Exotic materials with strongly correlated electron systems (SCES) are of considerable interest to researchers engaged in fundamental and practical studies, owing to their versatile electronic, magnetic, and optical properties. Examples of these properties include the high-temperature (high-T_c) superconductivity observed in cuprates and the colossal magnetoresistance (CMR) observed in manganites. In this chapter, we review the terahertz (THz) optics of SCES, whose studies can be broadly categorized as ones using the illumination of femtosecond optical pulses to produce THz radiation from high-T_c superconductors and CMR manganites and THz time-domain spectroscopic studies of these materials. We mainly concentrate on showing the salient optical properties at THz frequencies.

1 Introduction

Strongly correlated electron systems (SCES), in which the electron–electron and electron–phonon interactions play the dominant role, as typified by transition metal oxides, have attracted considerable attention because of their extraordinary electronic, magnetic, and optical properties [1]. In these systems, *correlated electrons* play the leading role, and their properties have still to be fully understood by researchers in condensed matter. The well-known example is the high-temperature (high-T_c) superconductivity that was discovered in copper oxides with a perovskite structure in 1986 [2]. Since then, a wide variety of experimental and theoretical efforts have concentrated on SCES and have found new characteristics that may be of crucial importance to future scientific and technological endeavors [1, 3]. One such discovery is the colossal magnetoresistance (CMR) of manganites, a gigantic decrease in resistance caused by an external magnetic field [4, 5]. The common aspect of these materials is their perovskite structure, where the transition metal ion is located at the center of oxygen octahedra and is connected to the nearest oxygen ions.

In SCES, the inherent attributes (charge, spin, and orbital) of electrons, which can be treated in the same way in the case of semiconductors, are thought to play separate roles. Therefore, "phase control" of SCES with external perturbations – i.e., the conductivity can be switched by a magnetic field and the magnetization can be switched by an electric field – is one of the central concerns of recent materials science [3]. These phase control schemes

K. Sakai (Ed.): Terahertz Optoelectronics, Topics Appl. Phys. **97**, 271–330 (2005)
© Springer-Verlag Berlin Heidelberg 2005

have employed physical attributes other than those exploited in current semi-conductor technology.

Recent advances in the use of visible optical pulses are seen as a new route for the exploration of phase control in SCES. Indeed, new revelations on the microscopic origin of the attributes of the nonequilibrium state and extraordinary optical properties exceeding the limitations of conventional materials have been reported; for example, *Kise* et al. [6] have reported that the magnetization of double-perovskite Sr_2FeMoO_6 ferromagnets, whose ground state is very similar to that of CMR manganites, can respond instantaneously to an illumination of femtosecond (fs) optical pulses, as revealed by pump-and-probe Kerr-effect measurements. The gigantic optical nonlinearity of Sr_2CuO_3, the analog of high-T_c superconductors, has been found in the transparent region by *Kishida* et al. [7] and *Ogasawara* et al. [8]. Ultrafast photoinduced insulator–metal transitions with a typical time scale of picoseconds have been observed in CMR manganites [9, 10, 11]. In particular, *Miyano* et al. [12] showed that a persistent photoinduced insulator–metal transition in CMR manganites could be achieved as long as the bias voltage is applied continuously. The recent experimental observations strongly suggest that the time has come to explore the new functionalities of SCES.

In light of the above context, we review here terahertz (THz) optics using fs optical pulses intended as a tool to produce variations in high-T_c superconductors and CMR manganites, as well as the use of THz pulses as sensitive probes to measure the optical properties of these materials at THz frequencies. We mainly focus on our recent experiments on THz optics in these materials.

A broad range of frequency, covering roughly $0.1\,THz$ to $10\,THz$ (corresponding to photon energies between $0.4\,meV$ to $40\,meV$ or $3\,cm^{-1}$ to $300\,cm^{-1}$ and wavelengths between $30\,\mu m$ to $3\,mm$), is at the boundary between microwaves and light waves, and are generated by electrical circuits and commercial light sources, respectively. At these frequencies, the performance of electrical circuits made of semiconductors reaches the critical limit and tunable light sources are not available. Therefore, this range has been referred to as the "smoked THz gap" [13]. During the last decade, rapid progress in THz pulse-generation and detection systems has been made and has involved the use of photoconductive (PC) semiconductor switches excited by fs optical pulses [14] (see the Chapter by *Sakai* and *Tani*). The generation mechanism of THz pulses phenomenologically involves natural optical conversion; light pulses are transformed into THz pulses through ultrafast modulation of current. This conversion process can be expressed by the following classical Maxwell's equation

$$\boldsymbol{E} \propto \frac{\partial \boldsymbol{J}}{\partial t}, \tag{1}$$

where $\boldsymbol{E}$ is the radiation field at the far-field and $\boldsymbol{J}$ the current. In other words, a transient current, typically of subpicosecond order, is created by ex-

(a) Semiconductors

(b) Superconductors

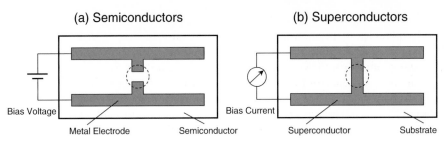

Bias Voltage

Bias Current

Metal Electrode   Semiconductor   Superconductor   Substrate

Fig. 1. Schematic illustration of the different antenna structures of photoswitch-ing devices when (**a**) semiconductors and (**b**) superconductors are used as parent materials. The femtosecond optical pulses (*dashed lines*) are irradiated on (**a**) the gap between metal electrodes while under the bias *voltage* and (**b**) the bridge made of superconductor while under the bias *current*. Under these conditions, the ul-trafast increase in photocurrent in the semiconductor or the ultrafast decrease in supercurrent in the superconductor is used as a source of terahertz radiation

citing carriers with ultrafast optical pulses, which in turn produce THz pulses propagating into the free space based on (1). Therefore, THz radiation from SCES is of great interest as a direct probe for photogenerated carriers in nonequilibrium states with subpicosecond time scales and may be associated with attractive new features of SCES, i.e., ultrafast modulation of the super-current and spin in response to light. SCES is a prospective source of THz radiation for various applications, including THz time-domain spectroscopy and T-ray tomography.

The first part of this chapter reviews the first findings of THz radiation from SCES by exciting charge carriers with fs optical pulses. We systemat-ically compare the THz radiation characteristics of photoswitching devices made of the well-known family of high-T_c cuprates: $YBa_2Cu_3O_{7-\delta}$ [15, 16, 17, 18, 19, 20, 21, 22, 23], $Bi_2Sr_2CaCu_2O_{8+\delta}$ [24, 25], and $Tl_2Ba_2CaCu_2O_{8+\delta}$ [26, 27, 28]. THz radiation experiments with semiconductors as the parent material have typically used the antenna structure shown in Fig. 1a; the gap between the metal electrodes under a constant *voltage* supply acts as an optical on/off switch triggered by the fs optical pulses. In contrast, the transient decrease of supercurrent is the source of THz radiation in high-T_c cuprates. There-fore, one can use an antenna without a gap (Fig. 1b) that is supplied with a constant *current*. The CMR manganites, $Pr_{0.7}Ca_{0.3}MnO_3$, were the first "magnetic" materials shown to radiate THz pulses [29]. Their characteristics are considerably different from the typical THz radiation characteristics of semiconductors, as manifested by the appearance of the anomalous tempera-ture dependence of THz radiation below the temperature when spin ordering sets in [29, 30, 31]. A reversible and bistable nature of THz radiation from $Pr_{0.7}Ca_{0.3}MnO_3$, as a result of the photoinduced insulator–metal transition, has also been discovered [32, 33].

In the THz frequency range, especially, in ranges below the characteristic frequencies of phonons, spectroscopic studies of even well-known materials have lagged behind, and therefore, THz frequencies are thought to be one of the final frontiers of optical spectroscopy. In high-T_c superconductors, the intrinsic Josephson plasma resonance in the c-axis optical spectrum was found around 1 THz for a superconducting state confined within the CuO_2 plane [34]. Recently, a finite frequency peak centered around 3 THz has been observed in the CuO_2 plane for various carrier doping levels and has been assigned to a collective excitation mode due to "charge stripes" [35], in which a regular arrangement of charge is separated by magnetic domains. These results generally support the idea that the unknown elemental excitation of SCES may be further down this frequency regime.

THz time-domain spectroscopy (THz-TDS) is the means of clarifying the overall optical properties of SCES in this frequency regime. As described in the Chapter by *Nishizawa* et al., THz-TDS is a powerful tool for revealing the low-energy charge response of materials as a function of frequency. Thus far, the optical properties of SCES in the THz region are not as well understood as those in the visible region. THz-TDS combined with pump-and-probe experiments on the underdoped superconducting $Bi_2Sr_2CaCu_2O_{8+\delta}$ has provided some insight in this regard [36, 37, 38, 39, 40]. We also describe THz-TDS experiments on two different CMR manganites, $La_{0.7}Ca_{0.3}MnO_3$ [41, 42, 43] and $Pr_{0.7}Ca_{0.3}MnO_3$ [44, 45], which exhibit antipodal properties; the ground state of the former is 100% spin-polarized *metallic* and that of the latter is charge-ordered *insulating*.

2 Terahertz Radiation from High-T_c Superconductors

Femtosecond laser technology has potential in a broad range of endeavors, from basic physics to applicational studies. One application is as an optical source for electromagnetic radiation from various materials. It is well known that a PC switch formed on low-temperature-grown GaAs (LT-GaAs) or semi-insulating GaAs (SI-GaAs) radiates electromagnetic pulses with frequency components extending up to several THz after irradiation by fs optical pulses. These devices have been effectively used as sources of THz radiation to study the high-frequency properties of a variety of materials. On the other hand, it was recently discovered that current-biased high-T_c superconductors (HTSCs) radiate THz pulses into free space after irradiation by fs optical pulses [15]. The phenomenon is due to ultrafast supercurrent modulation induced by fs optical pulses.

This section reviews the THz radiation properties and ultrafast carrier dynamics of typical HTSC materials.

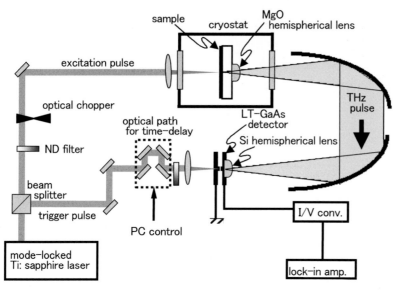

Fig. 2. Schematic illustration of photoconductive (PC) sampling technique using low-temperature-grown GaAs (LT-GaAs) as a detector, which is a general scheme for generation and detection of terahertz radiation. The light of the femtosecond optical pulses is divided by the beam splitter: one is used as the pump pulse, which is illuminated on the sample mounted in the temperature-controlled cryostat; the other is used as the trigger pulse, which is illuminated on the LT-GaAs PC switch after an appropriate time delay. To enhance the collection efficiency, the MgO and the Si hemispherical lens are attached to the back side of the sample and the LT-GaAs, respectively. To increase the signal-to-noise ratio, the pump is mechanically chopped and the signal is lock-in detected

2.1 Terahertz Radiation and Detection System

To measure THz radiation emitted from HTSC thin films, the film is usually patterned into a bow-tie or dipole antenna structure. The patterning is done with a conventional photolithographic technique followed by wet or dry etching. Figure 2 shows a THz time-domain optical setup utilizing HTSCs. Femtosecond optical pulses with a pulse width of about 100 fs are used as optical excitation sources. The optical pulses are generated by a mode-locked Ti:sapphire laser with a central wavelength of 780 nm to 800 nm at a repetition rate of 82 MHz. The pump laser light is chopped at several kHz for lock-in detection, and illuminated on the center of an HTSC antenna bridge with a bias current at a focal spot of several tens of µm in diameter at full width at half maximum (FWHM). The optical pulses produce a great number of quasiparticles at the excitation states of the superconducting gap by avalanche effects, followed by direct excitation of Cooper pairs. These excited quasiparticles relax to the superconducting condensed state within several picoseconds. Since the excitation and relaxation processes cause ul-

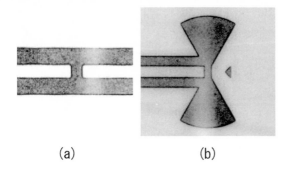

(a)   (b)

Fig. 3. Micrographs of the (a) dipole and (b) bow-tie antennas made of YBCO thin films

trafast modulations in the supercurrent, THz pulses are radiated according to the classical formula of electrodynamics given by (1). In the case of superconductors, the right-hand side of (1) is expressed as follows [15]

$$\frac{\partial \boldsymbol{J}_s}{\partial t} = e n_s \frac{\partial \boldsymbol{v}}{\partial t} + e \boldsymbol{v} \frac{\partial n_s}{\partial t} , \tag{2}$$

where $\boldsymbol{J}_s$ is the current density, e the elementary electric charge, n_s the supercarrier density carrying the supercurrent, and $\boldsymbol{v}$ the velocity of carriers. The THz pulses are radiated into free space through a substrate backed by a MgO hemispherical lens, and collimated and focused by a pair of off-axis paraboloidal mirrors onto an LT-GaAs PC switching detector after passing through a Si hemispherical lens. To observe the time-domain waveform, the time interval between the arrival times of the THz pulse and the trigger pulse at the detector are varied by changing the length of the optical path of trigger pulse (this is accomplished with computer-controlled sliding mirrors). The photocurrent induced in the LT-GaAs detector is lock-in detected after amplification.

2.2 Terahertz Radiation from YBa$_2$Cu$_3$O$_{7-\delta}$

Figure 3 shows dipole and bow-tie antennas made of YBa$_2$Cu$_3$O$_{7-\delta}$ (YBCO) thin film. The dipole antenna is set between two 6 mm long and 30 μm wide transmission lines that are separated by 30 μm. The angle and radius of the bow-tie structure are 60° and 240 μm, respectively. Figure 4 shows the waveform of the THz pulse radiated from the YBCO thin-film dipole antenna at a bias current I_B of 40 mA, laser power P_P of 15 mW, and temperature of 16 K. A sharp negative pulse with a FWHM of 0.5 ps occurs around 6 ps (the time origin is arbitrary), and is immediately followed by a positive pulse. The spectrum can be interpreted as the first derivative of the supercurrent in the films, and the time integral of the observed amplitude corresponds to the current change in the time domain. Therefore, the negative pulse corresponds to an abrupt decrease in supercurrent after irradiation by the fs optical pulse, whereas the positive one corresponds to the recovery of the supercurrent.

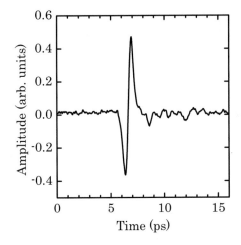

Fig. 4. Temporal detected electrical THz pulse of freely propagating electromagnetic wave radiated from a YBCO thin-film dipole antenna at 16 K

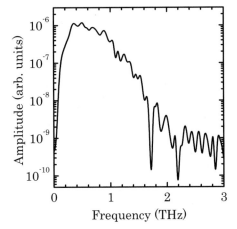

Fig. 5. Amplitude spectrum of the waveform in Fig. 4 on logarithmic scale

By Fourier transforming the detected pulse, one obtains an amplitude spectrum of the freely propagating THz pulse shown in Fig. 5. The spectrum extends from 0.1 THz to 2.0 THz with a peak intensity at around 0.5 THz, and has a bandwidth at half-maximum of 0.8 THz, which is similar to that obtained from LT-GaAs PC switches with a dipole antenna.

The waveform in Fig. 4 suggests that the whole characteristic time period, from photon arrival to relaxation into the equilibrium state, is about 2 ps, and that the supercurrent-decrease process is predominant in the first 500 fs. This value coincides with the relaxation time constant obtained from pump-and-probe optical reflection measurements in the fs time domain that were made above the critical temperature T_c [46]. Thus, the intraband hot-electron relaxation process, including avalanche pair breaking, governs the supercurrent decrease.

If we assume that all photons are absorbed in the 100 nm thick film and assume a laser beam diameter of 25 μm, an excitation energy of 1.56 eV, P_P of 1 mW, and an optical reflectance of 0.5, the excitation roughly corresponds to a photon injection density of 1×10^{18} cm^{-3}, which is low enough to maintain the superconductivity. Actually, the excitation at P_P above 150 mW sometimes induces a phase transition to the normal state at I_B of 100 mA, which accompanies the slow drift affecting the radiation properties and conspicuous deviation from those.

In the normal state, regardless of what causes the phase transition, no radiation is observable, because the bridge cannot carry a large current.

According to classical electrodynamics, the far-field electric field E of the THz radiation is proportional to the time derivative of the current. Using the simple two-fluid model, neglecting carrier acceleration, the change in the supercurrent ΔJ_s due to the irradiation of optical pulses is given by

$$\Delta J_s = e\Delta n_s v_s \propto e\left(\frac{n_s}{n}\right) P_P v_s = \frac{J_s P_P}{n}, \tag{3}$$

where n is the total carrier density, n_s is the pair density, and v_s is the drift velocity. Namely, the THz radiation amplitude is expected to increase linearly with I_B and P_P.

The minimum amplitudes of the first negative pulse for several samples are plotted as functions of I_B and P_P, respectively, in Figs. 6 and 7. Although the radiation properties strongly depend on the film thickness and quality, linear relationships for both parameters are obtained under the condition that I_B and P_P are relatively small in comparison with those inducing the phase transition [15, 19]. The characteristics agree with (3), which suggests that the THz radiation mechanism from HTSCs is attributable to the optical supercurrent modulation.

The linear relationship between the current density and THz amplitude has an application. One can develop a supercurrent distribution imaging system utilizing laser THz generation and a detection system with scanning laser pulses on sample surfaces. Up to the present, an imaging system with a spatial resolution of 7 μm has been used to observe the supercurrent distributions in YBCO thin-film striplines. It has even been used to observe vortices trapped in the films without destroying them. The details of this imaging system have been published elsewhere [47, 48, 49, 50, 51, 52, 53, 54].

Under an external magnetic field B_{ext}, which induces the shielding current of superconductors, bridges disconnected from the current source can also generate THz pulses in a similar way [17, 18]. Figure 8 shows the waveforms of THz radiation from a YBCO thin-film bow-tie antenna under a B_{ext} of 15 mT (middle curve). This figure shows that the THz radiation occurs by modulating the shielding current. Furthermore, THz radiation was still emitted after removal of the field. The waveforms were identical except for their sign of polarity and intensity. This indicates that the magnetic flux is trapped at the bridge by the fs optical excitation in the presence of B_{ext}. The above

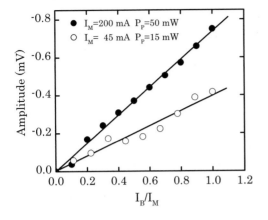

Fig. 6. Peak amplitude of the first negative pulses as a function of bias current for samples with thicknesses of (•) 350 nm and (∘) 200 nm that were excited with laser powers of 50 mW and 15 mW, respectively

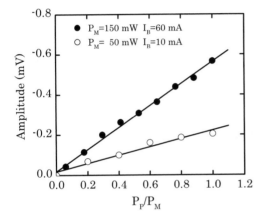

Fig. 7. Peak amplitude of the first negative pulses as a function of laser power for samples with thicknesses of (•) 350 nm and (∘) 85 nm that were excited with bias currents of 60 mA and 10 mA, respectively

phenomena gave the authors the impetus to start research on optical-vortex generation in HTSCs [55, 56, 57, 58, 59, 60, 61, 62, 63].

A proper antenna structure is crucial to producing strong THz radiation. The authors have examined three types of superconducting thin-film antenna; dipole, bow-tie, and log-periodic [64, 65, 66, 67, 68]. Relatively strong THz radiation has been detected from the log-periodic antennas.

Figure 9 shows a self-complementary log-periodic toothed planar antenna made of a 100 nm thick YBCO thin film. The antenna is based on the bow-tie antenna with teeth spread on both sides. The central angles of each tooth are 45 degrees. The n-th tooth is characterized by an outer radius R_n and inner radius r_n where $r_n/R_n = 0.7$, $R_n - 1/R_n = 0.49$, the maximum and minimum outer radius being 700 m and 82 m, respectively. The antennas are coupled to a 30 µm wide stripline at the center. The film on an MgO substrate is patterned into the antenna by using conventional photolithography and an ion-milling process. The radiation power was evaluated with a 4.2 K InSb

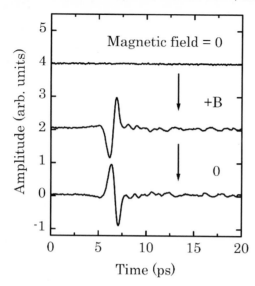

Fig. 8. Detected THz radiation waveforms: signal before applying the magnetic field (*upper curve*), signal under B_{ext} (*middle curve*), and signal after removal of the field (*lower curve*). The center bridge connected to the bow-tie antenna was excited with P_P of 50 mW

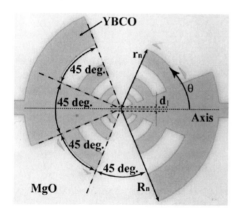

Fig. 9. Micrograph of the prepared YBCO thin-film log-periodic antenna

hot-electron bolometer, which was carefully calibrated using a high-pressure mercury lamp as the radiation standard. The MgO lens is attached to the back side of the substrate to increase the THz radiation collection efficiency.

Figure 10 shows the waveforms of radiation from the YBCO log-periodic antenna and bow-tie antenna excited under the same conditions: I_B of 80 mA, P_P of 10 mW, and probe power of 2 mW. The waveforms were measured with a bow-tie antenna detector. The amplitude of the radiation from the log-periodic antenna is twice that from the bow-tie antenna. When the spectrum was measured with a dipole-antenna detector, there appeared a frequency component over 2 THz with a center frequency of 500 GHz, which coincided with that radiated from a dipole antenna. The waveform differs from that of the log-periodic antennas because the MgO lens enhances the collection

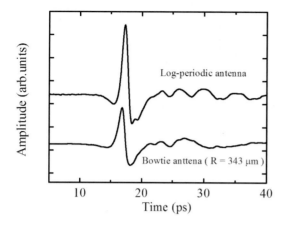

Fig. 10. Temporal THz electrical field radiated from YBCO log-periodic antenna and bow-tie antenna

efficiency only in the center of the antenna, i.e., within a diameter of less than $300\,\mu$m [49], which results in waveforms similar to those from the dipole antenna. The waveforms monitored without the lens have complicated structures resonating at each tooth of the antenna [66, 67]. They are strongly polarized along the direction of 90°, whereas those measured with the MgO lens are polarized mainly at 45° or 135°.

Figure 11 shows the detected radiation power as a function of I_B for various laser powers. The amplitude increases linearly with I_B. The radiation power from the log-periodic antenna is about one order of magnitude larger than that from the bow-tie antenna. The maximum power exceeds $22\,\mu$W on average. InAs is known to be a strong THz radiation emitter, with a radiation power of $12\,\mu$W at a P_P of $140\,$mW, B_{ext} of $8\,$T, and temperature of $170\,$K [69]. In our present system, the reflection at the window of the cryostat reduces the detectable radiation to about 70% compared with the radiated radiation, and heating at the Au/YBCO interface limits the maximum I_B. Thus, despite these losses, we can conclude that the YBCO log-periodic antenna is a potentially good THz radiation emitter.

In addition to elucidating the properties of THz radiation from YBCO thin-film antennas, other experimental research on THz pulse generation in HTSCs has been conducted, perhaps the most interesting being on ultrafast carrier dynamics in optically excited systems. Since the THz pulse-generation mechanism completely differs from the one in semiconductors, one can expect to discover many new ultrafast phenomena, such as optical vortex generation [55, 56, 57, 58, 59, 60, 61, 62, 63]. The authors have been studying the carrier dynamics based on THz radiation waveforms [70, 71, 72, 73] and carrier-doping effects [74, 75, 76], as well as studying their potential applications [77, 78, 79]. These efforts are still in progress, and will be reported in future articles.

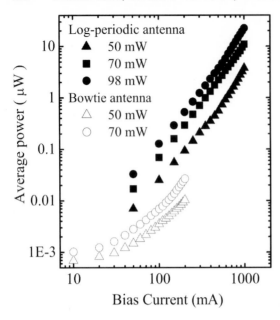

Fig. 11. Comparison of absolute powers of radiation from YBCO log-periodic antenna and bow-tie antenna as a function of bias current. The laser power was used as a parameter

2.3 Terahertz Radiation from Highly Anistropic High-T_c Superconductors

In this section, we review in detail the properties of THz radiation emitted from highly anisotropic HTSCs, $Bi_2Sr_2CaCu_2O_{8+\delta}$ (BSCCO) and $Tl_2Ba_2CaCu_2O_{8+\delta}$ (TBCCO) and discuss their fundamental properties, such as bias current, laser power, and temperature dependence of the radiated THz pulses.

2.3.1 General Properties of Terahertz Radiation from $Bi_2Sr_2CaCu_2O_{8+\delta}$ and $Tl_2Ba_2CaCu_2O_{8+\delta}$

Figures 12a and b show the time-domain waveforms of radiation from under-doped BSCCO (superconducting transition temperature $T_c \approx 76\,K$, critical current $I_c \approx 6.4\,mA$ at $9.6\,K$) and slightly underdoped TBCCO ($T_c \approx 99\,K$, $I_c \approx 90\,mA$ at $77\,K$) as a function of the delay times at different DC-bias currents [24, 26, 27, 80, 81]. It can be seen that the time-domain waveform keeps its shape independently of the bias current, while its amplitude increases with bias current. The maximum signal of the waveform for BSCCO is displayed in Fig. 13 as a function of bias current at different laser powers, and the inset shows its dependence on the laser power at a bias current of $I = 4\,mA$. The maximum amplitude of the signal increases almost linearly with bias current up to near the critical current. On the other hand, the amplitude tends to saturate even at laser powers as low as about $15\,mW$, as an effect that is peculiar to the underdoped HTSCs [24, 28]. For optimal and overdoped HTSCs,

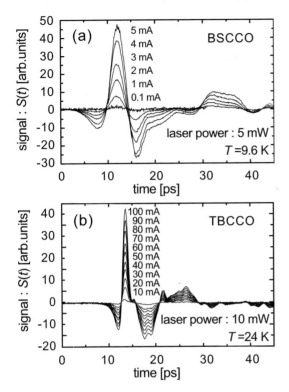

Fig. 12. (a) The time-domain waveforms of a BSCCO bow-tie antenna device under different DC-bias current conditions from 0.1 mA to 5 mA at $T = 9.6$ K and at a laser power of 5 mW. (b) The time-domain waveforms of a TBCCO bow-tie antenna device under different DC-bias current conditions from 10 mA to 100 mA at $T = 24$ K and at a laser power of 10 mW. It can be seen that the amplitude increases almost linearly with bias current up to the respective critical currents I_c, according to the classical formula of (2). The critical currents along the bridge are $I_c = 6.43$ mA for BSCCO at 9.6 K and 90 mA for TBCCO at 77 K

an almost linear increase with the laser power is generally observed up to 100 mW [16]. It is considered that in the underdoped samples with low I_c, a complete elimination of local supercurrent readily takes place near the center of the bridge, because the central region of the bridge corresponds to the minimum and maximum regions for the local supercurrent density and local laser power density, as shown in the next section.

The maximum amplitude of a THz pulse can be simply estimated on the assumption that the radiation amplitude is proportional to the number of photoexcited supercarriers $n_{ex}(x)$. Since the $n_{ex}(x)$ is proportional to the convolution of the local laser power density $p(x)$ and the local current density $j(x)$, a simple one-dimensional approximation suffices for $j(x)$ and $p(x)$ across the BSCCO bridge [61, 82], as follows,

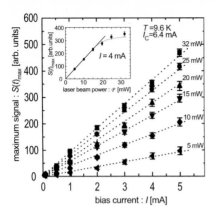

Fig. 13. Maximum signal near 12 ps in the time-domain waveform of BSCCO bow-tie antenna as a function of bias current for different laser powers. The temperature is 10 K. The *dotted lines* are drawn to indicate the trend. The *inset* shows the laser-power dependence for a bias current of $I = 4\,\text{mA}$. The *solid lines* are drawn to indicate the trend. The *dotted line* shows the result calculated from the convolution model using the local current density and local beam power density

$$
j(x) = \begin{cases} \dfrac{2j_c}{\pi} \arctan\left(\sqrt{\dfrac{W^2 - a^2}{a^2 - x^2}}\right), & -a < x < a\,, \\[2mm] j_c\,, & a < |x| < W\,. \end{cases} \tag{4}
$$

Here, $W = 15\,\mu\text{m}$ is half the width of the bridge, and a is a parameter depending on the transport current I_T determined by

$$
a = W\sqrt{1 - \left(\dfrac{I_T}{I_c}\right)^2}\,. \tag{5}
$$

On the other hand,

$$
p(x) = \mathcal{A}\exp(-0.693x^2)\,. \tag{6}
$$

The normalized distributions of $j(x)$ and $p(x)$ are displayed in Fig. 14a, where $I_T = 4\,\text{mA}$ has been used. To explain the nonlinear dependence of the maximum amplitude on the laser power, the minimum laser power density at $x = 0$ that can excite all the supercarriers $n_s(0)$ is defined as $\mathcal{A}_C$. Figure 14b shows the convolution of $j(x)$ and $p(x)$ for various $\mathcal{A}/\mathcal{A}_C$. It can be clearly seen that the most effective excitation of n_s takes place around the central region ($\mathcal{A}/\mathcal{A}_C < 1$) and extends towards the edge regions ($1 < \mathcal{A}/\mathcal{A}_C$). On the assumption that the maximum amplitude of THz radiation is directly proportional to the total number of optically excited supercarriers, the total number can be estimated by integrating the excited supercarriers along the x-axis. The dotted line in the inset of Fig. 13 shows the results obtained from

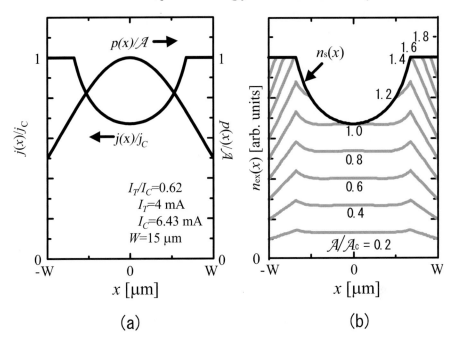

Fig. 14. (a) Normalized local current distribution $J(x)$ and local power density $p(x)$. (b) Excited supercarriers calculated by taking the convolution of $J(x)$ and $p(x)$ at several laser powers, from $\mathcal{P}/\mathcal{P}_{\mathrm{C}} = 0$ to 2

a one-dimensional analysis given a critical laser power of 16 mW and bias current of $I = 4\,\mathrm{mA}$. It can be seen that the nonlinearity in the experimental data is in qualitative agreement with the numerical result, taking account of the excitation ratio of supercarriers across the bridge.

2.3.2 Temperature Dependence of Terahertz Radiation from $\mathrm{Bi_2Sr_2CaCu_2O_{8+\delta}}$ and $\mathrm{Tl_2Ba_2CaCu_2O_{8+\delta}}$

Figures 15a and b show the time-domain waveforms at several temperatures. The waveforms do not have the same universality as observed in the measured bias-current dependence (see Fig. 12). This indicates that the transient non-equilibrium superconductivity also changes with temperature, probably due to changes in the relaxation process, magnetic flux flow effects, etc. [17, 36].

Figure 16a shows the temperature dependence of the maximum amplitude for BSCCO at several bias currents. The maximum amplitude increases with temperature and saturates above T_{S}. The increase can be qualitatively explained by the temperature dependence of the refractive index value of the BSCCO thin film [36], i.e., the decrease and increase in the imaginary and real parts of the refractive index, respectively. Another possibility is the effect due to magnetic flux flow, as mentioned above. Since the pinning force to the

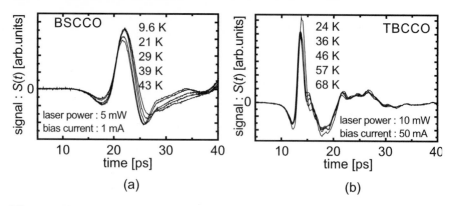

Fig. 15. Temperature dependence of the time-domain waveforms of (**a**) BSCCO and (**b**) TBCCO bow-tie antenna devices. The laser powers and bias currents are listed in the figures

fluxes weakens with increasing temperature, more radiation can be expected due to the flux flow effects.

The saturation in the temperature dependence of the maximum amplitude can be explained by the temperature dependence of the critical current, as follows. Since n_s decreases with increasing temperature, the ability of optical pulses to excite the supercarriers, $n_{ex}(x)$, exceeds n_s near the central region of the bridge. Under this condition, the radiation amplitude should saturate, as shown above. This relationship is clearly seen in Fig. 16b, which shows the close relationship between the temperature T vs. critical supercurrent I_c and the saturation temperature T_S vs. bias-current I_{bias} [24]. The bias-current ($< 1\,mA$) dependence of T_S at the laser power of $5\,mW$ shows almost the same temperature dependence of I_c.

2.3.3 Fourier Components of Terahertz Pulses

Since the time-domain waveform is strongly related to the transient-current state modulated by fs optical pulses, we can approximately ascertain the quasiparticle excitation and relaxation processes from the waveform. The difference in pulse width is clearly seen by comparing the waveforms of BSCCO and TBCCO in Fig. 12; BSCCO's pulse width is considerably wider than TBCCO's. In these experiments, however, the signals were measured with bow-tie antennas for the purpose of enhancing the output signal, which leads to a time integration of the transient electric field. Therefore, to survey the transient processes, one needs to differentiate the bow-tie antenna signal. The upper figures of Figs. 17a and b show the time derivative of the bow-tie antenna signals of Fig. 12, and the bottom figures show their Fourier components obtained through a fast Fourier transformation. The Fourier-transformed spectra show that the central components of $0.1\,THz$ to $0.2\,THz$

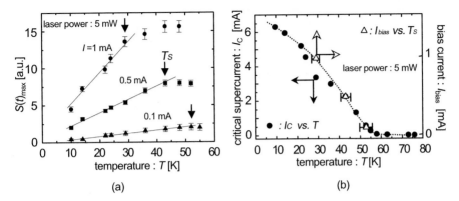

Fig. 16. (a) Temperature dependence of the maximum signals radiated from a BSCCO bow-tie antenna at several bias currents. The *solid lines* are drawn to indicate the trend. The *arrows* show the saturation temperatures T_S. (b) Relationship between the temperature T vs. critical supercurrent I_c (*closed circles*) and T_S vs. bias current (*open triangles*). The *dotted line* indicates the trend of the variation

for BSCCO and ≈ 0.33 THz for TBCCO. These central frequencies are much lower than ≈ 0.5 THz for YBCO, and the effective maximum frequencies, ≈ 0.2 THz for BSCCO and ≈ 0.7 THz for TBCCO, are much lower than the ≈ 2 THz for YBCO [15, 16]. The origin of the difference in the transient process can be seen more clearly from time-resolved pump-and-probe data, which yields the detailed relaxation processes of photoexcited carriers to a time resolution of ≈ 0.1 ps (see Sect. 4.1). Figure 18 shows representative time-resolved pump-and-probe reflectivity data for BSCCO, TBCCO and YBCO. These show that the relaxation times of the optically excited quasiparticles are as long as 4.2 ps to 5.8 ps for BSCCO, 3.3 ps to 4.1 ps for TBCCO, and 1.7 ps to 2.1 ps for YBCO [37, 39, 40, 83]. Thus, the main component in the Fourier-transformed spectrum mostly depends on the second term, $\partial n_s/\partial t$, of (2), which is the relaxation time of quasiparticles.

The maximum frequency may be limited by intrinsic physical properties such as Josephson plasma resonance (JPR), which has been observed in high-T_c cuprates. Microwave absorption studies of BSCCO revealed that the resonance frequency of JPR is below ≈ 160 GHz and ≈ 700 GHz for BSCCO and TBCCO, which exactly correspond to the respective maximum frequencies [84, 85]. The relationship between the maximum frequency and the JPR phenomenon is discussed in the next sections.

2.3.4 Coherent Terahertz-Wave Radiation from $Tl_2Ba_2CaCu_2O_{8+\delta}$ Caused by Josephson Plasma Resonance

It has been theoretically predicted that strong coherent THz waves could be radiated from HTSCs by the collective excitation of JPR [86]. A lot of effort

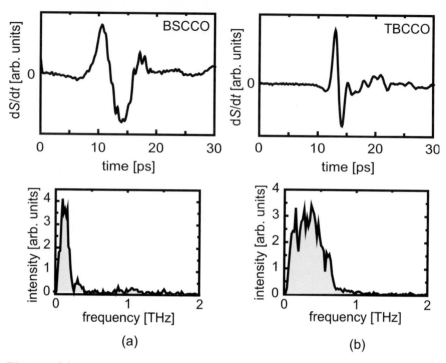

Fig. 17. (a) Time derivative of the bow-tie antenna signal of BSCCO for the bias current of 5 mA and its Fourier-transformed spectrum at $T = 9.6$ K. (b) Time derivative of the bow-tie antenna signal of TBCCO for the bias current of 100 mA and its Fourier-transformed spectrum at $T = 24$ K. Differences in pulse width and Fourier components between BSCCO and TBCCO are clearly seen, reflecting the difference in the ultrafast optical response property of these high-T_c material

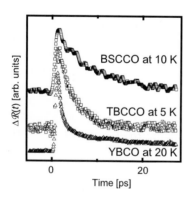

Fig. 18. Representative data of the temporal evolution of reflection change $\Delta\mathcal{R}(t)$ on the photoexcitation of BSCCO, TBCCO, and YBCO observed using the optical system of the pump-and-probe measurement with a reflection configuration. These data are well fitted with single exponential decay functions with a relaxation time of 5.8 ps for BSCCO, 3.5 ps for TBCCO, and 2.1 ps for YBCO

has gone into achieving free-space radiation of coherent THz waves by using JPR phenomena [34,87,88,89,90]. As mentioned above, the different effective

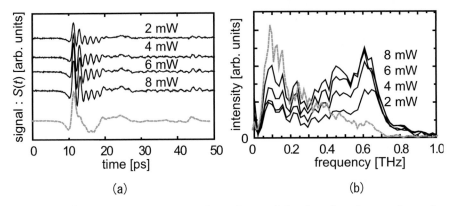

(a)   (b)

Fig. 19. (a) Excitation laser-power dependence of the time-domain waveform of a TBCCO thin film under a 100 Oe magnetic field applied along the c-axis. The measured temperature is 24 K. The *gray dashed line* shows the time-domain waveform under the bias-current condition without magnetic field. (b) Fourier-transformed spectra obtained from the time-domain waveforms in (a)

maximum frequencies of BSCCO, TBCCO and YBCO may be attributed to their different JPR frequencies.

TBCCO is a good candidate to investigate the relationship between optically radiated THz pulses and JPR, because the resonant frequency of JPR for TBCCO is about 700 GHz at low enough temperatures [91], and corresponds to the most sensitive frequency range of PC switching devices. Here, the THz radiation properties of c-axis-oriented TBCCO thin films under a magnetic field are compared with those under a bias current.

Figure 19a shows waveforms of THz radiation from a slightly underdoped TBCCO thin film by fs optical pulses at several laser powers. A TBCCO bow-tie antenna was used for the measurements in which a permanent magnet ring was used to apply a radial magnetic field of about 100 Oe nearly parallel to the c-axis. The leading pulse around 11 ps followed by the resonant oscillations can be seen. The waveform under the magnetic field is quite different from those under the bias current without a magnetic field, as shown by the gray line in Fig. 19. Namely, under the magnetic field, the strong oscillation lasts for about 10 ps, which is exactly the period of a single THz pulse from the current-biased TBCCO. This indicates the possibility that the resonant oscillation is excited by the THz pulse radiated due to the ultrafast modulation of the eddy-currents around the vortices. Furthermore, the positive sign around ≈ 600 GHz in the Fourier-transformed spectrum of Fig. 19b shows that there is substantial spatial radiation of resonant THz waves into free space.

Figure 20a shows the differential waveform for the laser power of 8 mW in Fig. 19a. This differential mode more clearly reveals the coherent radiation emitted after 50 ps, although their intensity is weaker than that of the initial

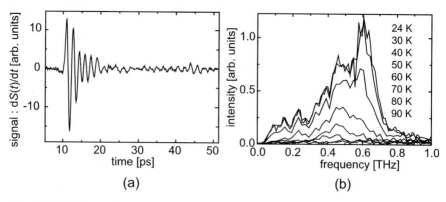

Fig. 20. (a) Time derivative of the time-domain waveform of Fig. 19's TBCCO thin-film excited with the laser power of 8 mW. (b) Fourier-transformed spectra at several temperatures obtained from the time derivative of waveforms measured under a magnetic field

oscillation around 10 ps. Figure 20b shows the Fourier-transformed spectra obtained from the differential of the waveform. The intensity of the lower-frequency peaks around ≈ 100 GHz is weaker compared with those of Fig. 19b due to the high-pass filtering of the signal. Notably, the oscillation frequency (≈ 600 GHz at 24 K) is temperature dependent, shifting towards ≈ 400 GHz with increasing temperature up to 80 K and vanishing above T_c, as would be expected for the JPR phenomenon.

As shown above, the resonant THz wave under a magnetic field is apparently different from that observed under a bias current. Under a small radial magnetic field, the vortices should penetrate the film and form a vortex lattice with a stairs structure due to the tilted elements of the radial magnetic field. However, the vortex solid state would be easily affected by the illumination of fs optical pulses, as the ultrafast modulation occurs in the eddy-currents around the vortices. Under this transient condition, a strong correlation could be expected between the electric field due to the eddy-current modulation and the fluctuating inplane vortices, and a resonant THz wave due to the collective excitation of the JPR would be radiated. In fact, strong resonant oscillations occur during the period of the supercurrent modulation.

On the other hand, Fig. 21 shows normalized Fourier-transformed spectra of the bias current in Fig. 15. It can be seen that the amplitude in the high-frequency regions from 0.4 THz to 0.7 THz decreases with increasing temperature, while the amplitude in the low-frequency regions from 0.2 THz to 0.4 THz increase. The observed spectral weight shifts towards the lower-frequency regions that corresponds to the resonant frequency shift of JPR, and strongly suggests that the maximum frequency of the biased current is determined under the influence of the JPR absorption.

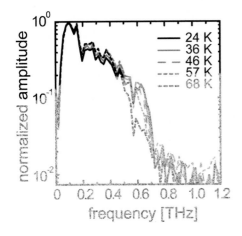

Fig. 21. Normalized Fourier-transformed spectra of the THz pulses in Fig. 15. It can be seen that the spectral weight in the higher-frequency regions shifts towards the lower-frequency region with increasing temperature

Assuming that the observed coherent radiation originated from the JPR, the London penetration depth along the c-axis λ_c can be estimated from the following equation;

$$\lambda_c = \frac{c}{\omega\sqrt{\epsilon}},$$ (7)

where c is the speed of light, and ϵ is the optical dielectric constant in the c-axis direction of TBCCO. Using $\epsilon = 9.1$ [92], λ_c becomes $\approx 26\,\mu\text{m}$ at $24\,\text{K}$ and $\approx 53\,\mu\text{m}$ at $80\,\text{K}$ for a laser power of $6\,\text{mW}$. These values show good agreement with those obtained from the JPR absorption experiments [93].

3 Terahertz Radiation from Colossal Magnetoresistive Manganites

In this section, we comprehensively review the results of the discovery of THz radiation from "magnetic" material (CMR manganite) $Pr_{0.7}Ca_{0.3}MnO_3$ thin films by exciting carriers with fs optical pulses [29, 30, 31]. We also describe a new class of THz radiation characteristics for this material [32, 33], i.e., reversible and bistable THz radiation, which has not been previously reported. These new characteristics arise from the instability of the metallic patches embedded in the charge-ordered insulating phase, which are created by illumination of visible light under an electric field, and that can be controlled by sweeping the temperature.

Hole-doped manganites with a perovskite structure, $A_{1-x}B_xMnO_3$, where A and B are rare-earth and alkali-earth elements, respectively, are exotic materials with useful electronic, magnetic, and optical properties [4, 5]. They are eliciting surprise from many researchers in condensed matter; the appearance of the CMR effect – a gigantic decrease in resistance caused by an external magnetic field around the ferromagnetic transition temperature T_C, is one such surprising property. So far, a wide variety of methods have been applied to study their unique features and have revealed the presence of two competitive phases, i.e., *metallic* and *insulating* phases. The typical example of the metallic phase appears in the ground state of $La_{0.7}Ca_{0.3}MnO_3$. This phase is an almost 100% spin-polarized ferromagnetic one, which is thought to be the driving force of the CMR effect near T_C. On the other hand, with a decrease of one electron bandwidth W (cationic substitution of La with Pr at the A-site), the insulating phase appears by overcoming the competitive ferromagnetic metallic phase. This phase occurs in $Pr_{0.7}Ca_{0.3}MnO_3$ and results from the real-space charge ordering of Mn^{3+} and Mn^{4+} ions with a 1:1 ratio in the underlying lattice [94]. However, the subtle balance between these two competitive phases is easily broken by external perturbations, e.g., changes in nominal hole concentration x, W, pressure P, temperature T, and magnetic field H (CMR effect) [4,5]. This dynamic could be called the "seesaw" effect in manganites; T and/or H act as weights on the seesaw, whose fulcrum could be x, W, and/or P (Fig. 22).

In $Pr_{0.7}Ca_{0.3}MnO_3$ as compared with other charge-ordered (CO) manganites, the ferromagnetic metallic state that accompanies the resistance reduction of over ten orders of magnitude (the melting of the CO insulating state) can be induced by not only small H [95,96] but also by other external perturbations such as X-ray irradiation [97] or illumination by visible light under an electric field [12] or H [98]. Due to deviation of x from $1/2$, at which the charge ordering is most stable, the extra electrons ($x < 1/2$) occupy Mn^{4+} sites in $Pr_{0.7}Ca_{0.3}MnO_3$. This reduces the strength of the charge ordering and therefore small external perturbations can induce the insulator–metal transition due to the melting of the CO insulating state [96]. Moreover, the CO insulating state can be easily switched to the ferromagnetic metallic one within a picosecond time scale by illumination by fs optical pulses [9]. Therefore, $Pr_{0.7}Ca_{0.3}MnO_3$ is a good SCES for examining exotic phase controls [3]. Further details on the physics and applications of CMR manganites are available in [3, 4, 5, 99, 100].

3.1 Methods

Pulsed laser deposition (PLD) was used to grow the thin films of $Pr_{0.7}Ca_{0.3}MnO_3$ on MgO(100) substrates at various temperatures. The films were characterized using room-temperature X-ray diffraction profiles and with magnetization measurements using a superconducting quantum interference device (SQUID) magnetometer. The films were single phase and nearly a-axis oriented, and their magnetic properties, except for those of the 950 °C deposition temperature case, were consistent with those of the reported previously single-crystalline and ceramic samples of $Pr_{0.7}Ca_{0.3}MnO_3$ [101].

The bow-tie and dipole-type antennas were patterned on $Pr_{0.7}Ca_{0.3}MnO_3$ thin film by photolithography and etching. The antennas were made by sput-

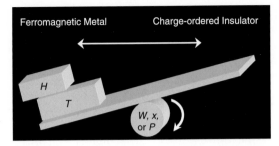

Fig. 22. How ferromagnetic metallic and charge-ordered insulating phases in manganites compete with each other. In this "seesaw" example temperature T and magnetic field H are regarded as weights and the seesaw resultantly leans to one side. The seesaw is sustained and controlled by fulcrums (hole concentration x, one electron bandwidth W, or pressure P)

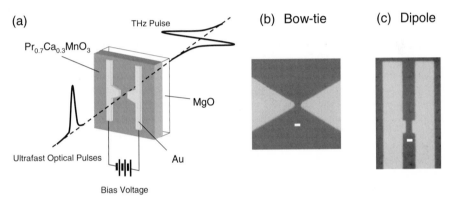

Fig. 23. Setup for terahertz (THz) radiation experiments and device structure used here. (**a**) Schematic illustration of the photoswitching device on $Pr_{0.7}Ca_{0.3}MnO_3$ thin film: Irradiating the fs optical pulses on the gap between the coplanar strip lines, which also act as electrodes of the constant-voltage supply, the THz pulse is generated and radiated from the back side of the MgO substrate into free space. Photographs of the photoswitching structure coupled with (**b**) the bow-tie-type and (**c**) dipole-type antennas made of Au. The scale of the *horizontal white bars* is $5\,\mu m$

tering Au onto the films. Figures 23b and c show optical photographs of the patterned bow-tie and dipole-type antennas, respectively. The antennas' gaps are $5\,\mu m$ wide.

We used the PC sampling technique with a bow-tie antenna made on LT-GaAs as a detector, which can directly measure the temporal waveform of the radiated THz pulse (Fig. 23a). The fs optical pulses (center wavelength of 780 nm to 800 nm, pulse width of $\approx 100\,fs$, and repetition rate of 82 MHz), from a mode-locked Ti:sapphire laser were illuminated on the gap between the coplanar strip lines. The focal diameter was estimated to be $25\,\mu m$ to

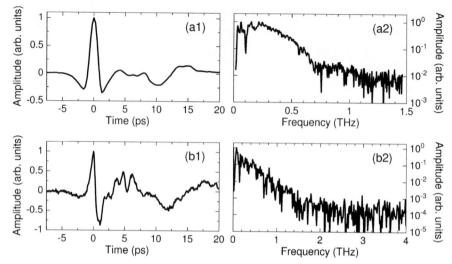

Fig. 24. Terahertz (THz) radiation characteristics of $Pr_{0.7}Ca_{0.3}MnO_3$ thin films with different antenna structures, measured at low temperature, by using the photoconductive sampling technique with the bow-tie-type low-temperature-grown GaAs antenna as a detector. *Upper* and *lower* panels show the THz radiation from (**a**) bow-tie- and (**b**) dipole-type photoswitching devices fabricated on $Pr_{0.7}Ca_{0.3}MnO_3$ thin films. Left (**1**) and right (**2**) panels show the propagated THz pulse in the time domain and their Fourier-transformed spectra on logarithmic plots in the frequency domain, respectively. All data is normalized by the respective maximum amplitude. The time delay at which the amplitude reaches the maximum is offset to 0 ps for comparison. Note that the scales of the frequency axes in (**a2**) and (**b2**) are different

50 µm, which is the FWHM of the distribution of the THz pulse calculated using the THz radiation imaging technique.

3.2 Terahertz Radiation from Magnetoresistive $Pr_{0.7}Ca_{0.3}MnO_3$

Figure 24 shows typical examples of THz radiation from the bow-tie (upper panels) and dipole (lower panels) antennas on $Pr_{0.7}Ca_{0.3}MnO_3$ thin films. The propagated waveform in the time domain and its Fourier-transformed spectrum in the frequency domain are shown in the left and right panels, respectively. The measurements for the bow-tie (dipole) antenna were made at 23 K (10 K) under the bias voltage V_{bias} of 32 V (20 V) and laser power P_{pump} of 184 mW (65 mW).

In the propagated waveform of the bow-tie antenna, a sharp positive peak, whose amplitude rapidly increases and decreases within a few ps, appears at 0 ps (Fig. 24a1). It consists of bilateral negative peaks centered around ±1.5 ps. The FWHM of the sharp positive peak is estimated to be ≈ 1 ps. Figure 24a2 shows the corresponding frequency spectrum on a logarithmic

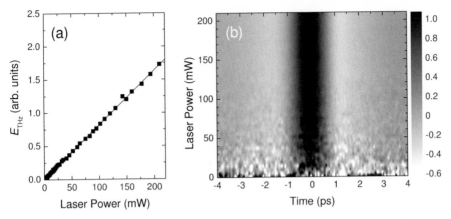

Fig. 25. Terahertz (THz) radiation characteristics of $Pr_{0.7}Ca_{0.3}MnO_3$ as a function of laser power P_{pump} at a fixed bias voltage of 15 V, measured at 23 K. (a) Main peak amplitude E_{THz} at 0 ps in the time-domain waveform as a function of P_{pump}. Data are normalized by the value of P_{pump} at 122 mW. The *solid line* is a least-squares fit to the data (*solid squares*) assuming a linear function. (b) Contour plot of the propagated THz pulse in the time domain as a function of P_{pump}. Data are normalized by the value of E_{THz} at respective P_{pump}s. The zero position of the time delay is offset to 0 ps for comparison

intensity scale as derived from the Fourier transformation of the waveform in Fig. 24a1; it has the center frequency of ≈ 0.2 THz and expands to ≈ 1 THz.

Figures 24b1 and b2 show the time-domain waveform of the dipole antenna and its frequency spectrum, respectively. We can see the nearly single cycle pulse, whose amplitude rapidly increases at 0 ps, followed by the reduction of the amplitude with the negative peak at 1 ps. The corresponding frequency spectrum expands by 200% (≈ 2 THz), as compared with the case of the bow-tie antenna.

3.2.1 Laser-Power and Bias-Voltage Dependences

Figure 25 shows the change in the propagated waveform under a fixed V_{bias} of 15 V as a function of P_{pump}, measured at 23 K. The main peak amplitude E_{THz} at 0 ps in the time-domain waveform (see Fig. 24a1) is found to increase linearly with increasing P_{pump} irrespective of the polarity of V_{bias}, as can be seen by the closed squares in Fig. 25a (we only show the data for V_{bias} of 15 V). E_{THz} is normalized by the value at P_{pump} of 122 mW. The solid line shows the least-squares fit to the data, assuming a linear response of E_{THz} with P_{pump}. Such a linear increase in E_{THz} is widely observed in THz radiation characteristics of semiconductor antennas with small gaps ($\approx 5 \mu m$), i.e., gaps that are nearly the width of that used in $Pr_{0.7}Ca_{0.3}MnO_3$ [102]. It has also been observed that both V_{bias} and fs optical pulses are needed to generate

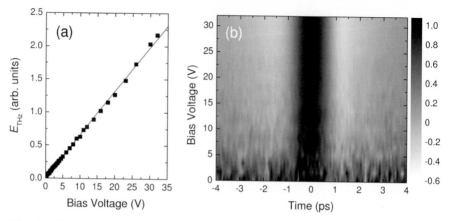

Fig. 26. Terahertz (THz) radiation characteristics of $Pr_{0.7}Ca_{0.3}MnO_3$ as a function of bias voltage V_{bias} at a fixed laser power of 122 mW, measured at 23 K. (a) Main peak amplitude E_{THz} at 0 ps in the time-domain waveform as a function of V_{bias}. Data are normalized by the value of V_{bias} at 15 V. The *solid line* is a least-squares fit to the data (*solid squares*) assuming a linear function. (b) Contour plot of the propagated THz pulse in the time domain as a function of V_{bias}. Data are normalized by the value of E_{THz} at respective V_{bias}s. The zero position of the time delay is offset to 0 ps for comparison

the THz pulse (see Figs. 25b and 26b). In addition, the change in the polarity of V_{bias} causes the reversal of the time-domain waveform. These properties exclude an optical nonlinear mechanism as a source of THz radiation from $Pr_{0.7}Ca_{0.3}MnO_3$. E_{THz} at the fixed P_{pump} of 122 mW also depends linearly on V_{bias}, as shown in Fig. 26a (E_{THz} is normalized by the value at V_{bias} of 15 V).

Figures 25b and 26b show the time-domain waveforms as a function of P_{pump} at fixed V_{bias} and as a function of V_{bias} at fixed P_{pump}, respectively. To see how the waveform varies with P_{pump} and V_{bias}, all data are normalized by the value of E_{THz} in the respective time-domain waveforms. In the case of semiconductors such as LT-GaAs, photoexcited carriers have a strong influence on various kinds of interactions, such as intervalley scattering and screening, when the THz pulse radiates into free space [103]. These interactions' strengths can be changed by varying P_{pump} and V_{bias}. In particular, the width of the pulse from LT-GaAs is mainly determined by the electron-trapping time of the shallow donor state; therefore, the precipitation of As by changing the growth conditions plays a crucial role in controlling the pulse width [104]. As can be seen in Figs. 25b and 26b, no influence on the pulse width of the main peak is observed after changing P_{pump} and V_{bias}. In addition, the pulse width of $Pr_{0.7}Ca_{0.3}MnO_3$ is not affected by the deposition temperature [30]; the pulse widths of the samples grown at 790 °C and 950 °C are estimated to be ≈ 1 ps and ≈ 1.3 ps, respectively, indicating that no efficient trapping center exists within the charge gap of $Pr_{0.7}Ca_{0.3}MnO_3$. Moreover, even under P_{pump} of 200 mW (Fig. 25a), E_{THz} continues to increase linearly without the saturation due to the screening effect of the photoexcited charge carriers, as is usually observed in the THz radiation characteristics of semiconductors.

By using pump-and-probe reflection spectroscopy, *Fiebig* et al. [9] have revealed that the metallic state resulting from the melting of the CO state exists for less than 1 ps; the photogenerated metallic patch acts as the seed of the persistent photoinduced insulator–metal transition as long as the electric field is maintained. Accordingly, this can explain the THz radiation charac-teristics described above, i.e., that the pulse width does not strongly depend on growth temperature and the negligible effect on the waveform by varying the strength of V_{bias} and P_{pump}. In particular, the reversible and bistable na-ture of THz radiation, as will be described in Sect. 3.3, can be thought to be one aspect of the occurrence of the photoinduced insulator–metal transition. However, as described below, this seems to be incapable of explaining the T dependence of E_{THz} below the temperature where magnetic order sets in.

3.2.2 Temperature Dependence

This experiment used two film samples, which were deposited at 790 °C (sam-ple A) and 950 °C (sample B) using the PLD technique. X-ray diffraction profiles showed that sample A was nearly a-axis oriented. Sample A had an intense (100) peak, whereas sample B had a (110) peak. In addition, the X-ray diffraction intensity of sample B was much smaller than that of sample A. These results indicate that sample B did not undergo the charge ordering transition. Figure 27 shows the T dependences of E_{THz} (upper panels) and the magnetization M with field cooling (FC; open circles) and zero field cool-ing (ZFC; open squares) under an H of 500 Oe (lower panels), respectively; the results for samples A and B are shown in the left and right panels of Fig. 27, respectively. The E_{THz} of sample A (sample B) was measured under a fixed V_{bias} of 15 V (10 V) and P_{pump} of 100 mW (80 mW). The respective $M(T)$s were calibrated by subtracting the contribution of $M(T)$ of the MgO substrate, which was measured using the same experimental procedure.

With increasing T, the E_{THz}s of both samples decrease but then increase around $T_C \approx 130$ K (the point at which M steeply increases with decreasing T). However, we observed different $E_{THz}(T)$ behaviors in the two samples above T_C; E_{THz} of sample A exhibits a broad maximum around 230 K, which corresponds to $T_{CO/OO}$ of the bulk $Pr_{0.7}Ca_{0.3}MnO_3$, while E_{THz} of sample B shows less T dependence and no correlation with $T_{CO/OO}$. As mentioned above, sample B seems not to exhibit charge ordering. Therefore, sample A's change in $E_{THz}(T)$ around $T_{CO/OO}$ can be ascribed to melting of the CO state (the creation of the metallic patches within the CO insulating phase) by external perturbations; the melting of the CO state attenuates the THz radiation efficiency, which is consistent with sample B's decrease in E_{THz}

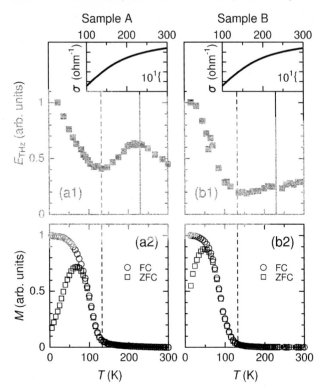

Fig. 27. Terahertz THz radiation and magnetic characteristics of $Pr_{0.7}Ca_{0.3}MnO_3$ grown at different temperatures. Temperature T dependences of the main peak amplitude E_{THz} at 0 ps in the time-domain waveform [*upper panels*; (**1**)] and the magnetization M under a magnetic field of 500 Oe [*lower panels*; (**2**)] for samples A [*left panels*; (**a**)] and B [*right panels*; (**b**)]. *Open circles and squares* represent M taken from the field cooling (FC) and zero field cooling (ZFC), respectively. M of $Pr_{0.7}Ca_{0.3}MnO_3$ was obtained by calibrating it to M of MgO substrate, which was measured separately. *Inset of upper panel* shows T dependence of DC conductivity σ on a logarithmic scale, measured by the conventional four-probe method with a constant current of $1\,\mu A$. The *vertical solid* and *dashed lines* indicate the charge/orbital ordering T of the bulk $Pr_{0.7}Ca_{0.3}MnO_3$ as reported by *Jirák* et al. [94] and the spin ordering T as derived form the $M(T)$ curves of the *lower panels*

around $T_{CO/OO}$, the details of which are presented in the following section (Sect. 3.3).

To date, the T dependence of THz radiation has been widely investigated not only for semiconductor surfaces [105, 106, 107, 108] but also semiconductor photoswitches [109]. In contrast to the intuitive view, it was found that the T dependence of THz radiation, which is independent of the magnitude of the bandgap or type of charge carrier, follows the T dependence of the static mobility μ. This is especially the case with narrow-gap semiconductors whose bandgap ($\approx 0.2\,\mathrm{eV}$) is comparable to the charge gap ($\approx 0.5\,\mathrm{eV}$) of $Pr_{0.7}Ca_{0.3}MnO_3$ [110], and can be phenomenogically explained as follows [111]. After illumination by optical pulses, the charge redistributes as a result of the large difference between the velocities of the electrons and holes, giving rise to an electric field. This is a major source of surge current (photo-Dember effect). Therefore, the change in μ with T is thought to strongly affect the THz radiation characteristics [108, 112].

The μ data for CO manganites are not available at present. *Yamada et al.* [113] have reported the measurement of the T dependence of the static carrier density n in single crystals of $Pr_{1-x}Ca_xMnO_3$ ($0.33 < x < 0.48$) on the thermopower Q between $100\,\mathrm{K}$ and $300\,\mathrm{K}$. They revealed that the absolute magnitude of dQ/dT, which is sensitive to the change in n, does not strongly depend on T; it does not exhibit the clear anomaly even in the vicinity of $T_{CO/OO}$. Therefore, in the framework of the photo-Dember description given above, it can be anticipated that the T dependence of THz radiation from $Pr_{0.7}Ca_{0.3}MnO_3$ resembles that of σ through the relation $\sigma = en\mu$. The insets of Figs. 27a1 and b1 show the T dependence of σ of samples A and B on a logarithmic scale, respectively, measured by the conventional four-probe method with a constant current of $1\,\mu\mathrm{A}$. Both sets of σ exponentially increase with increasing T and show no clear correlation with T_C or $T_{CO/OO}$. Apparently, $E_{THz}(T)$ does not follow $\sigma(T)$. In particular, $E_{THz}(T)$ below T_C has the opposite signature as compared to $\sigma(T)$; E_{THz} increases, while σ decreases exponentially. This leads to the conclusion that the T dependence of THz radiation from $Pr_{0.7}Ca_{0.3}MnO_3$ cannot be explained by the change in μ, which is the widely accepted picture for understanding the THz radiation from narrow-gap semiconductors.

We also noted that $E_{THz}(T)$ below T_C of both samples shows the same tendency and increases with decreasing T. This cannot be explained by the melting of the CO state, as described before (see also Sect. 3.3), because sample B seems not to exhibit charge ordering as far as X-ray diffraction measurements and the behavior of E_{THz} around $T_{CO/OO}$ indicate.

3.3 Reversible and Bistable Terahertz Radiation

In this section, we describe new THz radiation characteristics that are a result of the photoinduced insulator–metal (IM) transition, that is, the reversible and bistable nature of THz radiation, which can be controlled by external perturbations (V_{bias}, P_{pump}, and T). These characteristics could be employed to overcome the weaknesses of PC switches made on semiconductors such as LT-GaAs, which is frequently used as the parent material of efficient THz emitters.

To confirm the reversible and bistable nature of THz radiation, the following experimental procedures [(1)–(4)] had to be carefully implemented. The photocurrent I_{photo} of $Pr_{0.7}Ca_{0.3}MnO_3$ in the steady state was also measured during these experiments. The thin film (sample C) was prepared with the

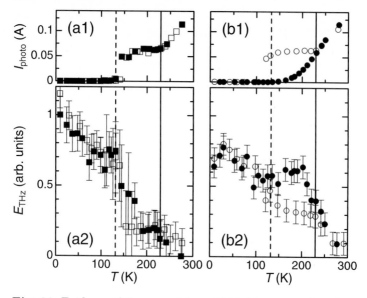

Fig. 28. Evidence of the reversible and bistable nature of terahertz radiation from $Pr_{0.7}Ca_{0.3}MnO_3$ thin film under bias voltage of 18 V and laser power of 65 mW. *Upper* and *lower panels* show the temperature T dependences of the photocurrent I_{photo} of $Pr_{0.7}Ca_{0.3}MnO_3$ and the main peak amplitude E_{THz} at 0 ps in the time-domain waveform, respectively. The *closed squares, open squares, closed circles,* and *open circles* indicate the data obtained for (1) first and (2) second warming, (3) cooling after the measurement of (1) or (2), and (4) warming after the measurement of (3), respectively. All data in (a2) and (b2) are normalized by the value (*closed square*) in (a2) at 10 K

PLD technique under nearly the same conditions as used to make sample A. A dipole-type antenna structure was patterned on it.

1. THz radiation was detected under a V_{bias} of 18 V and P_{pump} of 80 mW after the sample was cooled from room T to 10 K without external perturbations. The closed squares in Fig. 28a2 show E_{THz} as a function of T; the data were gathered in the T warming run. All data described below are normalized by the value at 10 K. With increasing T, E_{THz} decreases and seems to exhibit a broad minimum around 130 K (indicated by the dashed line), the T at which M steeply increases with decreasing T. This behavior is consistent with the previous result shown in Fig. 27a1 under V_{bias} of 15 V and P_{pump} of 100 mW, including the decreasing rate of E_{THz} and its magnitude. When T is further increased, contrary to the expectation that E_{THz} should increase and exhibit a broad maximum around $T_{CO/OO} \approx 230$ K (indicated by the solid line) as in the previous case (Fig. 27a1), one can see a sudden reduction in E_{THz} around 130 K. Concurrently, a transient increase in I_{photo} (closed squares) oc-

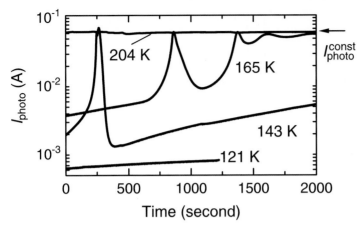

Fig. 29. The elapsed-time dependence of the photocurrent I_{photo} of $Pr_{0.7}Ca_{0.3}MnO_3$ at various temperatures. No elapsed-time dependence of I_{photo} is observed below 130 K and above 180 K. Note that the I_{photo}s seem to converge to the constant value of about 60 mA(I_{photo}^{const}), as indicated by the *arrow*

curs, as seen in Fig. 28a1. Above 130 K, we can see that both E_{THz} and I_{photo} are less sensitive to T, while I_{photo} continues to increase linearly above $T_{CO/OO}$. At room T, I_{photo} reaches 150 mA, which corresponds to a photoconductivity of the order of $10^3\,\Omega^{-1}\cdot cm^{-1}$, which is the typical magnitude of the metallic state of manganites. Figure 29 shows the time dependence of I_{photo} of $Pr_{0.7}Ca_{0.3}MnO_3$ at various temperatures. We can see a significant change in I_{photo} of up to two orders of magnitude between 140 K and 170 K, while no time dependence of I_{photo} is observed below 130 K and above 180 K. Another noticeable signature is that I_{photo} at various temperatures seems to converge to a constant value of about 60 mA. This simply indicates that the illumination of the optical pulse under V_{bias} can induce the metallic patches embedded with the CO insulating matrix. Therefore, the rapid reduction of E_{THz} with the concomitant increase of I_{photo} above 130 K (Fig. 28) can be phenomenogically understood by the occurrence of the photoinduced IM transition. This characteristic makes it possible to create reversible THz radiation and so overcome the problems of PC switches made on semiconductors such as LT-GaAs. One well-known problem is dielectric breakdown due to impact ionization of impurities [114]. In the case of LT-GaAs, the limiting *steady* current has roughly been estimated to be ≈ 1 mA (the breakdown field is ≈ 250 kV/cm, which is independent of the annealing temperature of LT-GaAs) [115, 116]. Once dielectric breakdown occurs above the limiting value of V_{bias} under constant P_{pump} (a few kV/cm under tens of mW in our measurements of LT-GaAs), the THz radiation characteristics do not completely recover. However, as the following experimental procedure shows, we can confirm the reversible nature of THz radiation from $Pr_{0.7}Ca_{0.3}MnO_3$.

2. After the measurement of (1), the illumination and V_{bias} were switched off at room T. The sample was then cooled from room T to 10 K and the THz radiation characteristics measured while warming under the same conditions as in (1); the results for $I_{photo}(T)$ and $E_{THz}(T)$ are shown by the open squares in Figs. 28a1 and a2, respectively. As can be seen, both the open and closed squares coincide within error bars defined as the maximum variation of E_{THz} with time, providing evidence of reversible THz radiation from $Pr_{0.7}Ca_{0.3}MnO_3$ after the current exceeded 100 mA. Another unique THz radiation characteristic is bistability, whose evidence is shown by the closed and open circles in Figs. 28b1 and b2.

3. After the measurements of (1) or (2), external perturbations were left unchanged, and E_{THz} and I_{photo} were measured while cooling; the data are represented by closed circles in the figures. E_{THz} and I_{photo} show a clear hysteresis behavior, which begins around $T_{CO/OO}$ and finishes around T_C. As T decreases from room T, E_{THz} increases until it reaches a maximum at 200 K; E_{THz} then decreases until T falls below T_C, after which it increases again. Such a hysteresis behavior can be ascribed to the T instability of the photogenerated metallic patches created by external perturbations described before. However, around 40 K, we can clearly see a significant change in E_{THz}; the values begin to *decrease* with decreasing T, whereas the E_{THz} values of measurement (1) or (2) *increase*. This low-T bistable phase was sustained even when V_{bias} and P_{pump} were switched off simultaneously at 10 K, which is inconsistent with the results of the photoinduced IM transition phenomena reported previously, since the initial insulating state before the occurrence of the photoinduced IM transition recovers when V_{bias} is switched off [9, 12, 117, 118, 119].

4. To gain an understanding of the stability of this low-T phase, the sample was kept at 10 K for 24 h without external perturbations, after which the THz radiation experiment under increasing T was performed again. The results for I_{photo} and E_{THz} are shown by the open circles in Figs. 28b1 and b2, respectively. On warming, E_{THz} *increases* and reaches a maximum value around 40 K; E_{THz} well reproduces the data (closed circles) of measurement (3). As T increases further, the behaviors of I_{photo} and E_{THz} appear to be quite similar to those (closed and open squares shown in Fig. 28a2) of measurement (1) or (2). All of which described above strongly indicates that the low-T bistable phase does not have the same origin as the hysteresis behavior observed near $T_{CO/OO}$, which results from the photogenerated metallic patches embedded in the CO insulating state [117, 118, 119]. Moreover, in addition to the removal of V_{bias} and P_{pump}, a sweep of T is necessary to go from the low-T bistable phase to the initial insulating state. Further study is needed to make detailed measurements of the V_{bias} and P_{pump} dependences at constant T after

the photoinduced IM transition occurs and in so doing assess their stability with T. The reversible and bistable nature of THz radiation from $Pr_{0.7}Ca_{0.3}MnO_3$ may be exploitable for scientific and technological purposes, but the radiated power must be improved; the total power of THz radiation from $Pr_{0.7}Ca_{0.3}MnO_3$ is approximately three orders of magnitude smaller than that from LT-GaAs at room T. However, E_{THz} under a fixed P_{pump} of 65 mW linearly increased as a function of V_{bias} up to ≈ 400 kV/cm (the limit of our voltage source) at 10 K (only an I_{photo} of 2 μA flows under these conditions). This behavior can be simply ascribed to the CO *insulating* ground state in $Pr_{0.7}Ca_{0.3}MnO_3$. Therefore, we should investigate the possibility of intense THz radiation from other CO manganites having strong charge ordering; e.g., half-doped manganite $Pr_{1/2}Ca_{1/2}MnO_3$ [120]. This is because an H of ≈ 20 T (≈ 6 T in $Pr_{0.7}Ca_{0.3}MnO_3$ [98]) is needed to melt the CO insulating state at low T in $Pr_{1/2}Ca_{1/2}MnO_3$. The ultrafast photoresponse should thus appear in the same manner as it does in $Pr_{0.7}Ca_{0.3}MnO_3$ [9].

4 Terahertz Time-Domain Spectroscopy of Strongly Correlated Electron Systems

This section describes the optical properties at THz frequencies of high-T_c superconducting cuprates and CMR manganites, in connection with THz optics. One line of study combines time-resolved pump-and-probe and THz time-domain spectroscopies of the underdoped high-T_c superconductor, $Bi_2Sr_2CaCu_2O_8$ [36, 39]. The other includes THz time-domain spectroscopic studies of CMR manganites with $x = 0.3$: $La_{0.7}Ca_{0.3}MnO_3$ [41, 42, 43] and $Pr_{0.7}Ca_{0.3}MnO_3$ [44, 45]. These are contrastive compounds in the absence of H; the former is a 100% spin-polarized ferromagnetic metal with an IM transition temperature of 230 K and the latter is a CO insulator without an IM transition.

4.1 Ultrafast Carrier Dynamics in Underdoped $Bi_2Sr_2CaCu_2O_{8+\delta}$

Pseudogap (PG) evolution is a peculiar phenomena of HTSCs, occurring particularly in the underdoped regime. It appears as a substantial reduction in the electronic density of states (DOS) around the Fermi level below a certain temperature T^* much higher than T_c. Although a great deal of experimental and theoretical efforts have been made to disclose its relation to macroscopic phase-coherent superconductivity [83, 121, 122, 123, 124, 125, 126, 127, 128, 129], much remains to be resolved. The greatest concern is whether the order parameter amplitude persists up to T^*, although the macroscopic superconductivity properties – perfect DC conductivity and diamagnetism – are lost at T_c. Another concern is whether the PG coexists with a superconducting gap (SG) below T_c.

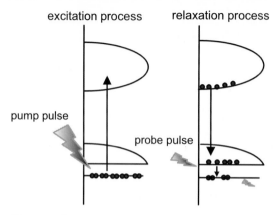

Fig. 30. Schematic illustration of excitation and relaxation processes of carriers. Many quasiparticles are excited just above the low-energy superconducting gap or pseudogap through the initial electron–electron and electron–phonon scatterings of photoexcited carriers directly excited by the pump-pulse illumination

The time-resolved pump-and-probe (TRP) technique using fs optical pulses is very useful to investigate such a complicated system where the SG (Δ_s) and the PG (Δ_p) may coexist [46, 83, 130, 131, 132, 133]. It has the potential to decompose the relaxation processes of photoexcited carriers at the respective low-energy gaps into distinct relaxation dynamics and disclose the low-lying electronic structure around the Fermi level [130].

On the other hand, if the pair amplitude persists above T_c, the correlation times among the pairs should be extremely short because of the lack of macroscopic phase coherence. To detect such a partial (or short-range) phase coherence, one can measure the high-frequency conductivity and use THz-TDS to capture the partial phase coherence with ultrashort correlation times of subpicosecond to picosecond order [134, 135].

This section describes TRP and THz-TDS measurements of an underdoped BSCCO thin film, which yielded experimental evidence for the coexistence of the SG and PG below T_c, as well as evidence for the appearance of superconducting fluctuations with ultrashort lifetimes that arise simultaneously with the PG evolution.

4.1.1 Coexistence of Superconducting Gap and Pseudogap in Underdoped $Bi_2Sr_2CaCu_2O_{8+\delta}$

TRP has revealed a multiple excitation state from the relaxation dynamics of photoexcited carriers with a time resolution of about 0.1 ps. Figure 30 shows a schematic illustration of the excitation and relaxation processes of the photoexcited carriers. In the TRP measurements made on HTSCs, pump pulses (1.5 eV for an 800 nm wavelength) can directly excite the Cooper pairs

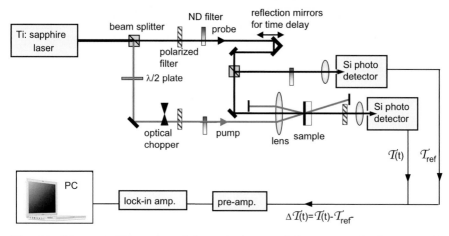

Fig. 31. Schematic illustration of the optical setup of the time-resolved pump-and-probe transmittance measurement

to the upper band, and the excited hot carriers immediately relax to an excited state just above the SG and PG, accompanied by electron–electron and electron–phonon scatterings. Through the relaxation processes of hot carriers, a great number of quasiparticles are excited by avalanche effects in the nonequilibrium state of superconductivity. Since the initial excitation and relaxation processes cease within $\approx 0.1\,\mathrm{ps}$ [83, 130], we can analyze in detail the relaxation processes of quasiparticles at the respective gaps by observing the transient changes in the transmittance of the probe pulse.

Figure 31 shows the optical setup for TRP measurement. The pump and probe pulses are separated from the main beam delivered from a mode-locked Ti:sapphire laser with a beam splitter, and both are illuminated on the same spot on the sample surface (a focal spot size of several tens of µm in diameter). To observe time-resolved transmittance data from the probe pulse, the probe pulse is time delayed by varying the length of the optical path with a multiple-reflection mirror system sliding under computer control. To detect the changes in the transmittance, the intensities of the transmitted probe pulse, $\mathcal{T}(t)$, and the reference pulse, $\mathcal{T}_{\mathrm{ref}}$, are detected by photodiodes, and the difference, $\Delta\mathcal{T}(t) = \mathcal{T}(t) - \mathcal{T}_{\mathrm{ref}}$, is lock-in detected. Here, $\mathcal{T}_{\mathrm{ref}}$ is adjusted to the value of $\mathcal{T}(t)$ just before the pump-pulse excitation.

Typical TRP transmittance data as a function of delay time, $\Delta\mathcal{T}(t)/\mathcal{T}_{\mathrm{ref}}$, are plotted in Fig. 32. The sign of $\Delta\mathcal{T}(t)$ is negative, in contrast to the positive sign of the reflectivity of Fig. 18. We can see a sudden increase in $|\Delta\mathcal{T}(t)|$ at $t \simeq 0\,\mathrm{ps}$ due to the irradiation of fs optical pulses. Also revealed are fast and slow relaxation processes below T_{c}, and a fast one above T_{c}. Therefore, on the assumption of two excitation gaps of Δ_{s} and Δ_{p} below T_{c}, a two-component fit can be applied to the data with use of an exponential decay function,

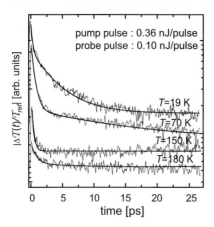

Fig. 32. Time-resolved transmittance as a function of the delay time of the probe pulse with respect to the pump pulse at several temperatures below and above T_c. The *lines* show the fits using the exponential decay function given by (8). It can be seen that there are two relaxation processes (fast and slow) below T_c, reflecting the coexistence of the SG and PG below T_c. The fast relaxation process above T_c is the relaxation process of photoexcited carriers at the PG

$$|\Delta T(t)/T_{\mathrm{ref}}| = A(T)\mathrm{exp}(-t/\tau_s) + B(T)\mathrm{exp}(-t/\tau_p). \qquad (8)$$

Here, τ_s and τ_p are the relaxation times of photoexcited carriers at Δ_s and Δ_p, respectively, and $A(T)[A(T) = 0$ at $T_c < T]$ and $B(T)$ are characteristic amplitudes closely related to the photoexcited carriers at the SG and PG, respectively. The solid lines in Fig. 32 are the fitting curves, which show a fairly good agreement with the relaxation data.

The amplitudes, $A(T)$ and $B(T)$, are temperature dependent and closely related to both the magnitude of the gap and the number of photoexcited carriers. Figure 33a plots $A(T)$ and $B(T)$ as functions of temperature. $B(T)$ gradually increases below $T^* \approx 210\,\mathrm{K}$ and begins to saturate near $T_S \approx 95\,\mathrm{K}$, whereas $A(T)$ rapidly increases below T_c, reaching an almost constant value below $\approx 60\,\mathrm{K}$ and almost corresponds to the increase in J_c.

The estimated relaxation times calculated from the fitting process are summarized in Fig. 33b. τ_s has an almost constant value of $4.2\,\mathrm{ps} \pm 0.3\,\mathrm{ps}$ below $40\,\mathrm{K}$, but diverges near T_c. On the other hand, τ_p is almost T independent below $150\,\mathrm{K}$ with a value of $0.7\,\mathrm{ps} \pm 0.3\,\mathrm{ps}$, and it diverges near T^*. The temperature dependence of τ_p indicates that the magnitude of PG is T independent at least below $150\,\mathrm{K}$ and approaches zero near T^*. The temperature dependence of these relaxation times apparently shows that the PG and SG coexist below T_c, because it is unlikely that there are any other temperature-dependent relaxation processes other than the SG and PG. Furthermore, from the fact that the amplitude $B(T)$ substantially increases, while τ_p remains at a constant value of about $0.7\,\mathrm{ps}$, one may consider that the PG does not evolve over all the sample but only in local domains below T^*, and that the

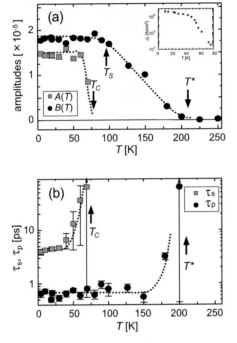

Fig. 33. (a) Temperature dependence of the amplitudes, $A(T)$ and $B(T)$, utilized for the best fit to the observed data using the exponential decay function given by (8). The fitting used the relationship, $A(T) + B(T) = C(T)$ for $T_c > T$, and $B(T) = C(T)$ for $T_c < T$. Here, the amplitude $C(T)$ is the experimental value, estimated as $\Delta T(0)/T_{\text{ref}}$. The *inset* shows the temperature dependence of the critical supercurrent density J_c along the *ab*-plane for the same underdoped BSCCO thin film. (b) Temperature dependence of relaxation times, τ_s and τ_p, obtained as the best parameters. The *solid lines* show the error bars of the fitting. The *dashed lines* indicate the trend

number (or area) of the PG domains increases and saturates with decreasing temperature from T^* to T_S. It is worthwhile noting that macroscopic superconductivity occurs a little below T_S at T_c.

4.1.2 Ultrafast Superconductivity Fluctuation in Underdoped $Bi_2Sr_2CaCu_2O_{8+\delta}$

THz-TDS measurements on the identical BSCCO thin film also reveal a normal state anomaly in the high-frequency conductivity below T^*. The advantage of the phase-sensitive THz-TDS technique is that it can directly measure the electric field of the THz pulse transmitted through the thin film, thus gaining information about the phase and amplitude. Therefore, we can directly obtain the real and imaginary parts of the complex conductivity as a func-

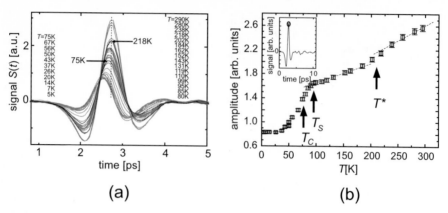

Fig. 34. (a) Time-domain signals observed at temperatures from $T = 290$ K to 5 K. Here, only the first pulses are displayed for clarity. It can be seen that only a slight phase shift occurs below the characteristic temperature $T^* \approx 210$ K, whereas one clearly occurs below T_c. The *dotted line* is drawn to see the shift clearly. (b) Temperature dependence of the maximum amplitude of the transmitted THz pulse. The amplitude clearly decreases below T_c, as well as near T^*

tion of frequency from the complex transmission without Kramers–Kronig analysis.

Figure 34a shows time-domain signals of a THz pulse transmitted through the sample in the normal and superconducting states. In this kind of phase-sensitive method, the phase shift takes place below T_c because of the super-conducting transmission line acting as a high-pass filter [134]. It can be seen that the phase shift takes place below the two characteristic temperatures, T^* and T_c. With decreasing temperature, the peak position shifts slightly towards the left below $\approx T^*$, and then drastically shifts to the left below $\approx T_c$. The phase shift near T^* shows the possibility of partial phase coherence even at T^* in the normal state. Figure 34b shows the temperature dependence of the maximum amplitude of the THz waveform around ≈ 2.7 ps. The decrease in the amplitude can be seen not only below T_S, but also near T^*, probably due to the appearance of partial phase coherence. To investigate this possibility, it is useful to estimate the imaginary part of the complex conductivity σ_2, which is closely related to the pair density.

Figure 35 shows the temperature dependence of σ_2 at several frequencies on a semilogarithm scale. It is remarkable that σ_2 at the higher frequencies ($f = 1.08$ THz and 1.41 THz) shows a definite increase below T^* corresponding to the PG evolution, and then a rapid increase below T_c. The selective increase observed only at the higher frequency σ_2 (1 THz $< f$) suggests the appearance of partial phase coherence with an ultrashort lifetime of less than ≈ 1 ps, as predicted by time-dependent Ginzburg–Landau theory [136]. Furthermore, as the frequency is lowered, the onset temperature of the superconductivity fluctuations shifts towards lower temperature, $\approx T_S$

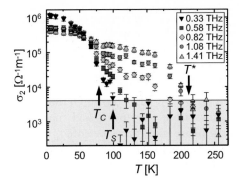

Fig. 35. Temperature dependence of the imaginary part of the complex conductivity σ_2, at several frequencies. σ_2 at $f = 1.08$ THz and 1.41 THz show a definite increase below the characteristic temperature $T^* \simeq 210$ K, and a rapid increase below T_c. The onset temperature of the increase decreases as the frequency decreases. The *gray zone* corresponds to data scattering caused by numerical error

at $f = 0.33$ THz. It is worth noting that by utilizing THz-TDS, the superconducting fluctuations could be observed near T^*, which is much higher than T_c.

Taking account of all the results obtained from the TRP and THz-TDS measurements, the simultaneous appearance of partial phase coherence and a PG at $T \approx T^*$ suggests that a PG domain is itself in a fluctuating superconducting state with a lifetime of less than 1 ps. Furthermore, the decrease of the onset temperature of σ_2 also supports the local domain picture drawn from the TRP data, because the overlapping of the order parameters of the partially coherent domains, which is caused by the evolution of local domains, extends the superconductivity fluctuation lifetime. As a result of the extension, a macroscopic superconductivity state could be realized at T_c.

4.2 Low-Energy Charge Dynamics in Half-Metallic Ferromagnets

As mentioned in Sect. 3, the low-temperature phase of $La_{0.7}(Sr,Ca)_{0.3}MnO_3$ is a 100% spin-polarized ferromagnetic metallic phase. This was first suggested by early infrared to far-infrared spectroscopy [137] and was confirmed by spin-polarized photoemission spectroscopy [138]. Since then, various optical studies have clarified that the electronic structure is distorted with the typical order of eV with T and H (see a review of the optical spectroscopy of manganites [139]). However, low-energy charge dynamics, which is sensitive to the dynamical motion of the charge carriers, is not well understood at present. This is due to the difficulty of obtaining a light source in this low-energy region. Here, we describe the THz-TDS of the 100% spin-polarized ferromagnetic metallic phase of $La_{0.7}Ca_{0.3}MnO_3$ thin film, focusing on its T dependence [41, 42]. THz-TDS was also used to study thin films of the oxygen-deficient (δ) compound; $La_{0.7}Ca_{0.3}MnO_{3-\delta}$. We compared the charge dynamics in these films within the framework of the simple Drude model [43].

4.2.1 Methods

The thin films of $La_{0.7}Ca_{0.3}MnO_3$ were grown by PLD on MgO(100) substrates and characterized with X-ray diffraction, transport, and magnetic measurements. The films were of a single phase and a-axis oriented, with a lattice constant of 3.86 Å. The conventional four-probe method was used to meaure the T dependence of the DC resistivity ρ_{DC}. Figure 36c shows that the film undergoes the IM transition around T_{IM} of ≈ 230 K, with the resistance falling by over two orders of magnitude. The same figure shows that the magnetization M in the field-cooling run under an H of 10 kOe steeply increases and that the ferromagnetic order sets in at this temperature T_C [140]. The oxygen-deficient $La_{0.7}Ca_{0.3}MnO_{3-\delta}$ sample was prepared by changing the oxygen pressure during the deposition. The film had a T_{IM} of around 190 K. According to the phase diagram (ρ_{DC} vs. T) of $La_{1-x}Ca_xMnO_3$ [141], δ should be ≈ 0.04. Note that the oxygen-deficient sample is located near the compositional IM transition boundary.

The THz-TDS experiments employed the PC sampling technique in a transmission configuration. The light source was THz radiation from the surface of InAs or from a biased dipole-type antenna on LT-GaAs, which was generated after illumination by fs optical pulses from a mode-locked Ti:sapphire laser (center wavelength of 800 nm and pulse width of 100 fs). The transmitted THz pulse was detected by the biased bow-tie-type LT-GaAs. A complex optical spectrum in the frequency-domain was numerically calculated from raw data in the time domain. Multiple reflections within the MgO substrate were ignored in this calculation, the details of which are found in [41].

4.2.2 100% Spin-Polarized Ferromagnetic Metallic Phase

Figures 36a and b respectively, show the real n and imaginary κ parts of the complex refractive index spectrum of the frequency $\omega/2\pi$ plot of $La_{0.7}Ca_{0.3}MnO_3$ thin film, measured at 16 K. The closed squares are experimental data. Reflecting the metallic character at low temperature (see, T dependence of ρ_{DC} shown in Fig. 36c), each spectrum nearly holds to $n(\omega) \approx \kappa(\omega)$ within the measured ω range. Both n and κ show a steep increase with decreasing $\omega/2\pi$ and form a Drude-like peak centered at $\omega \approx 0$ at 16 K. With increasing T, the intensity of the peak decreases, while its FWHM increases. Finally, the peak diminishes around the characteristic temperature $T^* \approx 170$ K. Note that $T^*(\approx 0.7T_{IM})$ is well below T_{IM} and T_C as determined by the $\rho_{DC}(T)$ and $M(T)$ measurements, respectively (Fig. 36c). T^* for $La_{0.7}Ca_{0.3}MnO_{2.96}$ thin film is estimated to be 95 K, which corresponds to $\approx 0.5T_{IM}$ ($T_{IM} \approx 190$ K).

The simple Drude model can be used to analyze the data as a function of T. Both $n(\omega)$ and $\kappa(\omega)$ can be expressed using the real $\epsilon_1(\omega)$ and imaginary $\epsilon_1(\omega)$ parts of the complex dielectric constant:

$$
n(\omega) = \sqrt{\frac{\epsilon_1(\omega) + \sqrt{\epsilon_1(\omega)^2 + \epsilon_2(\omega)^2}}{2}} \, ,
$$

$$
\kappa(\omega) = \sqrt{\frac{-\epsilon_1(\omega) + \sqrt{\epsilon_1(\omega)^2 + \epsilon_2(\omega)^2}}{2}} \, . \tag{9}
$$

Given the scattering rate γ and the plasma frequency ω_p as parameters, $\epsilon_1(\omega)$ and $\epsilon_2(\omega)$ become

$$
\epsilon_1(\omega) = \epsilon_\infty - \frac{\omega_\mathrm{p}^2}{\omega^2 + \gamma^2}, \qquad \epsilon_2(\omega) = \frac{\omega_\mathrm{p}^2 \gamma}{\omega(\omega^2 + \gamma^2)}, \tag{10}
$$

where ϵ_∞ is the dielectric constant of the bound electrons. The solid lines in Figs. 36a and b are the least-squares fits to the data using (9) and (10). The data can be well reproduced with $\omega_\mathrm{p} \approx 1.6\,\mathrm{eV}$ and $\gamma \approx 0.1\,\mathrm{eV}$ at 16 K. The Drude model can also be applied to analyze the data of $La_{0.7}Ca_{0.3}MnO_{2.96}$ thin film, yielding $\omega_\mathrm{p} \approx 1.46\,\mathrm{eV}$ and $\gamma \approx 0.1\,\mathrm{eV}$ at 20 K. These estimations are reasonable for raw time-domain data, which show a negligible phase shift of the transmitted THz pulse through the MgO substrate and sample/MgO substrate, because $\omega/\gamma < 1$ holds.

To clarify the features of the charge carriers at low temperature, the effective mass m^*_optics of $La_{0.7}Ca_{0.3}MnO_3$ thin film was estimated by assuming a carrier density n of 1.6 holes per Mn site, according to Hall-effect measurements of single crystals of $La_{0.7}Ca_{0.3}MnO_3$ made by *Chun* et al. [142]; by using the following relation, m^*_optics is about ≈ 15 in units of bare electron mass m_0.

$$
m^*_\mathrm{optics}/m_0 = \frac{e^2 n}{\epsilon_0 \omega_\mathrm{p}^2 m_0}, \tag{11}
$$

where ϵ_0 is the permittivity of the vacuum and e is the electric charge. The magnitude of m^*_optics is a few times larger than that of the effective mass m^*_heat estimated from specific heat measurements of single crystals of $La_{0.7}Ca_{0.3}MnO_3$ made by *Okuda* et al. ($m^*_\mathrm{heat} \approx 2 - 3m_0$) [140]. This simply implies that the charge carrier has an incoherent nature even at low temperature (even the 100% spin-polarized ferromagnet metallic phase). Such a discrepancy as manifested by the anomalous suppression of the Drude term has been reported by *Okimoto* et al. [137, 143] in their early optical studies of $La_{1-x}Sr_xMnO_3$ single crystals. Using the above equation, m^*_optics of $La_{0.7}Ca_{0.3}MnO_{2.96}$ thin film is estimated to be $\approx 23m_0$ by assuming $n \approx 1.6$ holes per Mn site [144].

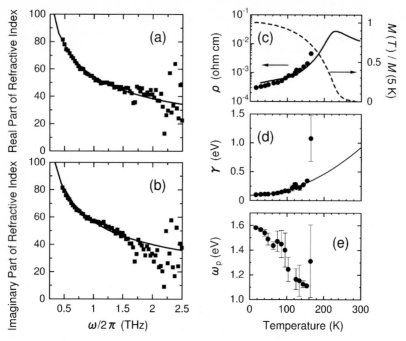

Fig. 36. Charge dynamics at terahertz (THz) frequencies of the ferromagnetic metallic phase of $La_{0.7}Ca_{0.3}MnO_3$ thin film as revealed with THz time-domain spectroscopy in a transmission configuration. (**a**) Real and (**b**) imaginary parts of the complex refractive index spectra in units of frequency $\omega/2\pi$, measured at 16 K. *Closed squares* are numerically estimated from the transmitted THz waveform. *Solid lines* are least-squares fits to data using the Drude model with the scattering rate γ and the plasma frequency ω_p as fitting parameters in (9) and (10). Temperature T dependence of (**c**) the resistivity ρ, (**d**) γ, and (**e**) ω_p, all of which (*closed circles*) are estimated using (10) and (14). Also shown in (**c**) are T dependences of ρ (*left axis; solid line*) measured by the conventional four-probe method and the magnetization M (*right axis; dashed line*) normalized by M at 5 K in a field-cooling run under a magnetic field of 10 kOe. M of $La_{0.7}Ca_{0.3}MnO_3$ thin film was obtained by calibrating it to M of the MgO substrate, which was measured separately under the same experimental conditions. The *solid line* in (**d**) is a least-squares fit to data assuming a T^2 dependence

Other important physical quantities at low temperature are the mean free path l and the Fermi wavelength λ_F. Assuming a degeneracy of two in the conduction band (the degree of occupancy of the electron is two in e_g orbitals), l can be expressed as:

$$l = \frac{\hbar}{\gamma m^*}(3\pi^2 n)^{1/3}, \tag{12}$$

where $\hbar$ is Planck's constant divided by 2π. For a spherical Fermi surface, λ_F can be obtained as

$$\lambda_F = (8\pi/3n)^{\frac{1}{3}}. \tag{13}$$

We calculated l and λ_F of $La_{0.7}Ca_{0.3}MnO_3$ thin film to be $\approx 4.7\,\text{Å}$ and $\approx 6.7\,\text{Å}$, respectively. Therefore, the condition $\lambda_F > l > a$ (a is the lattice constant) holds even in the low-temperature ferromagnetic phase, which is in contrast to the case of a conventional metal, in which the condition $\lambda_F < l$ is satisfied. The physical characteristics of $La_{0.7}Ca_{0.3}MnO_3$ thin film are listed in Table 1, together with those of $La_{0.7}Ca_{0.3}MnO_{2.96}$ thin film.

Table 1. The physical quantities of the ferromagnetic metallic phase of $La_{0.7}Ca_{0.3}MnO_3$ and $La_{0.7}Ca_{0.3}MnO_{2.96}$ thin films at low temperature. The values in the table were obtained from the fitting results by using the simple Drude model (10) and were estimated from (11), (12), (13). T_{IM} is the insulator–metal transition temperature determined from the DC-resistivity measurements, T^* is the temperature at which the Drude term diminishes, normalized by T_{IM}, ω_p is the plasma frequency, γ is the scattering rate, m^* is the effective mass in units of the bare electron mass m_0, l is the mean free path, and λ_F is the Fermi wavelength

Sample	T_{IM}	T^*	ω_p	γ	m^*	l	λ_F
	K	T_{IM}	eV	eV	m_0	Å	Å
$La_{0.7}Ca_{0.3}MnO_3$	230	≈ 0.7	≈ 1.6	≈ 0.1	≈ 15	≈ 4.7	≈ 6.8
$La_{0.7}Ca_{0.3}MnO_{2.96}$	190	≈ 0.5	≈ 1.46	≈ 0.1	≈ 20	≈ 3.4	≈ 6.8

The $\rho(T)$ measurements indicate that $La_{0.7}Ca_{0.3}MnO_{2.96}$ thin film is located at the compositional IM phase boundary. On decreasing the carrier concentration (increasing δ) toward the IM phase boundary, ω_p decreases by 90%. On the other hand, γ does not change within the experimental accuracy. In conventional metals, γ at low temperature is mainly determined by the amount of impurity scattering. Therefore, the magnitude of γ is related to the residual resistivity ρ_0; ρ_0 of $La_{0.7}Ca_{0.3}MnO_3$ thin film is one order of magnitude smaller than that of $La_{0.7}Ca_{0.3}MnO_{2.96}$ thin film, while the γs of the two samples are comparable. This clearly shows that the impurity scattering has no great influence on the motion of the charge carriers in the ferromagnetic metallic phase of manganites. In fact, m^*_{optics} is a few times larger than m^*_{heat} [140]. These results indicate that the motion of charge carriers has an incoherent nature even in the ferromagnetic metallic ground state. *Okuda* et al. [140] revealed that ρ_0 of $La_{0.7}Ca_{0.3}MnO_3$ single crystals shows a strong pressure-dependent behavior and concluded that an anomalous scattering channel for charge carriers exists in the ferromagnetic metallic phase, a finding that is consistent with the THz-TDS experiments.

It has been shown that the electronic specific heat of $La_{1-x}Ca_xMnO_3$ single crystals linearly depending on m^*_{heat}, shows less mass renormalization near the IM critical point ($x_c \approx 0.22$) [140], compared with other hole-doped transition metal oxides that show a large mass renormalization. Despite the fact that the magnitude of m^*_{optics} is considerably different from that of m^*_{heat}, no significant enhancement of m^*_{optics} near the IM critical point was observed in the THz-TDS experiments (Table 1).

4.2.3 Temperature Dependence of the Ferromagnetic Metallic Phase

Figures 36d and e show the temperature dependences of γ and ω_p of $La_{0.7}Ca_{0.3}MnO_3$ thin film, respectively. The measurements were performed while the film was being warmed. The closed circles are data obtained from the fittings to the respective spectra using (9) and (10). The error bars are estimated from the uncertainty of the fitting results. With increasing T, γ increases in proportion to T^2, as indicated by the solid line in Fig. 36d. Around $T^* \approx 160$ K, the temperature at which the Drude term disappears, a clear deviation from the T^2 term can be seen. The same tendency of $\gamma(T)$ is also observed in $La_{0.7}Ca_{0.3}MnO_{2.96}$ thin film. Such a T^2-dependent γ below T^* has been reported by *Simpson* et al. [145], who used conventional Fourier-transform spectroscopy to study $La_{0.7}(Sr,Ca)_{0.3}MnO_3$ thin films on $LaAlO_3$ substrates. ω_p of both compounds also has the T dependence; it decreases with increasing T, which is in contrast to the behavior of conventional metals.

We also calculated the T-dependent resistivity ρ_{optics} by using the measured $\omega_p(T)$ and $\gamma(T)$ in the following equation:

$$\rho_{optics}(T) = \frac{\gamma(T)}{\omega_p(T)^2 \epsilon_0}. \tag{14}$$

Figure 36c plots the T dependence of ρ_{optics} (closed circles) and ρ_{DC} (the solid line). The T dependences of ρ_{optics} and ρ_{DC} are quite similar, indicating the validity of the simple Drude model within the measured ω range.

In $La_{0.7}Ca_{0.3}MnO_3$ and $La_{0.7}Ca_{0.3}MnO_{2.96}$ thin films, the condition $\lambda_F > l > a$ is satisfied even in the ferromagnetic metallic ground state (Table 1). As T increases toward T^*, γ increases, while ω_p decreases. Namely, the hopping amplitude of the charge carriers decreases. In the vicinity of T^*, one obtains $l < a$ and $l < \lambda_F$ in both compounds, indicating that Boltzmann transport is not valid in this regime. A suitable explanation for this behavior can be found by considering the phase-separation scenario. Here, the ferromagnetic metallic phase may serve as a metallic domain in the neighboring insulating phase. Indeed, a moderate mixture of the metallic and nonmetallic phase coexists in the vicinity of T_C, as observed by scanning tunneling spectroscopy [146].

Sawaki et al. [147] have systematically investigated the effect of Al-substitution on transport properties in $La_{1-x}Sr_xMnO_3$ single crystals. They revealed that coherent–incoherent crossover occurs when l reaches λ_F at

T^* [147] and categorized the metallic state into a low-temperature coherent metallic phase (LCM) or a high-temperature incoherent metallic phase (HIM) [147, 148]. In their classification, the condition $l < \lambda_F$ is violated in LCM but not in HCM. $La_{0.7}Ca_{0.3}MnO_3$ thin film, only HIM characteristics can be clearly observed by THz-TDS; no LCM characterisitics are seen in the measured T range. *Sawaki* et al. also reported that the critical resistivity ρ_c of $La_{1-x}Sr_xMnO_3$ is about $300\,\mu\Omega \cdot cm$ at T^*. The magnitude of ρ_c is comparable with that of ρ_0 for $La_{0.7}Ca_{0.3}MnO_3$ thin film. This may be the reason for the lack of LCM in $La_{0.7}Ca_{0.3}MnO_3$ and $La_{0.7}Ca_{0.3}MnO_{2.96}$ thin films.

4.3 Observation of a Collective Excitation Mode in Charge-Ordered Manganites

In contrast to $La_{0.7}Ca_{0.3}MnO_3$, $Pr_{0.7}Ca_{0.3}MnO_3$ exhibits an insulating behavior with no IM transition in the whole T range. This insulating state comes from the real-space charge ordering of Mn^{3+} and Mn^{4+} ions in a 1:1 ratio, which is the so-called charge-exchange (CE)-type structure. Such a real-space charge ordering is characterized by a single-particle excitation across the charge gap 2Δ. Therefore, optical spectroscopy has been extensively employed to investigate the single-particle excitation spectrum of CO manganites. These studies have mainly focused on the manganites' T and H dependences [110, 149, 150, 151, 152]; for example, *Okimoto* et al. [110, 149] identified from polarized reflectivity measurements the presence of a 2Δ of $\approx 75\,THz$ for the CO manganite $Pr_{0.6}Ca_{0.4}MnO_3$ and clarified that the electronic structure is dramatically changed (on the order of eV) by varying T and H. Using transmission measurements, *Calvani* et al. [150] found that $2\Delta(T)$ of another CO manganite, $La_{1-x}Ca_xMnO_3$ with $x = 0.5$ and 0.67, can be well described by the Bardeen–Cooper–Schrieffer (BCS) relation.

CE-type charge ordering can be viewed as a quasi-one-dimensional process by taking the orbital degrees of freedom of Mn^{3+} ions into account. In fact, recent transmission electron microscopic studies support this view [153]; the Mn^{3+} ion is surrounded by three Mn^{4+} ions rather than a regular arrangement of Mn^{3+} and Mn^{4+} ions along the c-axis. *Mori* et al. [154] revealed the presence of "paired charge stripes" – a new periodic form composed of a pairing of diagonal Mn^{3+} ions with Mn^{4+} ions – in another CO manganite, $La_{1-x}Ca_xMnO_3(x \geq 1/2)$, whose period is proportional to $1/x$. This distinctive pattern of the charge is like an *electron river* of Mn^{3+} ions separated by a *levee* of Mn^{4+} ions in one direction. This modifies the charge density's uniformity, leading to the charge-density-wave (CDW) condensate, an analog of which can be seen in Japanese temple gardens (Fig. 37). In this case, the charge ordering is also characterized by the collective excitation below 2Δ. However, no evidence for the collective excitation in CO manganites has been reported so far. By the illumination of light, the collective excitation can make coupled oscillations at a finite frequency and appears as a mode in

Fig. 37. Beauty of Tofukuji temple, Japan. The dry garden (Karesansui in Japanese) is a traditional art of revealing "nature" in a small garden, the flow of a river can be expressed with sand (so-called Sunamon in Japanese); the waves shown here seem to be analogous to the charge-density-wave condensate in condensed matter. The photo of Sunamon was taken by *Y. M.*

the optical spectrum (see also Fig. 42). As in the case of the low-dimensional CDW system, this mode is located in the millimeter frequency range, where it is difficult to employ the probe light of conventional spectroscopy. Here we review the optical properties at THz frequencies of the typical CO manganite; $Pr_{0.7}Ca_{0.3}MnO_3$ and give the first identification of the collective excitation mode by using THz-TDS [44, 45]. We alo examine the possibility that the CDW condensate is the ground state in CO manganites.

4.3.1 Observation of the Finite Frequency Peak

Figures 38a and b show a logarithmic frequency $\omega/2\pi$ plot of the real part of the complex optical conductivity spectrum $\sigma_1(\omega)$ of samples E and F, respectively, recorded at low T. Both samples were grown by PLD on MgO(100) substrates under the same growth conditions and are of a single phase, as characterized by their X-ray diffraction patterns recorded at room T. The experimental setup for THz-TDS and the calculation procedure of the complex optical spectra are described in Sect. 4.2.1. The open circles are experimental data including error bars, which are relatively large above 0.6 THz due to the poor sensitivity of the light source. As can be clearly seen, finite frequency peaks centered at ≈ 0.7 THz and ≈ 0.5 THz appear in $\sigma_1(\omega)$ of samples E and F, respectively. The peak frequency ω_0 depends on whether the growth condition of the two samples are identical, while $\sigma_1(\omega)$ at ω_0 does not. In

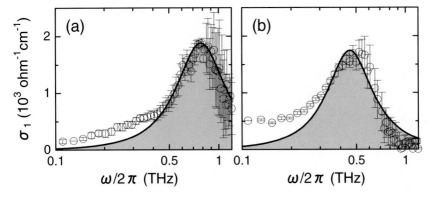

Fig. 38. Observation of the finite frequency peak at terahertz (THz) frequencies in $Pr_{0.7}Ca_{0.3}MnO_3$ thin film by using THz time-domain spectroscopy in a transmission configuration. Real part of the complex optical conductivity spectrum $\sigma_1(\omega)$ on the logarithmic frequency $\omega/2\pi$ plot of two [(**a**) and (**b**)] $Pr_{0.7}Ca_{0.3}MnO_3$ thin films (grown under the same conditions), and measured at low T [(**a**) 8.6 K and (**b**) 4 K]. The *open circles* including the error bars are experimental data. The *solid lines* are the results of least-squares fits to data using (15)

addition, the magnitude ($\approx 2000\,\Omega^{-1}\cdot cm^{-1}$) of $\sigma_1(\omega)$ at ω_0 is comparable to that of the single-particle excitation at 2Δ [110, 149, 152].

Figure 39 shows the T dependence of $\sigma_1(\omega)$ at ω_0 (closed circles) and at 0.2 THz (open squares) for sample E. The respective data are normalized by the value at 280 K for comparison. The measurements were performed while the sample was being warmed. Solid and dashed lines indicate $T_{CO/OO}$ of the bulk $Pr_{0.7}Ca_{0.3}MnO_3$ reported by *Jirák* et al. [94] and T_C, where the magnetization steeply increases with decreasing T under an H of 500 Oe (see Fig. 27a2), respectively. As can be clearly seen, neither $\sigma_1(\omega)$ at ω_0 nor $\sigma_1(\omega)$ at 0.2 THz depends on T above $T_{CO/OO}$. However, $\sigma_1(\omega)$ at ω_0 increases, while $\sigma_1(\omega)$ at 0.2 THz decreases on the border of $T_{CO/OO}$ without the change around T_C, indicating that the peak is related to the charge ordering and not spin ordering.

Note that there is a discrepancy between the T dependence of the DC conductivity σ_{DC} and that of $\sigma_1(\omega)$ at 0.2 THz. The samples used in the THz-TDS experiments exhibit an insulating behavior, as reported previously; σ_{DC} falls by ten orders of magnitude as T is decreased from room T. $\sigma_1(\omega)$ at 0.2 THz only falls to half of its room T value with decreasing T (Fig. 39). This discrepancy originates from the presence of the T-dependent Debye-like dispersion in the optical spectra of the kHz–MHz frequency range (the relaxation time increases with decreasing T), as revealed by dielectric measurements using an LCR meter [155].

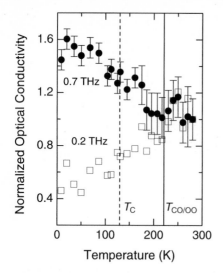

Fig. 39. Temperature dependence of the real part of the complex optical conductivity spectrum at 0.7 THz (the finite frequency peak) (*closed circles*) and at 0.2 THz (*open squares*) of $Pr_{0.7}Ca_{0.3}MnO_3$ thin film. The respective data are normalized by the value at 280 K for comparison. The measurements were performed while warming the sample. The *vertical dashed and solid lines* represent the critical temperature T_C at which the magnetization steeply increases with decreasing temperature (see Fig. 27a2) and the charge/orbital ordering temperature $T_{CO/OO}$ of the bulk $Pr_{0.7}Ca_{0.3}MnO_3$ reported by *Jirák* et al. [94], respectively

4.3.2 Origin of the Finite Frequency Peak

To determine the origin of the optical conductivity peak at THz frequencies in $Pr_{0.7}Ca_{0.3}MnO_3$, we first compare the results of the time dependence with the typical phase-separation characteristics of CO manganites [100]. This leads us to discuss the possibility of thermal fluctuations as the origin of the finite frequency peak. Finally, the origin of the peak is discussed in comparison with various characteristics of low-dimensional materials showing the CDW condensate.

Phase separation (PS) – the volume fraction of the ferromagnetic domain embedded in CO insulating phase – is thought to be key to understanding the anomalous magnetotransport properties of CO manganites [100]. PS, as well as the disorder commonly seen in thin films using MgO substrates, is known to produce a finite frequency peak as a result of the Drude peak centered at $\omega = 0$ shifting. This is the case with impurity doping on the Cu site in copper-oxide superconductors; *Basov* et al. [156] reported that the finite frequency peak structure emerges around 2.5 THz in Zn-doped $YBa_2Cu_4O_8$ as a consequence of the destruction of superconductivity.

In the $Pr_{0.7}Ca_{0.3}MnO_3$ of the THz-TDS experiments, the ferromagnetic order sets in at low T; there is a significant hysteresis in the $M–H$ curve at

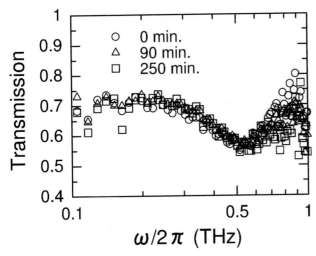

Fig. 40. The elapsed-time dependence of transmission spectra of $Pr_{0.7}Ca_{0.3}MnO_3$ thin film shown on a logarithmic frequency $\omega/2\pi$ scale. The measurements at the elapsed time of 0 (*circles*), 90 (*triangles*), and 250 (*squares*) minutes, were performed at 4 K after the sample was rapidly cooled from room temperature to 4 K. The respective transmission spectra are normalized by the value of the transmission spectrum at 0.5 THz of the elapsed time at 0 min

30 K with a coercive force of $\approx$ 500 Oe, which may imply the presence of the ferromagnetic domain in the CO insulating phase. Therefore, we should compare these experimental results with the typical PS characteristics of CO manganites. One such characteristic is the slow relaxation effect [157,158], in which the physical quantities (i.e., resistance and magnetization) are strongly affected by the history of the external perturbations.

The experimental sample was rapidly cooled from room T to 4 K at a cooling rate of $\approx$ 8 K per minute and the time dependence of the transmission spectra was evaluated at 4 K. Figure 40 shows the transmission spectra on a logarithmic ω scale after 0 (circles), 90 (triangles), and 250 (squares) min normalized by the value of the transmission spectrum at 0.5 THz at 0 min. There is no slow relaxation effect as far as the finite frequency peak structure is concerned; both the peak frequency $\approx$ 0.5 THz and the spectral shape show no variation after $\approx$ 250 min. This result provides clear evidence that the observed structure does not originate from the shifting of the Drude peak even if the PS occurs in the samples of the THz-TDS experiments.

In the measured ω range, thermal fluctuations of the order of $k_B T$ ($\approx$ 1 meV at 10 K) and the presence of the impurity band play an important role; they excite the charge carriers within 2Δ and the thermally excited charge carriers give rise to the finite frequency peak structure [159]. In this case, the scattering rate Γ of the charge carriers should follow an exponential T dependence as observed in semiconductors with shallow impurities: Si-doped

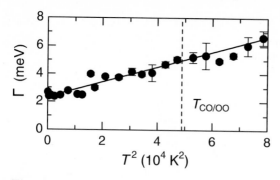

Fig. 41. Temperature T squared plot of the scattering rate Γ (*solid circles*) as derived from the fitting curve using (15). The *vertical dashed line* represents the charge/orbital ordering temperature $T_{CO/OO}$ of the bulk $Pr_{0.7}Ca_{0.3}MnO_3$ reported by *Jirák* et al. [94]. The *solid line* is a least-squares fit to the data below $T_{CO/OO}$ using (16)

GaAs near the compositional IM boundary [160]. To examine this scenario, $\sigma_1(\omega)$ can be analyzed by a single Lorentz oscillator;

$$\sigma_1(\omega) = \frac{\sigma_0 \omega^2 \Gamma^2}{(\omega_0^2 - \omega^2)^2 + \omega^2 \Gamma^2},\tag{15}$$

where ω_0 is the peak frequency and σ_0 is $\sigma_1(\omega)$ at ω_0. The solid lines with shared area in Figs. 38 denote least-squares fits to the experimental data; $\sigma_1(\omega)$ of sample E (F) can be well reproduced by (15) with $\omega_0 = 3.3\,\mathrm{meV}$ (1.9 meV) and $\Gamma = 2.6\,\mathrm{meV}$ (1.9 meV). Figure 41 shows the T-squared plot of Γ (closed circles) for sample E. The solid line is the least-squares fit to the data below $T_{CO/OO}$ by using the following phenomenogical relation:

$$\Gamma(T) = \Gamma_0 + A_2 T^2,\tag{16}$$

where Γ_0 is Γ at $0\,\mathrm{K}$ and A_2 is a coefficient. As can be clearly seen, Γ has a T^2 dependence, not an exponential one, providing the evidence that $\Gamma(T)$ cannot be accounted for by the behavior of the thermally excited charge carriers within 2Δ.

4.3.3 Collective Excitation Mode due to Charge-Density-Wave Condensate

As is well known, low-dimensional materials like $K_{0.3}MoO_3$ have a one-dimensional character due to their antistropic electronic structure, which leads to a strong electron–lattice interaction that separates the electrons and lattice. Therefore, a gap tends to open in the q-space at low T, which is referred to as the CDW condensate. Optical properties of the materials showing a CDW condensate are quite different from conventional band insulators, especially their optical properties below 2Δ. The remarkable signature of the CDW condensate is the appearance of the collective excitation modes at a finite frequency below 2Δ. Positionally and temporally dependent modes are referred to as amplitudons and phasons, respectively [161]. According to Lee et al. [162], the dispersion relation of phasons is given by

$$\omega^2 \propto |q|^2 , \tag{17}$$

where q is the wave number. The dispersion relation is shown in Fig. 42a as a thick dashed line. By illumination of light, the collective excitation can make a coupled oscillation (solid lines in Fig. 42b), such that

$$\frac{q^2}{\omega^2 \epsilon_0 \mu_0} \propto \frac{\omega_{LO}^2 - \omega^2}{\omega_{TO}^2 - \omega^2} , \tag{18}$$

where μ_0 is the magnetic permeability of the vacuum and ω_{LO} and ω_{TO} are the longitudinal and transverse optical frequencies, respectively (horizontal dashed lines in Fig. 42). The change in the dielectric constant at $q = 0$ is thus proportional to $(\omega_{LO}^2 - \omega^2)/(\omega_{TO}^2 - \omega^2)$. Therefore, this coupled mode appears as a collective excitation mode in the complex optical spectrum. Figures 42b and c show the calculated dielectric constant and transmission spectra of the phasons, respectively. Impurities and/or lattice imperfections also have a striking influence on the CDW dynamics, because the CDW condensate is pinned by impurities, and consequently, the peak position of the collective excitation mode shifts to a finite frequency [161]. The simplest representation of this is to assume that its response in one direction x can be expressed by the following equation of motion:

$$\frac{d^2 x}{dt^2} + \frac{1}{\tau}\frac{dx}{dt} + \omega_0^2 x = \frac{eE}{m^*} , \tag{19}$$

where τ is the scattering time and ω_0 is the pinning frequency. By solving (19) on the assumption that the electric field of light E and x are proportional to $\exp(i\omega t)$, $\sigma_1(\omega)$ becomes the same as in (15). This causes the finite frequency peak structure centered at ω_0 (Figs. 42b and c).

So far, the properties of the collective excitation mode of the low-dimensional materials have been investigated by using the cavity perturbation technique. The following characteristics have been revealed; (i) the pinning frequency is generally of the order of GHz [161, 163, 164, 165, 166, 167]. (ii) The spectral weight of the collective excitation mode is typically two orders of magnitude smaller than that of the single-particle excitation [161, 163, 164, 165, 166, 167]. This is due to the relatively large 2Δ of ≈ 75 THz, as estimated by optical spectroscopy, which is in strong contrast to the small spin gap ≈ 2 THz observed in the spin-density-wave (SDW) system [168]. (iii) The pinning frequency in $(Ta_{1-z}Nb_z Se_4)_2 I$ alloys linearly shifts to higher ω,

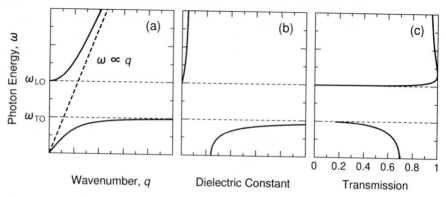

Fig. 42. Dispersion relation of phasons given by (17). (**a**) Dispersion relation of phasons indicated by the *thick dashed line* in the photon energy ω vs. wavenumber q plot. *Solid lines* indicate the dispersion of the coupled mode given by (18), which is created when light is illuminated on the light. ω_{LO} and ω_{TO} are longitudinal and transverse optical frequencies, respectively. Their peak frequencies are indicated by the *horizontal thin dashed lines*. Calculated (**b**) dielectric constant and (**c**) transmission spectra of the phasons

and $\sigma_1(\omega)$ at the pinning frequency does not change with increasing impurity concentration z [165]. The same tendencies are seen in $K_{0.3}Mo_{1-z}W_zO_3$ alloys [167]. (iv) The swelling structure on the low-energy side of the collective excitation mode in $K_{0.3}MoO_3$ has been identified as an internal deformation on the basis of generalized Debye analysis [164].

The characteristics of the finite frequency peak structure of $Pr_{0.7}Ca_{0.3}MnO_3$ as described in Sect. 4.3.1 (Fig. 38) are remarkably similar to those of the collective excitation mode described above [161, 163, 164, 165, 166, 167]: (i) ω_0 appears around 1 THz below the peak frequency (≈ 5 THz) of the lowest optical phonon mode arising from the perovskite structure. (ii) σ_0 ($\approx 2000\,\Omega^{-1}cm^{-1}$) bears comparison with that at 2Δ ($\approx 1000\,\Omega^{-1}cm^{-1}$), meaning that the spectral weight of the finite peak structure is $\approx 1\%$ of that at 2Δ because 2Δ is located around ≈ 75 THz. (iii) ω_0 changes even though the growth conditions of the two films are the same, while σ_0 does not. (iv) an increase in $\sigma_1(\omega)$ below ω_0 occurs that cannot be reproduced by (15). Moreover, the CDW system shows a T^2 dependence of Γ (Fig. 41) [161, 163], and (15) can be derived from the equation of motion for CDW dynamics given by (19) [161]. These similarities provide a clear indication that the finite frequency peak at THz frequencies in $Pr_{0.7}Ca_{0.3}MnO_3$ should be assigned to the collective excitation mode arising from the CDW condensate.

It is well known that the collective excitation mode originates from two different states; one is a "pinned" state due to the pinning of the CDW condensate by impurities and/or lattice imperfections and the other is a "bound" state, which is created by coupling of the pinned state with optical phonons or impurities near the pinned state [166, 167]. Despite the fact that Figs. 38

shows a swelling of $\sigma_1(\omega)$ below the low-energy side of the finite structure, which is usually ascribed to an internal deformation of the pinned collective excitation mode [164, 166], we cannot clearly say whether the observed mode should be assigned to the pinned state or to the bound state of CDW. Further spectroscopic measurements in the GHz ω range are necessary to identify the exact state.

The collective excitation mode is still visible above $T_{CO/OO}$, but its spectral shape is blurred; Γ continues to follow the T^2 dependence (Fig. 41). Within the framework of the mean-field theory, the magnitude of η estimated by the mean-field temperature T^{MF} divided by $T_{CO/OO}$, where fluctuation is neglected, is a good measure of the strength of the fluctuation. T^{MF} has been estimated to be $\approx 1000\,K$ using the value of $2\Delta \approx 75\,THz$ obtained by optical spectroscopy [110, 149]. This temperature is nearly four times higher than $T_{CO/OO}$. Such an anomalous difference has also been reported not only in other CO manganites (La$_{1-x}$Ca$_x$MnO$_3$ with $x = 0.5$ and 0.67 [150] and Bi$_{1-x}$Ca$_x$MnO$_3$ with $x = 0.74$ and 0.82 [151]) but also in CO transition metal oxides (e.g., La$_{1.67}$Sr$_{0.33}$NiO$_4$ [169] and La$_{1/3}$Sr$_{2/3}$FeO$_3$ [170]).

Here, we briefly discuss the dynamics of the charge fluctuation above $T_{CO/OO}$ in Pr$_{0.7}$Ca$_{0.3}$MnO$_3$ from a phenomenogical viewpoint. Although, we could not conclude whether the observed collective excitation mode arises from pinned or bound states of CDW, the following discussion will not be affected qualitatively. The simplest way of representing σ_0 is by $n_{CDW}e^2\tau_{CDW}/m^*_{CDW}$, where n_{CDW} and τ_{CDW} are the carrier number participating in the charge fluctuation and the relaxation time, respectively. The effective mass m^*_{CDW} is simply related to the amplitude of the distortion u in terms of the conservation of kinetic energy E_k [171, 172]; E_k of the ions is equal to E_k of CDW condensate. This leads to the following relation:

$$u \propto \sqrt{m^*_{CDW}}. \tag{20}$$

As we have shown in Fig. 39, σ_0 does not depend on T above $T_{CO/OO}$, while Γ increases with T^2 (Fig. 41). According to thermoelectric measurements of Pr$_{1-x}$Ca$_x$MnO$_3$ by *Yamada* et al. [113], $n \approx n_{CDW}$ is less T dependent above $T_{CO/OO}$. This clearly accounts for the large reduction of m^*_{CDW} following a T^2 dependence. Therefore, even with our simplifications, u in (20) is expected to fall linearly with increasing T. Local distortions arising from the Jahn–Teller-type Mn^{3+}O$_6$ octahedron have been observed above $T_{CO/OO}$ by using an X-ray scattering technique [173]. However, a detailed measurement of u in Pr$_{0.7}$Ca$_{0.3}$MnO$_3$ has not been made although low-T data is available from a neutron-diffraction study by *Louca* et al. [174]. A direct comparison between T dependencies of m^*_{CDW} and u can determine the direction of CDW condensate.

In summary, η cannot be explained in the framework of the mean-field approximation. In general, 2Δ is estimated from the linear extrapolation of the rising part of $\sigma_1(\omega)$ around the fundamental absorption edge [110, 149,

150, 151, 152, 169, 170]. With increasing T, the fundamental absorption edge becomes blurred. In other words, 2Δ fills below $T_{CO/OO}$. Above $T_{CO/OO}$, the peak structure around 2Δ can still be observed [110, 149, 150, 151, 152, 169, 170]. Therefore, it is difficult to determine the contribution of the charge fluctuation by only measuring the remnant part around 2Δ because of the presence of the strong absorption above $T_{CO/OO}$. In our case, the collective excitation mode cannot be subject to other contributions, e.g., the thermal fluctuation, which affects the overall spectral shape of $\sigma_1(\omega)$ around 2Δ [175]. Γ follows a T^2 dependence, and m^*_{CDW} decreases above $T_{CO/OO}$. In addition, the peak frequency of the collective excitation mode does not depend on T. These results can be ascribed to successive changes in 2Δ above $T_{CO/OO}$. Therefore, we may infer that the persistence of the peak structure around 2Δ above $T_{CO/OO}$ is a consequence of the continuous formation of the gap.

References

[1] M. Imada, A. Fujimori, Y. Tokura: Rev. Mod. Phys. **70**, 1039 (1998)
[2] J. Bednortz, K. Müller: Z. Phys. **64**, 189 (1986)
[3] Y. Tokura: Jpn. Soc. Appl. Phys. Int. **2**, 12 (2000)
[4] Y. Tokura (Ed.): *Colossal Magnetoresistive Oxides* (Gordon Breach, New York 2000)
[5] M. Salamon, M. Jaime: Rev. Mod. Phys. **73**, 583 (2001)
[6] T. Kise, T. Ogasawara, M. Ashida, Y. Tomioka, Y. Tokura, M. Kuwata-Gonokami: Phys. Rev. Lett. **85**, 1986 (2000)
[7] H. Kishida, H. Matsuzaki, H. Okamoto, T. Manabe, M. Yamashita, Y. Taguchi, Y. Tokura: Nature **405**, 929 (2000)
[8] T. Ogasawara, M. Ashida, N. Motoyama, H. Eisaki, S. Uchida, Y. Tokura, H. Ghosh, A. Shukla, S. Mazumdar, M. Kuwata-Gonokami: Phys. Rev. Lett. **85**, 2204 (2000)
[9] M. Fiebig, K. Miyano, Y. Tomioka, Y. Tokura: Appl. Phys. B **71**, 211 (2000)
[10] T. Ogasawara, T. Kimura, T. Ishikawa, M. Kuwata-Gonokami, Y. Tokura: Phys. Rev. B **63**, 113105 (2001)
[11] T. Ogasawara, K. Tobe, T. Kimura, H. Okamoto, Y. Tokura: J. Phys. Soc. Jpn. **71**, 2380 (2002)
[12] K. Miyano, T. Tanaka, Y. Tomioka, Y. Tokura: Phys. Rev. Lett. **78**, 4257 (1997)
[13] C. Sirtori: Nature **417**, 132 (2002)
[14] D. Auston, K. Cheung, P. Smith: Appl. Phys. Lett. **45**, 284 (1984)
[15] M. Tonouchi, M. Tani, Z. Wang, K. Sakai, S. Tomozawa, M. Hangyo, Y. Murakami, S. Nakashima: Jpn. J. Appl. Phys. **35**, 2624 (1996)
[16] M. Hangyo, S. Tomozawa, Y. Murakami, M. Tonouchi, M. Tani, Z. Wang, K. Sakai, S. Nakashima: Appl. Phys. Lett. **69**, 2122 (1996)
[17] M. Tonouchi, M. Tani, Z. Wang, K. Sakai, N. Wada, M. Hangyo: Jpn. J. Appl. Phys. **36**, L93 (1997)
[18] M. Tonouchi, N. Wada, S. Sikii, M. Hangyo, M. Tani, K. Sakai: Physica C **293**, 82 (1997)

[19] M. Tani, M. Tonouchi, M. Hangyo, Z. Wang, N. Onodera, K. Sakai: Jpn. J. Appl. Phys. **36**, 1984 (1997)

[20] N. Tanichi, N. Wada, T. Nagashima, M. Tonouchi, M. Hangyo, M. Tani, K. Sakai: Physica C **293**, 229 (1997)

[21] T. Kondo, M. Hangyo, M. Tonouchi: IEEE Trans. Appl. Supercond. **11**, 3166 (2001)

[22] T. Yoshimura, T. Kiwa, M. Tonouchi: Physica C **362**, 329 (2001)

[23] T. Kiwa, Y. Yoshimura, H. Murakami, M. Tonouchi: Sin. J. Phys. **18**, 161 (2002)

[24] H. Murakami, T. Kiwa, M. Misa, M. Tonouchi, T. Uchiyama, I. Iguchi, Z. Wang: Jpn. J. Appl. Phys. **41**, 1992 (2002)

[25] H. Murakami, T. Fukui, Y. Tominari, M. Tonouchi, T. Uchiyama, I. Iguchi, Z. Wang: Physica C **367**, 317 (2002)

[26] Y. Tominari, H. Murakami, M. Tonouchi, H. Wald, P. Seidel, H. Schneidewind: Appl. Phys. Lett. **80**, 3147 (2002)

[27] H. Wald, P. Seidel, H. Schneidewind, Y. Tominari, H. Murakami, M. Tonouchi: Physica C **378–381**, 372 (2002)

[28] Y. Tominari, T. Kiwa, H. Murakami, M. Tonouchi, H. Wald, P. Seidel, H. Schneidewind: IEEE Trans. Appl. Supercond. in press

[29] N. Kida, M. Tonouchi: Appl. Phys. Lett. **78**, 4115 (2001)

[30] N. Kida, M. Tonouchi: Physica B **329–333**, 773 (2003)

[31] N. Kida, K. Takahashi, M. Tonouchi: Opt. Lett. **29**, 2554 (2004)

[32] N. Kida, M. Tonouchi: Appl. Phys. Lett. **82**, 3412 (2003)

[33] K. Takahashi, N. Kida, M. Tonouchi: J. Magn. Magn. Mater. **415**, 272–276 (2004)

[34] K. Tamasaku, Y. Nakamura, S. Uchida: Phys. Rev. Lett. **69**, 1455 (1992)

[35] A. Lucarelli, S. Lupi, M. Ortolani, P. Calvani, P. Maselli, M. Capizzi, P. Giura, H. Eisaki, N. Kikugawa, T. Fujita, M. Fujita, K. Yamada: Phys. Rev. Lett. **90**, 037002 (2003)

[36] H. Murakami, T. Kiwa, M. Tonouchi, T. Uchiyama, I. Iguchi, Z. Wang: Physica C **367**, 322 (2002)

[37] Y. Tominari, T. Kiwa, H. Murakami, M. Tonouchi, T. Arai, T. Aomine: Physica C **367**, 332 (2002)

[38] H. Murakami, M. Tonouchi, T. Uchiyama, I. Iguchi, Z. Wang: J. Supercond. **15**, 345 (2002)

[39] H. Murakami, T. Kiwa, N. Kida, M. Tonouchi, T. Uchiyama, I. Iguchi, Z. Wang: Europhys. Lett. **60**, 288 (2002)

[40] H. Murakami, T. Yasuda, Y. Tominari, T. Kiwa, M. Tonouchi, T. Uchiyama, I. Iguchi, Z. Wang: Physica C **378–381**, 320 (2002)

[41] N. Kida, M. Hangyo, M. Tonouchi: Phys. Rev. B **62**, R11965 (2000)

[42] N. Kida, M. Hangyo, M. Tonouchi: J. Magn. Magn. Mater. **226–230**, 818 (2001)

[43] N. Kida, M. Tonouchi: *Physics in Local Lattice Distortions* (Am. Inst. Phys., New York 2001) p. 366

[44] N. Kida, M. Tonouchi: Phys. Rev. B **66**, 024401 (2002)

[45] N. Kida, M. Tonouchi: Physica B **329–333**, 842 (2003)

[46] S. Nashima, M. Tonouchi, M. Hangyo, K.-U. Barholz, F. Schmidl, P. Seidel: Jpn. J. Appl. Phys. **39**, L663 (2000)

[47] S. Shikii, T. Kondo, M. Yamashita, M. Tonouchi, M. Hangyo: Appl. Phys. Lett. **74**, 1317 (1999)
[48] O. Morikawa, M. Yamashita, H. Saijo, M. Morimoto, M. Tonouchi, M. Hangyo: Appl. Phys. Lett. **75**, 3387 (1999)
[49] M. Tonouchi, M. Yamashita, M. Hangyo: J. Appl. Phys. **87**, 7366 (2000)
[50] A. Moto, M. Hangyo, M. Tonouchi: IEICE Trans. Electron. **E84-C**, 67 (2001)
[51] M. Tonouchi, A. Moto, M. Yamashita, M. Hangyo: IEEE Trans. Appl. Supercond. **11**, 3230 (2001)
[52] A. Moto, M. Tonouchi: Physica C **357–360**, 1603 (2001)
[53] M. Tonouchi, A. Moto: Physica C **367**, 33 (2002)
[54] T. Kiwa, Y. Kamada, M. Misra, H. Murakami, M. Tonouchi: IEEE Trans. Appl. Supercond. in press
[55] M. Tonouchi, N. Wada, M. Hangyo, M. Tani, K. Sakai: Appl. Phys. Lett. **71**, 2364 (1997)
[56] M. Tonouchi, S. Shikii, M. Yamashita, K. Shikita, T. Kondo, O. Morikawa, M. Hangyo: Jpn. J. Appl. Phys. **37**, L1301 (1998)
[57] M. Tonouchi, S. Shikii, M. Yamashita, K. Shikita, M. Hangyo: IEEE Trans. Appl. Supercond. **9**, 4467 (1999)
[58] M. Tonouchi: Jpn. J. Appl. Phys. **40**, L542 (2001)
[59] T. Fukui, A. Moto, H. Murakami, M. Tonouchi: Physica C **357–360**, 454 (2001)
[60] M. Tonouchi, K. Shikita: Physica C **367**, 37 (2002)
[61] T. Fukui, H. Murakami, M. Tonouchi: IEICE Trans. on Electron. **E85-C**, 818 (2002)
[62] M. Tonouchi: Sin. J. Phys. **18**, 153 (2002)
[63] M. Tonouchi: *Vortex Physics*, Studies of High Temperature Superconductors **42** (Nova Science 2002) p. 109
[64] M. Tani, M. Tonouchi, Z. Wang, K. Sakai, M. Hangyo, S. Tomozawa, Y. Murakami: Jpn. J. Appl. Phys. **35**, L1184 (1996)
[65] M. Tonouchi, M. Tani, Z. Wang, K. Sakai, M. Hangyo, N. Wada, Y. Murakami: IEEE Trans. Appl. Supercond. **7**, 2913 (1997)
[66] M. Morimoto, H. Saijo, M. Yamashita, M. Tonouchi, M. Hangyo: *Advances in Superconductivity XII* (Springer, Tokyo 2000) p. 1111
[67] H. Saijo, M. Morimoto, T. Kiwa, M. Tonouchi: Physica C **362**, 319 (2001)
[68] M. Tonouchi, H. Saijo, M. Hangyo, O. Morikawa, P. Gu, M. Tani, K. Sakai: Physica C **357**, 1600 (2001)
[69] R. McLaughlin, A. Corchia, M. Johnston, Q. Chen, C. Ciesla, D. Arnone, G. Jones, E. Linfield, A. Davies, M. Pepper: Appl. Phys. Lett. **76**, 2038 (2000)
[70] M. Tonouchi, M. Tani, Z. Wang, K. Sakai, N. Wada, M. Hangyo: Jpn. J. Appl. Phys. **35**, L1578 (1996)
[71] H. Wald, P. Seidel, M. Tonouchi: Physica C **362**, 324 (2001)
[72] H. Wald, P. Seidel, M. Tonouchi: Physica C **357–360**, 146 (2001)
[73] H. Wald, P. Seidel, M. Tonouchi: Physica C **367**, 308 (2002)
[74] H. Wald, S. Nashima, M. Yamashita, M. Tonocuhi, P. Seidel, M. Hangyo: *Advances in Superconductivity XII* (Springer, Tokyo 2000) p. 224
[75] H. Wald, C. Steigmeier, P. Seidel, S. Nashima, M. Tonouchi, M. Hangyo: Physica C **341–348**, 1899 (2000)

[76] H. Wald, F. Schmidt, P. Seidel, M. Tonouchi: Supercond. Sci. Technol. **15**, 1494 (2002)

[77] M. Tonouchi, M. Hangyo, J. Ramos, R. Ijsselsteijn, V. Shultze, H.-G. Meyer, H. Hoening: *Advances in Superconductivity XI* (Springer, Tokyo 1999) p. 1297

[78] M. Morimoto, T. Yoshimura, M. Tonouchi: Physica C **357**, 1607 (2001)

[79] T. Kiwa, I. Kawashima, S. Nashima, M. Hangyo, M. Tonouchi: Jpn. J. Appl. Phys. **39**, 6304 (2000)

[80] T. Uchiyama, Z. Wang: IEEE Trans. Appl. Supercond. **11**, 3297 (2001)

[81] H. Schneidewind, M. Manzel, G. Bruchlos, K. Kirsch: Supercond. Sci. Technol. **14**, 200 (2001)

[82] E. Zeldov, J. Clem, M. McElfresh, M. Darwin: Phys. Rev. B **49**, 9802 (1994)

[83] J. Demsar, B. Podobnik, V. Kabanov, T. Wolf, D. Mihailovic: Phys. Rev. Lett. **82**, 4918 (1999)

[84] O. Tsui, N. Ong, J. Peterson: Phys. Rev. Lett. **76**, 819 (1995)

[85] M. Gaifullin, Y. Matsuda, N. Chikumoto, J. Shimokawa, K. Kishio, R. Yoshizakiy: Phys. Rev. Lett. **83**, 3928 (1999)

[86] M. Tachiki, T. Koyama, S. Takahashi: Phys. Rev. B **50**, 7065 (1994)

[87] C. Homes, T. Timusk, R. Liang, D. Bonn, W. Hardy: Phys. Rev. Lett. **71**, 1645 (1993)

[88] Y. Matsuda, M. Gaifullin, K. Kumagai, K. Kadowaki, T. Mochiku: Phys. Rev. Lett. **75**, 4512 (1995)

[89] I. Iguchi, K. Lee, E. Kume: Phys. Rev. B **61**, 689 (2000)

[90] H. Wang, P. Wu, T. Yamashita: Phys. Rev. Lett. **87**, 107002 (2001)

[91] A. Tsvetkov, D. van der Marel, K. Moler, J. Kirtley, J. de Boer, A. Meetsma, Z. Ren, N. Koleshnikov, D. Dulic, A. Damascelli, M. Gruninger, J. Schutzmann, J. van der Eb, H. Somal, J. Wang: Nature **395**, 360 (1998)

[92] D. Dulic, S. Hak, D. van der Marel, W. Hardy, A. K. R. Liang, D. Bonn, B. Willemsen: Phys. Rev. Lett. **86**, 4660 (2001)

[93] D. Dulic, D. van der Marel, A. Tsvetkov, W. Hardy, Z. Ren, J. Wang, B. Willemsen: Phys. Rev. B **60**, R15051 (1999)

[94] Z. Jirák, S. Krupička, Z. Šimša, M. Dlouhá, S. Vratislav: J. Magn. Magn. Mater. **53**, 153 (1985)

[95] Y. Tomioka, A. Asamitsu, Y. Moritomo, Y. Tokura: J. Phys. Soc. Jpn. **64**, 3626 (1995)

[96] Y. Tomioka, A. Asamitsu, H. Kuwahara, Y. Moritomo, Y. Tokura: Phys. Rev. B **53**, R1689 (1996)

[97] V. Kiryukhin, D. Casa, J. Hill, B. Keimer, A. Vigliante, Y. Tomioka, Y. Tokura: Nature **386**, 813 (1997)

[98] Y. Okimoto, Y. Tokura, Y. Tomioka, Y. Onose, Y. Otsuka, K. Miyano: Mol. Cryst. Liq. Cryst. **315**, 257 (1998)

[99] Y. Tokura, N. Nagaosa: Science **288**, 462 (2000)

[100] A. Moreo, S. Yunoki, E. Dagotto: Science **283**, 2034 (1999)

[101] I. Deac, J. Mitchell, P. Schiffer: Phys. Rev. B **63**, 172408 (2001)

[102] M. Tani, S. Matsuura, K. Sakai, S. Nakashima: Appl. Opt. **36**, 7853 (1997)

[103] C. Ludwig, J. Kuhl: Appl. Phys. Lett. **69**, 1194 (1996)

[104] P. Loukakos, C. Kalpouzos, I. Perakis, Z. Hatzopoulos, M. Logaki, C. Fotakis: Appl. Phys. Lett. **79**, 2883 (2001)

[105] X.-C. Zhang, D. Auston: J. Appl. Phys. **71**, 326 (1992)

[106] S. Howells, S. Herrera, L. Schlie: Appl. Phys. Lett. **65**, 2946 (1994)
[107] S. Howells, L. Schlie: Appl. Phys. Lett. **67**, 3688 (1995)
[108] S. Kono, P. Gu, M. Tani, K. Sakai: Appl. Phys. B **71**, 901 (2000)
[109] A. Markelz, E. Heiwell: Appl. Phys. Lett. **72**, 2229 (1998)
[110] Y. Okimoto, Y. Tomioka, Y. Onose, Y. Otsuka, Y. Tokura: Phys. Rev. B **57**, R9377 (1998)
[111] M. Johnston, D. Whittaker, A. Corchia, A. Davies, E. Linfield: Phys. Rev. B **65**, 165301 (2002)
[112] P. Gu, M. Tani, S. Kono, K. Sakai, X.-C. Zhang: J. Appl. Phys. **91**, 5533 (2002)
[113] S. Yamada, T. Arima, H. Ikeda, K. Takita: J. Phys. Soc. Jpn. **69**, 1278 (2000)
[114] E. Schöll: *Nonequilibrium Phase Transitions in Semiconductors* (Springer, Berlin, Heidelberg 1987) p. 40
[115] M. Frankel, J. Whitaker, G. Mourou, F. Smith, A. Calawa: IEEE Trans. Electron Devices **37**, 2493 (1990)
[116] J. Luo, H. Thomas, D. Morgan, D. Westwood: Appl. Phys. Lett. **64**, 3614 (1994)
[117] M. Fiebig, K. Miyano, Y. Tomioka, Y. Tokura: Science **280**, 1925 (1998)
[118] M. Fiebig, K. Miyano, Y. Tomioka, Y. Tokura: Appl. Phys. Lett. **74**, 2310 (1999)
[119] M. Fiebig, K. Miyano, T. Satoh, Y. Tomioka, Y. Tokura: Phys. Rev. B **60**, 7944 (1999)
[120] M. Tokunaga, N. Miura, Y. Tomioka, Y. Tokura: Phys. Rev. B **57**, 5259 (1998)
[121] H. Alloul, T. Ohno, P. Mendels: Phys. Rev. Lett. **63**, 1700 (1989)
[122] A. Puchkov, P. Fournier, D. Basov, T. Timusk, A. Kapitulnik, N. Kolesnikov: Phys. Rev. Lett. **77**, 3212 (1996)
[123] S. Uchida, Y. Fukuzumi, K. Takenaka, K. Tamasaku: Physica C **263**, 264 (1996)
[124] C. Renner, B. Revaz, J.-Y. Genoud, K. Kadowaki, O. Fischer: Phys. Rev. Lett. **80**, 149 (1997)
[125] V. M. Krasnov, A. Yurgens, D. Winkler, P. Delsing, T. Claeson: Phys. Rev. Lett. **84**, 5860 (2000)
[126] M. Suzuki, T. Watanabe: Phys. Rev. Lett. **85**, 4787 (2000)
[127] H. Ding, T. Yokoya, J. Campuzano, T. Takahashi, M. Randeria, M. Norman, T. Mochiku, K. Kadowaki, J. Giapintzakis: Nature **382**, 51 (1996)
[128] V. Emery, S. Kivelson, O. Zachar: Phys. Rev. B **56**, 6120 (1997)
[129] E. Demler, S.-C. Zhang: Nature **396**, 733 (1998)
[130] V. Kabanov, J. Demsar, B. Podobnik, D. Mihailovic: Phys. Rev. B **59**, 1497 (1999)
[131] S. Han, Z. Vardeny, K. Wong, O. Symko, G. Koren: Phys. Rev. Lett. **65**, 2708 (1990)
[132] G. Eesley, J. Heremans, M. Meyer, G. Doll, S. Liou: Phys. Rev. Lett. **65**, 3445 (1990)
[133] T. Gong, L. Zheng, W. Xiong, W. Kula, Y. Kostoulas, R. Sobolewski, P. Fauchet: Phys. Rev. B **47**, 14495 (1993)
[134] S. Brorson, R. Buhleier, J. White, I. Trofimov, H.-U. Habermeier, J. Kuhl: Phys. Rev. B **49**, 6185 (1994)

[135] I. Wilke, M. Khazan, C. Rieck, P. Kuzel, T. Kaiser, C. Jaekel, H. Kurz: J. Appl. Phys. **87**, 2984 (2000)

[136] C. Caroli, K. Maki: Phys. Rev. **159**, 306 (1967)

[137] Y. Okimoto, T. Katsufuji, T. Ishikawa, A. Urushibara, T. Arima, Y. Tokura: Phys. Rev. Lett. **75**, 109 (1995)

[138] J.-H. Park, E. Vescovo, H.-J. Kim, C. Kwon, R. Ramesh, T. Venkatesan: Nature **392**, 794 (1998)

[139] Y. Okimoto, Y. Tokura: J. Supercond. **13**, 271 (2000)

[140] T. Okuda, Y. Tomioka, A. Asamitsu, Y. Tokura: Phys. Rev. B **61**, 8009 (2000)

[141] P. Schiffer, A. Ramirez, W. Bao, S.-W. Cheong: Phys. Rev. Lett. **75**, 3336 (1995)

[142] S. Chun, M. Salamon, Y. Tomioka, Y. Tokura: Phys. Rev. B **61**, R9225 (2000)

[143] Y. Okimoto, T. Katsufuji, T. Ishikawa, T. Arima, Y. Tokura: Phys. Rev. B **55**, 4206 (1997)

[144] S. Chun, M. Salamon, P. Han: Phys. Rev. B **59**, 11155 (1999)

[145] J. Simpson, H. Drew, V. Smolyaninova, R. Greene, M. Robson, A. Biswas, M. Rajeswari: Phys. Rev. B **60**, R16263 (1999)

[146] M. Fäth, S. Freisem, A. Menovsky, Y. Tomioka, J. Aarts, J. Mydosh: Science **285**, 1540 (1999)

[147] Y. Sawaki, K. Takenaka, A. Osuka, R. Shiozaki, S. Sugai: Phys. Rev. B **61**, 11588 (2000)

[148] K. Takenaka, Y. Sawaki, S. Sugai: Phys. Rev. B **60**, 13011 (1999)

[149] Y. Okimoto, Y. Tomioka, Y. Onose, Y. Otsuka, Y. Tokura: Phys. Rev. B **59**, 7401 (1999)

[150] P. Calvani, G. D. Marzi, P. Dore, S. Lupi, P. Maselli, F. D'Amore, S. Gagliardi, S.-W. Cheong: Phys. Rev. Lett. **81**, 4504 (1998)

[151] H. Liu, S. Cooper, S.-W. Cheong: Phys. Rev. Lett. **81**, 4684 (1998)

[152] T. Tonogai, T. Satoh, K. Miyano, Y. Tomioka, Y. Tokura: Phys. Rev. B **62**, 13903 (2000)

[153] T. Asaka, S. Yamada, S. Tsutsumi, C. Tsuruta, K. Kimoto, T. Arima, Y. Matsui: Phys. Rev. Lett. **88**, 097201 (2002)

[154] S. Mori, C. Chen, S.-W. Cheong: Nature **392**, 473 (1998)

[155] S. Yamada, T. Arima, K. Takita: J. Phys. Soc. Jpn. **68**, 3701 (1999)

[156] D. Basov, B. Dabrowski, T. Timusk: Phys. Rev. Lett. **81**, 2132 (1998)

[157] A. Anane, J.-P. Renard, L. Reversat, C. Dupas, P. Veillet, M. Viret, L. Pinsard, A. Revcolevschi: Phys. Rev. B **59**, 77 (1999)

[158] R. Mathieu, P. Nordblad, A. Raju, C. Rao: Phys. Rev. B **65**, 132416 (2002)

[159] D. Romero, S. Liu, H. Drew, K. Ploog: Phys. Rev. B **42**, 3179 (1990)

[160] S. Liu, K. Karrai, F. Dunmore, H. Drew, R. Wilson, G. Thomas: Phys. Rev. B **48**, 11394 (1993)

[161] G. Grüner: Rev. Mod. Phys. **60**, 1129 (1988)

[162] P. Lee, T. Rice, P. Anderson: Solid State Commun. **14**, 703 (1974)

[163] T. Kim, D. Reagor, G. Grüner, K. Maki, A. Virosztek: Phys. Rev. B **40**, 5372 (1989)

[164] G. Mihály, T. Kim, G. Grüner: Phys. Rev. B **39**, 13 009 (1989)

[165] T. Kim, S. Donovan, G. Grüner, A. Philipp: Phys. Rev. B **43**, 6315 (1991)

[166] L. Degiorgi, B. Alavi, G. Mihály, G. Grüner: Phys. Rev. B **44**, 7808 (1991)

[167] L. Degiorgi, G. Grüner: Phys. Rev. B **44**, 7820 (1991)

[168] G. Grüner: Rev. Mod. Phys. **66**, 1 (1994)
[169] T. Katsufuji, T. Tanabe, T. Ishikawa, Y. Fukuda, T. Arima, Y. Tokura: Phys. Rev. B **54**, R14230 (1996)
[170] T. Ishikawa, S. Park, T. Katsufuji, T. Arima, Y. Tokura: Phys. Rev. B **58**, R13326 (1998)
[171] G. Grüner: *Density Waves in Solids* (Addison-Wesley, Reading 1994)
[172] P. Brüesch, S. Strässler, H. Zeller: Phys. Rev. B **12**, 219 (1975)
[173] S. Shimomura, T. Tonegawa, K. Tajima, N. Wakabayashi, N. Ikeda, T. Shobu, Y. Noda, Y. Tomioka, Y. Tokura: Phys. Rev. B **62**, 3875 (2000)
[174] D. Louca, T. Egami, W. Dmowski, J. Mitchell: Phys. Rev. B **64**, 180403(R) (2001)
[175] L. Degiorgi, S. Thieme, B. Alavi, G. Grüner, R. McKenzie, K. Kim, F. Levy: Phys. Rev. B **52**, 5603 (1995)

Terahertz Imaging

Michael Herrmann[1,2], Ryoichi Fukasawa[1,3], and Osamu Morikawa[4]

[1] National Institute of Information and Communications Technology
588-2 Iwaoka, Nishi-ku, Kobe 651-2492, Japan
[2] *Present address:* Institute for Scientific and Industrial Research
Osaka University
8-1 Mihogaoka, Ibaraki-shi, Osaka 567-0047, Japan
herrm32@sanken.osaka-u.ac.jp
[3] *Present address:* Terahertz Optoelectronics Laboratory
Tochigi Nikon Corporation
770 Midori, Ohtawara-shi, Tochigi 324-8625, Japan
[4] Department of Science, Japan Coast Guard Academy
Wakabacho 5-1, Kure, Hiroshima 737-8512, Japan

Abstract. This Chapter covers a survey of terahertz (THz) imaging stretching from hardware to applications. In the hardware section, we describe two THz-imaging setups, one of them based on photoconductive antennas and mechanical scanning, the other one offering real-time imaging with electro-optic sampling and a CCD camera. We go into details of some aspects of THz-imaging data processing and image interpretation. The application sections concentrate on the characterization of semiconductors and superconductors and the study of powders with a particular focus on real-time applications.

When it was discovered around 1990 that electromagnetic (EM) radiation in the THz range could be produced with femtosecond (fs) lasers at significant power levels, THz radiation developed rapidly into a valuable scientific tool. (This development, its background and implications have been described in detail in the Chapter by *Sakai* and *Tani* in this volume.) With THz radiation readily at hand, ideas for commercial, industrial and scientific applications were developed at a fast rate making use of, for example, the good transmittance of most insulators to THz radiation, or the excellent sensitivity of THz radiation to even small amounts of water. But it was soon found that most applications required not only probing the average properties of a large sample, but rather the two-dimensional mapping of sample properties with a camera-type instrument: i.e., THz imaging.

One set of THz applications can be derived from the good transparency of most insulators to THz radiation. Package inspection and quality control [1, 2, 3] were among the first applications suggested, as well as replacing X-ray baggage checks at airports by checks with THz radiation (then called "T-rays" [2]), followed by the screening of persons for weapons and other objects[1]. This necessarily incomplete list continues with counting and check-

---

[1] At least two systems are currently at the brink of commercial operation for security applications using passive imaging in the lower THz range. They

K. Sakai (Ed.): Terahertz Optoelectronics, Topics Appl. Phys. **97**, 331–381 (2005)
© Springer-Verlag Berlin Heidelberg 2005

ing banknotes [4] and screening letters for drugs and bioweapons [5] to the reproduction of the three-dimensional structure of objects with THz tomography [6]. Another set of applications is based on the excellent sensitivity of THz radiation even for small amounts of water. They stretch from studying the process of drying and refreshing of leaves [1, 2] to medical applications as the diagnosis of skin burns [2, 7], teeth [8] and cancer [9, 10, 11].[2]

This Chapter discusses the two prevailing types of THz imaging setups, elaborates some characteristics of THz images and then describes in detail a selected set of THz imaging applications: the study of powders, the characterization of semiconductors and superconductors, and real-time applications.

1 Variants of THz Imaging Equipment

The process of imaging corresponds to mapping points of an object onto a set of detectors of radiation that emerges from the object. For example, a CCD camera may simultaneously map visible light from 1600×1200 points of a real-world object with a lens on an array of 1600×1200 photodetectors. In a much simpler approach, there may be only one detector, and the data points will be mapped successively while the sample or detector is scanned. The former approach is certainly much faster, while the latter may have advantages in cost and image quality. For THz imaging both variants have been tried and developed simultaneously.

AT&T Bell Labs followed the method of THz time-domain spectroscopy (THz-TDS) based on photoconductive (PC) antennas, as described in the Chapter by *Nishizawa* et al. in this volume. Conventional THz-TDS has the sample in a wide and parallel beam. For imaging purposes, an additional focus was introduced in the center of this beam, where the sample was placed, and the sample was scanned mechanically in two directions [1]. In an alternative approach, a group at Rensselaer Polytechnic Institute used electro-optic (EO) signal conversion and could thus record THz images, with a commercial CCD camera [12]. These two methods have remained the prevailing methods to date, and both are discussed in this section in detail. The section is restricted to the hardware aspects of the topic and will not extend to data evaluation or any experimental results, which are left for later sections. Furthermore, we also restrict the discussion to the use of pulsed THz radiation neglecting continuous-wave equipment, which is treated in the Chapter by *Matsuura* and *Ito*.

---

are provided by QinetiQ Ltd (http://www.qinetiq.com/home/newsroom/news_releases_homepage/2001/3rd_quarter/qinetiq.html) and ThruVision Ltd (http://www.thruvision.com/).

[2] On recent advances in the field of THz imaging of skin and breast cancer, also see the web pages at Teraview Ltd (http://www.teraview.co.uk/ap_oncology.asp).

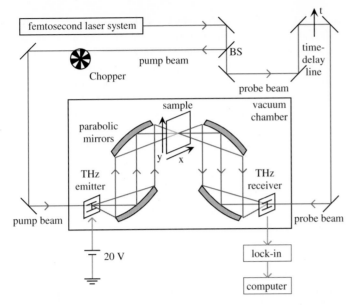

Fig. 1. THz imaging setup: The THz pump-probe setup is based on a fs laser system and on PC antennas as emitter and detector. The sample is scanned in x and y directions. With a delay stage, time-domain waveforms can be recorded at every image position

1.1 Scanning THz Imaging System Based on Photoconductive Antennas

In the most straightforward implementation of imaging with THz radiation, both emitter and detector are PC antennas, and the points of the object (sample) are mapped onto the detector successively while the sample is scanned. A corresponding THz imaging setup, which has been in use at the National Institute of Information and Communications Technology (NiCT), Japan, is shown in Fig. 1. It resembles a setup for THz-TDS as described in the Chapter by *Nishizawa* et al. but uses an additional pair of parabolic mirrors generating an intermediate focus at which the sample is placed. Some noteworthy points of the components of this setup are discussed in the following.

Laser: The fs laser system is often based on a Ti:sapphire laser with 50 fs to 100 fs pulse width. 10 fs lasers provide higher bandwidth and are necessary for a convenient signal-to-noise ratio (SNR) at frequencies higher than about 3 THz. The pulse repetition rate is typically of the order of 100 MHz, but an amplified laser system with a repetition rate of the order of 1 kHz can be used when high pulse energies are needed, for example in combination with EO sampling and CCD camera detection (see the next section).

Stages: The stages for sample movement and time delay typically employ stepping motors. The length of the time-delay line determines the frequency resolution of the setup. For example, with a $l = 15$ cm long delay stage, a frequency resolution of $\Delta f = c/(2l) = 1$ GHz can be achieved. The precision of the time-delay stage corresponds to the resolution on the time axis. For instance, for measuring a delay time of 10 fs, 1.5 µm stage precision is needed. The precision of the sample movement, in contrast, is noncrucial except in the case of a very good THz near-field imaging setup. The sample-movement stage should rather be chosen for a large range and high speed.

Antennas: THz emitters and detectors supply enough material for a book on their own. The chapter by *Sakai* et al. in this volume contains a detailed treatment of this topic. PC dipole antennas are often chosen for their bandwidth. Bow-tie or large-aperture antennas may supply higher power at lower bandwidth, spiral antennas can also supply higher power but emit circularly polarized THz radiation. Semiconductor surfaces without antenna structures can also emit THz radiation at significant power levels, particularly in combination with magnetic fields. EO pulse conversion can also be used on the emitter as well as the detector side. It requires a laser amplifier for generating intense THz pulses and an IR-sensitive camera and offers the potential for ultrahigh bandwidth. As the IR camera, commercial CCD cameras are typically used, but CMOS cameras can be operated more flexibly, which can be turned into an advantage for the SNR by using dynamic recording modes [13]. In the future, detector arrays of PC antennas may provide an interesting alternative [14]. PC antennas are usually combined with hemispherical Si lenses that have the purpose of delaying the pulse reflected at the second surface of the antenna chip and thus preventing oscillations in the THz spectrum. Hyperhemispherical Si lenses can be used at the emitter in order to direct more radiation into the forward direction, but they are more difficult to align and have unfavorable focal conditions.

Mirrors: For focusing THz radiation, off-axis parabolic mirrors are in widespread use. As opposed to lenses, mirrors have a great advantage by avoiding undesired reflections that can lead to radiation losses and particularly to fake structures in the waveform and spectrum. Mirrors do need great care in the alignment, though. Aligning parabolic mirrors and antennas in a conventional THz imaging setup amounts to an optimization problem in 30 dimensions (including rotation of the mirrors that may not always be glued to the mirror holders as precisely as one may desire), which means that a good starting position is indispensable. The alignment should start by placing the mirrors in the designed positions according to manufacturer specifications within sub-mm precision, then aligning them one by one as precisely as possible using laser light both from emitter to detector and back from detector to emitter, and finally by optimizing the THz waveform. For best spatial resolution, the mirrors should have a high numerical aperture, that is, a short focal length in relation to the mirror diameter. A set of parabolic mirrors will image one spot at the focus of the first mirror onto one spot at the focus of the second mirror, which is sufficient for use with simple PC antennas. But offcenter spots can suffer from considerable aberration. For use with detector arrays and cameras, parabolic mirrors are not suitable and should be replaced by lenses.

Chamber: It may be desirable to place the THz imaging equipment in a chamber where the environment can be controlled. This may even be necessary because of the high density of water-vapor absorption lines at frequencies > 1 THz. If the chamber is supposed to be pumped to vacuum, care should be taken that the mounting plate of the THz components is not subjected to mechanical stress from the pressure difference because such stress can substantially compromize the quality of the THz focus.

Amplifier: Since the configuration at the PC antenna gap is never exactly symmetrical, there is an average current even in the absence of THz radiation, so that it is necessary to chop the laser pump beam and use a lock-in amplifier. The chopper is typically operated at a frequency near 1 kHz. In the case of using an amplified laser system, the chopper may have to be locked to the laser frequency. In order to minimize stray pickup, a preamplifier should be situated as close as possible to the antenna. For minimizing noise, the preamplifier resistance should be high, and there is a noise advantage in separating the preamplifier from the antenna during the antenna-off times [15].

The waveform and spectral quality of a THz imaging system based on PC antennas is the same as for a THz-TDS system with the same equipment, or slightly worse because the additional set of mirrors makes the alignment more difficult. With an 80 fs laser and dipole antennas with a 5 μm gap, a center frequency of 0.3 THz to 0.5 THz can be achieved with an amplitude signal-to-noise ratio of 1000 or higher measured at 300 ms lock-in integration time. The data-recording time in the setup of Fig. 1 ranges from several minutes for a small image to several hours for an image containing long waveforms. It depends on the lock-in integration time, the data-transfer rate between lock-in and computer and the speed of the mechanical stages. Sub-mm spatial resolution was shown through a 0.3 mm aperture as demonstrated in Fig. 2. This is close to the 0.7 mm diffraction limit theoretically achievable for this setup at 1.0 THz and actually better than the refraction limit theoretically achievable for THz radiation with a center wavelength of 0.5 THz as generated with the THz emitter antenna. This is possible because the small aperture cuts off significant amounts of the long-wavelength radiation.

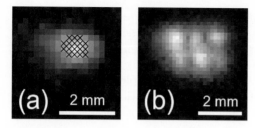

Fig. 2. (a) THz image of a 0.3 mm aperture demonstrating sub-mm spatial resolution. The pixel size in this image corresponds to 0.2 mm on the sample. The pixels marked "×" have an amplitude > 60% of the maximum and thus indicate the resolution limit. (b) THz image of a structure of 3 × 2 apertures at a minimum distance of 1 mm, also clearly demonstrating sub-mm resolution

1.2 Real-Time THz Imaging Based on Electro-Optic Sampling

While research into the utilization of THz radiation in imaging technology [1, 2, 16] and new spectroscopy [17] is underway around the world, THz radiation has been applied or suggested for application in fields as different as semiconductor characterization [18, 19, 20, 21], medical diagnoses [22], chemical analyses [23] and industrial process control [7]. There is a great potential for its application in a number of other fields as well. Industrial applications typically imply fast operation. Tochigi Nikon Corporation has developed a commercial real-time THz imaging system for industrial applications, in collaboration with the NiCT [24]. This system uses a CCD camera to capture a single image in a short time. To obtain real-time THz images, an array detector is desired. But it is technically difficult to make an array detector for the THz region. Scanning THz imaging systems [25], on the other hand, measuring one point at a time, require the sequential scanning of the sample, which is extremely time consuming, an image takes more than a minute to acquire. It is almost impossible to see moving objects using a scanning THz imaging system. Significant advances are required to allow real-time THz imaging. The two-dimensional (2D) EO sampling technique promises the possibility of real-time THz imaging without the need to spatially scan the object. The 2D EO sampling technique was first demonstrated by *Wu* et al. at the Rensselaer Polytechnic Institute [12]. The real-time imaging system developed by Tochigi Nikon Corporation can capture up to 30 images per second because the CCD image capture is precisely synchronized with optical pulse emission.

Figure 3 shows the schematic diagram of this real-time THz imaging system. It consists of a fs laser source, a source of THz radiation, the sample to be imaged, imaging optics, an EO crystal (ZnTe), a computer-controlled optical delay line and a CCD camera. The fs laser pulse is divided into pump and probe beams. The optical source is a Ti:sapphire regenerative amplifier with 130 fs pulse width, 800 nm center wavelength and 1 kHz repetition rate. The

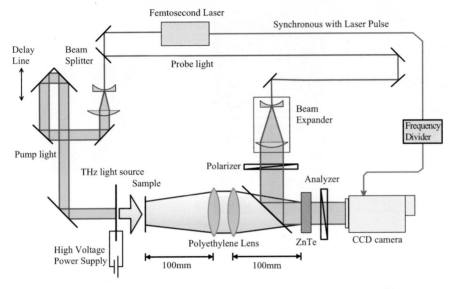

Fig. 3. Schematic diagram of the real-time THz imaging system

pump beam goes through the optical delay line, and drives the large-aperture PC antenna [26], which emits THz radiation.

The antenna consists of a 620 μm thick semi-insulating GaAs wafer with a 15 mm wide PC gap between gold electrodes. A bias voltage of 5 kV is applied between the electrodes. The fs laser pulse is illuminated onto the gap between the two electrodes. The current rises very rapidly after injection of photocarriers by the fs laser pulse, and then decays with a time constant given by the carrier lifetime of the semiconductor. The transient photocurrent radiates into free space according to Maxwell's equations. The THz radiation amplitude is proportional to the time derivative of this transient photocurrent. The radiation from the THz emitter passes through the sample and is focused onto the EO crystal by two polyethylene lenses to form an image of the sample. The probe beam is expanded at the same time, and the polarizer ensures that the probe beam is linearly polarized. The probe beam is then led into the same optical axis as the THz radiation by a pellicle beam splitter.

Since CCD cameras do not respond to THz radiation, the 2D EO sampling technique is employed for transfering the THz image into an intensity pattern in the 800 nm laser beam. At each point on the EO crystal, the refractive index is changed depending on the THz electric field within the EO crystal, and birefringence is induced. When the probe beam passes through the EO crystal, the birefringence changes the polarization of the probe beam. Only the light with changed polarization passes through a crossed polarizer positioned in front of the CCD camera, because the transmission axis of the polarizer is aligned perpendicular to the polarization of the incident probe

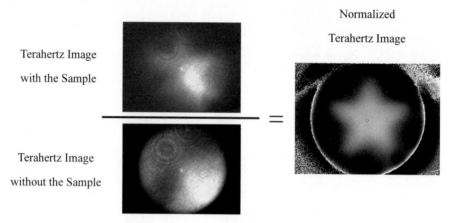

Fig. 4. Images with sample (upper left) and without sample (lower left) and processed image (right)

beam. Through these processes, the THz electric field distribution in the EO crystal is converted into an optical intensity distribution which can be recorded by the CCD camera. The CCD camera has a frame-transfer-type CCD with a pixel number of 384×288; the maximum frame rate is 39 frames per second (fps). However, a lower frame rate results in considerably better quality of the THz images. Therefore, this camera is used at rates under 30 fps. For the EO crystal, a 2 mm thick (110)-oriented ZnTe crystal is used. Because both the EO crystal and the probe beam are spatially nonuniform, the images obtained by the CCD camera are nonuniform even without THz radiation present. This background image is subtracted from the image obtained by the CCD camera, thereby detecting only the signals of the THz radiation. All the THz images presented in this section are background subtracted using this method.

Although we have observed several objects using the real-time THz imaging system, distinguishing the fine shape of objects was difficult. Figure 4 shows an example of the image processing to sharpen the images. The procedure for image processing is quite simple. The image data with a sample present is divided by the image data without a sample present. This means that the CCD output of each pixel is normalized by the CCD output in the absence of samples. The upper-left image with the sample present is divided by the lower-left image without the sample. The right-hand image is the result after this procedure. It is much clearer than the original image (the upper-left image in Fig. 4). The sample was a metallic film with a star-shaped aperture shown in Fig. 5a. A substrate transparent to the THz radiation was coated with a metal film, with a star-shaped cutout through which THz radiation was allowed to pass.

(a)

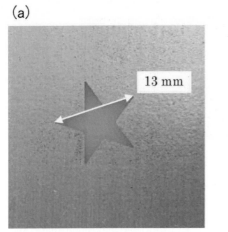

Spatial resolution
approximately
1.3 mm

(b)

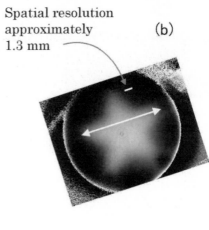

Fig. 5. (a) Optical image of a star-shaped test pattern. (b) THz transmission image

Figure 5 demonstrates the spatial resolution of the real-time THz imaging system. It takes only 0.2 s to capture a single image. The photographs are the star-shaped test pattern (left) and its image obtained through THz illumination (right). The wavelength of THz radiation is in the hundreds of micrometers. As can be seen from this photograph, the spatial resolution is roughly 1.3 mm, close to the diffraction limit for THz radiation.

2 Basic Properties of THz Images

The long wavelength is one point where imaging with THz radiation differs from imaging with visible light. Another point is the fact that THz imaging uses ultrafast light pulses. Let us now consider the effects of such differences on THz imaging and THz images.

2.1 Display Modes

In time-domain THz imaging the full data set is three-dimensional (3D) with two axes in space (along the sample, vertical to the THz beam) and one axis in time. This 3D data set can be arranged into a sequence of images, i.e., a movie on a picosecond time scale. Figure 6 shows such a movie representing the transision of a THz waveform through a sample consisting of the letters "THz" cut from a 0.52 mm thick Teflon plate. Zero THz electric field is displayed gray, while positive (negative) field is shown white (black). THz radiation that reaches the detector directly without transmitting the sample arrives 0.75 ps faster than transmitted radiation, corresponding to a refractive

0.00 ps   1.45 ps   1.75 ps   1.95 ps   2.05 ps

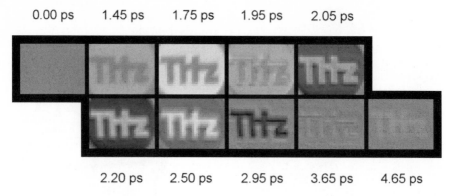

2.20 ps   2.50 ps   2.95 ps   3.65 ps   4.65 ps

Fig. 6. Images from a "THz movie" taken at various positions of the time-delay stage. The sample consists of the letters "THz" cut from a 0.52 mm thick Teflon plate. Gray corresponds to the zero level, white and black to positive or negative THz signal, respectively. For the radiation not transmitting the Teflon plate, the pulse maximum arrives in the third image at 1.75 ps, while the radiation transmitting the sample has its peak at 0.75 ps later in the seventh image. This time difference corresponds to a refractive index of Teflon of 1.43

index of Teflon of 1.43. An interesting point about this set of images is that there are images with very weak contrast, for example the fourth image at $t = 1.90$ ps. As a simple but important conclusion, THz images taken with the time-delay stage at constant position can have little quantitative significance. This is why THz imaging data usually starts with a 3D data set, moving the delay stage and recording waveforms at all sample positions. Images are then calculated based on certain features of the time-domain waveform as its amplitude or the delay time of the main peak. We call the ways of calculating such images "display modes".

Two straightforward examples of features used for display modes are the magnitude and the position (i.e., arrival time) of the main peak, resulting in "maximum-signal mode" and "maximum-position mode", respectively ((A) and (a) in Fig. 7). In a simple approximation, the maximum-signal mode maps sample transmittance and is thus related to the extinction coefficient of a sample, while the maximum-position mode maps the transmission time and corresponds to the refractive index. (For a more complex approach see Sects. 3.3 and 4.1.) Instead of the position of the main peak, one may want to use the position of the zero-crossing between the highest (positive) peak and the following negative peak ("zero position mode", (c) in Fig. 7). The latter has an advantage with respect to the maximum position mode because typically zero-crossings can be measured more accurately than maxima. However, when pulse shape and pulse width are strongly affected by the sample, maximum position and zero position change in different ways, and the maximum position may be more meaningful since it is the position of most pulse energy. Obviously, maximum-position mode and zero-position mode, although very

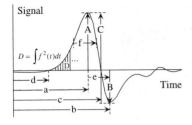

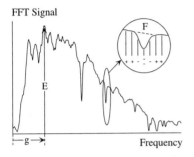

Fig. 7. display modes:
(A) maximum-signal mode
(B) minimum-signal mode
(C) amplitude mode
(D) total-energy mode
(E) FFT maximum mode
(F) absorption-line mode
(a) maximum-position mode
(b) minimum-position mode
(c) zero-position mode
(d) prezero-position mode
(e) maximum-to-minimum distance mode
(f) maximum (minimum)-width mode
(g) FFT maximum-position mode

simliar, can both be useful depending on the measurement situation. Other display modes include:

(D) Total-energy mode: Maps the total pulse energy by integrating the square of the signal over time.

(F) Absorption-line mode: Operates on the frequency axis after Fourier transformation by integrating the intensity drop over an absorption line and normalizing it against the background spectrum, see *Herrmann* et al. [25] for details. Collective absorption line modes probing several absorption lines simultaneously can be used for reliably detecting gases [25].

(f) Maximum-width mode: Maps the width of the main peak. This display mode emphasizes dispersion. But it can also very clearly outline objects because at the border of an object the waveform is an interference between passing and transmitting waveforms, and if these two have nonzero time delay, the main peak of the interference can appear as a very wide peak.

For examples of maximum-signal mode, maximum-position mode and maximum-width mode THz images see Sect. 3.2 of this Chapter.

2.2 Diffraction and Image Resolution

A THz imaging setup is an optical instrument, and as with other optical instruments, its resolution is limited by diffraction and determined by the wavelength of the radiation, the numerical aperture of the setup and the precision of the alignment. A typical resolution close to 1 mm is reached for a setup as described in Sect. 1.1 for a center wavelength in the range of 0.5 THz to 1 THz.

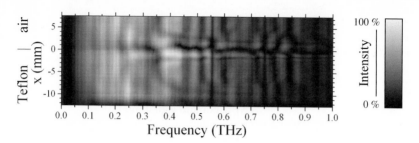

Fig. 8. Spectral intensity depending on frequency and sample position for a line scan through a 0.5 mm thick Teflon plate

THz imaging samples may often be as small as just a few centimeters and have features close to the diffraction limit in size. Diffraction and interference are then important for correctly interpreting THz images. Since THz pulses are short, interference patterns do not extend far from the edge of an object, but can be strong.

As an example that will be helpful for understanding effects on THz images of later sections, let us examine Fig. 8, where the frequency-dependent intensity is mapped into brightness levels as a function of the position on a line scan through a 0.5 mm thick Teflon plate. The plate extends from negative x to zero, positive x are off the plate. For very low frequencies $\ll 0.3$ THz the phase difference between the beam crossing the sample and the beam passing the sample is small, correspondingly, diffraction and interference are weak. For higher frequencies an intensity gradient occurs between the THz beams just inside and just outside the Teflon plate, so that diffraction becomes important and, as with visible light diffracted at an edge, the intensity at the sample edge can drop to zero. On both sides of the sample edge, a short pattern of bright and dark lines is seen. The width of these lines is proportional to the wavelength and decreases toward the right side of Fig. 8. The interference pattern is narrow because the THz pulses only contain a single cycle.

As this measurement was carried out in air, two vertical dark lines are seen due to water-vapor absorption of the THz beam at 0.557 THz and 0.752 THz.

3 THz Imaging Applications to Powders

With the wavelength of THz radiation and the grain size of many powders in the same order of magnitude, one can expect that the refractive indices and extinction coefficients of powders vary greatly within the THz range. THz radiation should thus be an excellent tool for the study of powders themselves as well as processes taking place in powders, such as the diffusion of humidity. This is particularly interesting and important because the study of powders has turned out to be surprisingly difficult not only theoretically but also experimentally using other methods. In addition to these academic applications, there may also be commercial applications in the alliance of powders with THz waves, ranging from tasks such as food-quality control (i.e., finding objects in powders) to drug detection (i.e., finding powders in objects).

3.1 Powders and THz Radiation

Granular materials, including powders, have long been unpopular materials both in theoretical and experimental physics. The disinclination of most theoreticians to granular materials is based on the fact that the properties of a large number of unordered particles are difficult to put into formulas – the situation being similar to that in the field of "chaos" up to the 1970s. Experimentalists hesitated to tackle the field partly because of a lack of practicable theories. Another restraint came from the experimental situation itself: A lack of convenient experimental techniques to probe granular materials. This remained true despite the fact that there is even commercial interest in this field. A sugar or cornflakes manufacturer, for example, may want to know how to avoid that the consumer finds packages only 80% full with his product although they were filled at his factory to 100%.

One problem of probing powders is that they are opaque in a wide range of the EM spectrum simply because of the multitude of grain surfaces, all of them serving as scattering centers. This argument does not hold for high-energy radiation (X-rays) because the refractive index of most materials in that region is close to 1 and scattering from the surfaces is weak, but for this very reason bulk interaction is also weak and the applications remain limited. Low-energy radiation, on the other hand, with a wavelength large compared to the grain size does not "feel" the surfaces any longer and can transmit through powders. But on the low-energy side, radio waves and even microwaves would have a wavelength larger than the sample size, which strongly limits applicability. Static electric fields were applied recently [27] and with a set of 4 capacitors this experiment even achieved some spatial resolution, but for a long time invasive, mechanical probing or ultrasonic measurements have remained the method of choice, the obvious disadvantage in both cases being that the process of probing is or can be destructive to the sample.

Obviously, a convenient way of probing many powders would use EM radiation with a wavelength larger than the grain size and smaller than the sample size. With a wavelength of 0.3 mm, THz radiation is this candidate for many commercially available powders both for studying average powder properties and also for probing samples with good spatial resolution. The latter may prove useful in quality control in the food industry by checking powder density and humidity or searching for impurities. X-rays could perform some of these tasks, for example, finding metal objects in a powder, but would hardly be able to detect insulator objects, particularly plastic.

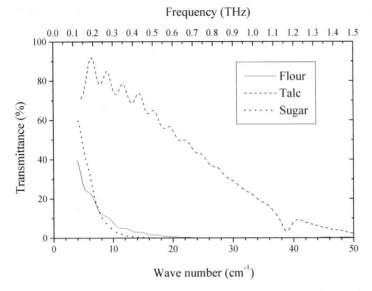

Fig. 9. THz transmission spectra of wheat flour (*solid line*), talc (*dashed line*) and sugar (*dotted line*)

Figure 9 shows the THz spectra of three commercially widely used powders: wheat flour, sugar and talc. While talc still transmits a considerable amount of radiation at 1 THz, wheat flour and sugar are opaque at this frequency. Lower-frequency radiation, however, is transmitted as expected and can be used for imaging purposes. The spectra were taken with a far-infrared spectrometer. The oscillations in the low-frequency region are due to multiple reflections in the walls of the plastic box that contained the powders.

The transmission spectra are certain to depend strongly on the optical properties of the grain material, the size of the grains and the size of the cavities between the grains. Other factors include the grain shape and possible contaminants, of which water is of particular interest because of its high absorption coefficient in the THz range and because of its abundance in the atmosphere.

1 THz corresponds to a wavelength of $\lambda = 300\,\mu m$ in vacuum or air. The grain size d of the above powders varies from $\approx 20\,\mu m$ in talc to $\approx 50\,\mu m$ in flour and $\approx 300\,\mu m$ in sugar. The size parameter is $d/\lambda < 1/10$, and thus in the range of Rayleigh scattering that predicts that the mean free path depends on the wavelength as $l^* \propto l^m$ with $m = 4$. However, Rayleigh scattering assumes that the scattering centers are weak or diluted. In particular, it neglects interference between different scattering centers. This approach need not work for a powder where scattering centers are closer to each other than the wavelength. So it is not surprising that the transmission in Fig. 9

does not follow the Rayleigh formula. We observe instead a proportionality with the exponents $m = 1$, 1.5 and 2 for flour, talc and sugar, respectively.

A more elaborate model for scattering in random media was presented by *Kawato* et al. [28] and shown to work well for a porous glass with pore sizes of $0.15\,\mu m$ to $0.5\,\mu m$ under visible and near-infrared light. We expect that it should work as well for our powders under THz radiation. More recently *Pearce* et al. have directly targeted specific scattering events in granular matter with time-resolved THz measurements [29].

3.2 Imaging Results

Since THz radiation is the shortest-wavelength EM radiation to transmit through many powders, it should be the radiation best suited for detecting and imaging structures in those powders. Such structures could be objects such as, for example, contaminations in food. THz radiation may thus be used for quality control in food production. In this section, we will show imaging results of various objects in powders and discuss the effects observed on those images.

Figure 10 is a set of four THz images showing the same sample in four different display modes. The sample consists of two objects, a piece of egg shell on the left and a piece of plastic on the right. The top left image (a) is a constant-time mode image taken without moving the time-delay stage. There is an inherent problem with this display mode when imaging objects in powders: Powders tend to be more dense at the bottom than at the top, which results in a considerable shift in pulse arrival time. For this reason the lower half of the image is white, which means strongly positive electric field at the main peak of the THz pulse. The top of the image is black because at the time of measurement the main peak of the THz pulse had already passed and the first strong minimum arrived. The phase or time difference between the beams through the top or bottom of the sample thus amounts to approximately $1\,ps$ or half a cycle of $0.5\,THz$ radiation. This means that for the purpose of imaging objects in powders constant-time mode images are difficult to interpret. Images calculated from time-domain waveforms, such as those following, can display objects much more clearly.

Figure 10b is an amplitude-mode display of the same measurement. It maps the peak signal of the waveform at each sample position, nonwithstanding the arrival time of the peak. Both objects can be seen with much better contrast than in (a). The piece of egg shell appears darker than the piece of plastic or than the surrounding powder. This means decreased transmittance due to stronger absorption or scattering. The dark edges around the objects are caused by diffraction due to the same phenomenon that causes a dark minimum and interference lines when visible light passes along a sharp edge (cf. Sect. 2.2). The black areas in the corners are shadows from the sample holder.

max- THz signal max+ 0   Amplitude   max.

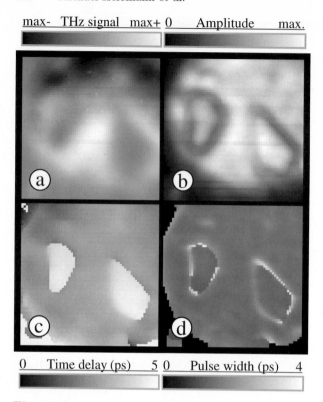

0   Time delay (ps)   5 0   Pulse width (ps)   4

Fig. 10. Images in various display modes showing a piece of egg shell (*left*) and a piece of plastic (*right*) hidden in powdered sugar. (**a**) Constant-time mode, (**b**) amplitude mode, (**c**) delay-time mode, (**d**) pulse-width mode. See the text for further explanations. In the corners of the images the THz beam was blocked by the sample holder

The bottom-left image (c) displays the arrival time of the pulse maximum. An offset plane was used on this image (and also on all subsequent images) in order to compensate for the vertical variation of powder density described above. Both objects produced similar pulse delay although the piece of plastic was considerably thicker than the egg shell (about 1.5 mm vs. 0.4 mm). This reflects the higher refractive index of the egg shell (approximately 2.7 vs. 1.6 for plastic). The piece of plastic in this display mode shows a sharp boundary on the right-hand side and a soft boundary on the left-hand side. The reason for this is the shape of the plastic that was broken off from a bigger disk. The right-hand edge was parallel to the THz beam, while the left-hand edge was at an angle of 45° so that the thickness on the left edge reduced gradually.

In Fig. 10d the width of the pulse maximum is displayed. Both objects display reduced brightness. The reasons for this are strong dispersion and the extinction of high-frequency components that increase the pulse width in the

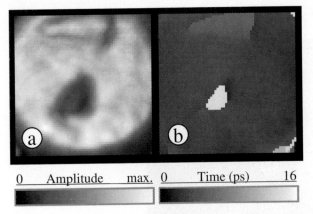

Fig. 11. Amplitude-mode display (**a**) and delay-time mode display (**b**) of a piece of plastic (*either top*) and a piece of glass (*either bottom*) hidden in talc. Glass produces strong contrast in these images because of its high refractive index and extinction coefficient

powder. A beam passing through the objects (and thus through less powder) correspondingly has a narrower peak. The boundaries of the objects appear bright. At these places, both the pulse through the object and the pulse around the object arrived in comparable intensity but at different times, and the resulting interfering waveform was a double pulse that was interpreted by the evaluation software as a very wide pulse. In the case of the piece of plastic, this effect was sharp on the right-hand side and soft on the left-hand side just as and for the same reason as the boundary effect in (c).

Figure 11 shows a talc sample containing a piece of plastic (top) and a piece of glass (bottom). Glass appears black in the amplitude mode (a) because it absorbs THz radiation strongly. There is also much contrast in the delay-time mode due to the high refractive index of glass. Materials with a high water content (meat, for example) also produce such strong signals. Comparing compact materials with their corresponding powders, as a rule, powders have a lower refractive index because of their air content, but a higher extinction coefficient due to scattering at the surfaces. For this reason, THz radiation can image a large variety of materials with good contrast.

Special interest was directed to imaging plastic objects in powders. Figure 12 shows images of a cable clamp in wheat flour. (a) is a sketch, while (b), (c) and (d) are THz images in amplitude mode, delay-time mode or pulse-width mode, respectively. The contrast in (b) and (c) is high where the THz beam crosses much plastic material. This is particularly true of the hook (right part) of the clamp that is made of solid material. The body of the clamp (left part) is hollow, so that strong contrast is produced only by its walls at the sides and the bottom. The clamp body is open to the top. In the bottom, there is a hole for a nail. It is also clearly identified, although

Sketch   0   Amplitude   max.

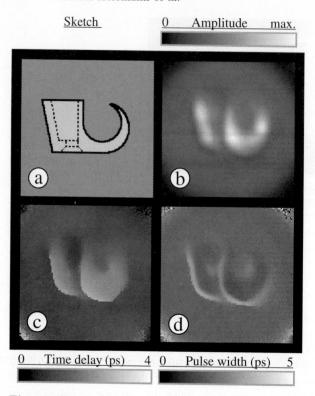

0   Time delay (ps)   4 0   Pulse width (ps)   5

Fig. 12. Sketch and THz images of a cable clamp in wheat flour. Such clamps are used for fixing cables to walls. The cable would be embraced by the hook on the right-hand part of the structure, the wall would be on top of it and the clamp fixed to the wall with a nail inserted from the bottom through a hole in the *left side*. (**a**) Sketch; (**b**) Amplitude mode; (**c**) Delay-time mode; (**d**) Pulse-width mode

with a diameter of 1.5 mm it is smaller than the optical resolution for most of the radiation transmitted through the powder. The bottom of the clamp appears thicker than it actually is, the likely reason being that the clamp was somewhat rotated with respect to the sketch (a), particularly around an axis horizontally aligned in the plane of the paper. This may have happened during the process of shaking the sample in order to distribute the powder in the box. A dark area in and on top of the body of the clamp in the delay-time mode (c) indicates that the body was probably not densely filled with flour. This is confirmed in the pulse-width mode (d), where the same region displays reduced pulse width.

3.3 Identifying Materials

The spectroscopic identification of materials can be simple in the case of gases that have specific absorption lines due to molecular rotation in the THz range, for example, or due to molecular vibrations typically at higher energies. Solid materials often lack such sharp features. They can be characterized by their refractive indices and extinction coefficients, which can be calculated from the intensity loss and delay time of the transmitted beam if the thickness of the sample is known. However, when detecting objects in a practical application, the sample thickness may often be unknown. A thickness-independent material parameter is desirable in that case, and such a material parameter can be derived from a THz transmission measurement.

Let us consider an object in a surrounding medium that may be, for example, air or a powder. The difference Δt in delay times between a pulse through the object (index x) as against a pulse that does not cross the object (index 0) is

$$\Delta t = t_x - t_0 = \Delta n d / c \,, \tag{1}$$

where $\Delta n = n_x - n_0$ is the difference in refractive indices of the object and the powder, d is the thickness of the object, and c the vacuum velocity of light. The amplitude reduces as

$$A_x / A_0 = \exp\left(-\Delta k \omega_c d / c\right) \,, \tag{2a}$$

where Δk is the difference in extinction coefficients of the object and the powder, and ω_c as a matter of convenience can be taken as the center angular frequency of the transmitted radiation, though here it serves only as a scaling constant. (Let us keep in mind that the refractive index or extinction coefficient of a powder of a material can be very different from the refractive index or extinction coefficient of the nongranular material.) Equation (2a) neglects the losses from reflection at the surface of the object, but this loss is $(n_x - n_0)^2 / (n + n_0)^2$ and may often be neglected. For example, typically $n_x = 1.6$ for various kinds of plastic, and $n_0 = 1.3$ to 1.4 for powders such as flour, sugar and powdered sugar (depending also on the density of the powder). The resulting error in amplitude here is less than 1%. For surrounding air the error reaches 5%.

Equation (2a) can be rewritten as

$$\ln\left(A_0 / A_x\right) = \Delta k \omega_c d / c, \tag{2b}$$

so that we arrive at two equations linear in d, and by dividing (2b) by (1) we obtain a material parameter $Q \equiv \frac{\ln(A_0/A_x)}{\omega_c \Delta t} = \frac{\Delta k}{\Delta n}$ that does not depend on the thickness. Multiplying (1) and (2b) yields a thickness parameter $P^2 \equiv \omega_c \Delta t \ln(A_0/A_x) = \Delta k \Delta n (\omega_c/c)^2 d^2$ that depends strongly on the thickness and less strongly on the materials.

Figure 13 shows the result of this evaluation for a sample with three plastic objects in wheat flour. The bottom object and the top-right object are small plates of polyoxymethylene (POM). The bottom plate is 2.8 mm thick, both top plates are 1.0 mm thick. The top-left plate is made of acrylnitrile-butadiene-styrole-copolymer (ABS). In (a) and (b) the amplitudes and delay times of the transmitted waveforms are displayed. The thick POM plate causes the strongest contrast in both images, but the two thin plates also differ in contrast. In (c) P^2 is displayed on a brightness scale, while at the same time Q is translated into a color scale ranging from red to green. The bottom plate still appears brightest according to its thickness. Both POM plates appear green, and the ABS plate red.

While the formulas presented above are simple, a more thorough calculation soon gets complicated because the extinction coefficients of the powders depend strongly on the wavelength, so this dependence must then be taken into account. Due to this, and other reasons, it is difficult to find a thickness-independent parameter in a stricter approximation.

3.4 Humidity in Powders

THz radiation is strongly absorbed by water, and water vapor also exhibits absorption lines in the THz frequency range. It is thus not surprising that the humidity of powders also strongly affects transmitted THz radiation. This can be used for measuring powder humidity with good spatial resolution, and even to study the diffusion of the water content in a powder.

As an example, in an experiment a drop of water was dropped on a sample of wheat flower in a plastic box. The box was closed, and the transmitted THz amplitude was observed along a lateral scan through the sample as a function of time for 4 h. The moistened part of the sample showed up with strong contrast due to absorption at the bottom of Fig. 14 early in the experiment, but absorption at that place weakened while the humidity spread through the sample. At the same time a slight increase of absorption could be observed at other parts of the sample, though this effect is weaker because the total sample volume is ~ 100 times the volume of the originally moistened part.

3.5 Powder-Density Relaxation and Local Variations of Powder Density

The refractive index n of a powder in air can be considered a weighted average of the refractive index n_0 of the solid material the powder was made of, and the refractive index of air (which is approximately 1). The weight is determined by the filling factor, i.e. by the volume fraction α of the respective material:

$$n(\alpha) = \alpha n_0 + (1 - \alpha) 1$$
$$= 1 + \alpha (n_0 - 1) . \tag{3}$$

This simple approach is not precise as a result of interference between multiply scattered waves [30], but works well enough for the purposes of this

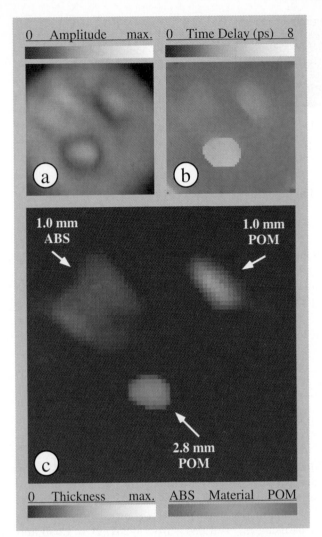

Fig. 13. Three objects in wheat powder: A 1.0 mm thick plate of ABS (*top left*), a 1.0 mm thick plate of POM (*top right*) and a 2.8 mm thick plate of POM (*bottom*). (a) and (b) show the amplitude and the delay time of the transmitted waveforms, respectively. In (c) the material parameter Q is translated into a color scale. POM appears green while ABS appears red. The brightness in this image corresponds to the thickness of the objects

section. It allows us to access the volume fraction, or density, of a powder by probing the refractive index with THz radiation.

The density or filling factor $\alpha(z)$ of a powder can be changed by shaking it z times. This process is called density relaxation and is well known, but an

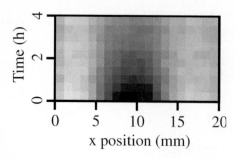

Fig. 14. Temporal development of a section of a powder sample that contains a water drop. The image displays THz amplitude, darkness corresponds to water absorption. The drying of the drop is observed very clearly, while at the same time the surrounding powder gets slightly more humid

underlying theory is still in its infancy. According to a heuristic approach [27],

$$\alpha\left(z\right) = \alpha_{\infty} - \frac{\Delta\alpha}{1 + B\ln\left(1 + \frac{z}{\zeta}\right)}, \tag{4}$$

where α_{∞} is the maximum achievable filling factor, $\alpha_{\infty} - \Delta\alpha$ is the initial filling factor, and B and ζ are constants that can depend on the way in which the powder is shaken.

When (4) is inserted into (3), one parameter can be dropped, arriving at

$$n\left(z\right) = n_{\infty} - \frac{\Delta n}{1 + B\ln\left(1 + \frac{z}{\zeta}\right)} \tag{5}$$

for the evolution of the refractive index after shaking z times. Here n_{∞} is the maximum achievable refractive index and Δn is the amount by which n can be compressed from its initial value.

A corresponding experiment was carried out after filling a plastic box with wheat flour and knocking it z times on a table from a height of 3 cm. The sample was then inserted into the THz chamber. The refractive index, calculated from the delay time of the main pulse, is shown in Fig. 15. The dashed line represents a fit using $n_{\infty} = 1.684$, $\Delta n = 0.257$, $B = 1.690$ and $\zeta = 10.15$.

It was mentioned in Sect. 3.2 that powders are usually more dense near the bottom than near the top of the samples. The reason for this is that during density relaxation the powder weight helps compress the subjacent material [27]. The additional weight influences the parameters B and ζ of (4). In THz imaging this effect manifests itself in a depth-dependent THz amplitude and pulse arrival time, and can be investigated nondestructively with a local resolution of approximately 2 mm.

For observing this effect quantitatively, a 6 mm thick wheat flour sample was prepared as above, and vertical scans were recorded with a THz imaging setup. The pulse arrival time and corresponding refractive index depending on the vertical position below the top surface is shown in Fig. 16 in various

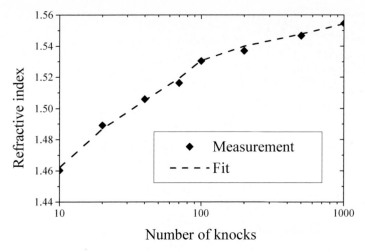

Fig. 15. Density relaxation of a wheat-flour sample observed by measuring the refractive index after a number of consecutive knocks. The *dashed line* is a fit (see text)

stages of relaxation (after 10 through 1000 knocks, see inset). For 10 knocks, the delay-time difference between top and bottom (18 mm vertical distance) amounts to 1.1 ps, while a temporal resolution better than 10 fs is achievable with the THz imaging setup. The delay time corresponds to a difference in refractive indices of 0.055. Figure 16 clearly shows that the refractive index near the top surface increases more strongly, and at greater depth increases less strongly. This could be expected because at great depths the powder density can be expected to approach the constant value of densely packed powder grains. With increased treatment the profiles become increasingly flat.

3.6 Powders in Envelopes

Package inspection for the purpose of quality control or baggage checks at airports were among the first applications proposed for THz imaging [1, 2]. With this in mind, it did not come as a surprise that at the BISAT conference [31] in Leeds at the end of November 2001, just 2 months after the deadly anthrax powder attacks in the U.S., several research groups independently demonstrated the potential of detecting powders in mail envelopes with THz radiation.

The example shown in Fig. 17 is one of the early demonstrations. No attempt was made to determine the material, but the experiment proves that very small amounts of a powder can be detected. The image is based on the delay-time of the transmitted THz beam and shows half a milligram of wheat powder with a volume of less than a cubic millimeter. In the case

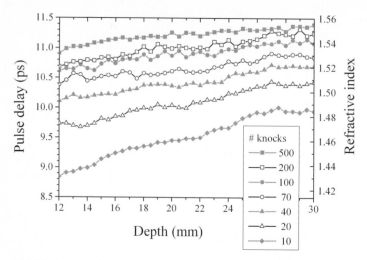

Fig. 16. Relaxation of the density profiles of wheat flour

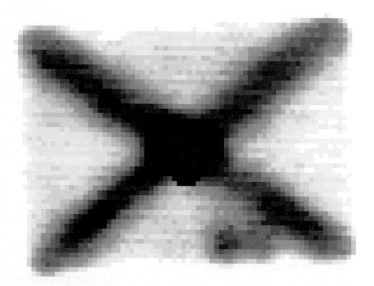

Fig. 17. THz transmission image (delay-time mode) of a small envelope (15 mm² × 11 mm²). The black "X" arises from the paper overlap. The small dark structure at the bottom is caused by a small amount of wheat flour, only 500 µg

of chemically less complex powders spectral fingerprints can be taken for detecting illicit drugs or explosives [5].

4 Characterization of Si Wafers with THz Imaging

Silicon is undoubtedly one of the technologically most relevant materials to-day. It is also one of the most cleanly produced materials, impurities can be controlled down to the level of 1 in 1 billion atoms. Consequently, the characterization of silicon is an essential part of the silicon production process securing the quality of a delicate product. With the plasma frequency of silicon in the THz range, THz radiation is efficient for optically determining the electrical properties of silicon and other semiconductors [18, 19, 32, 33, 34, 35, 36, 37] and THz imaging may, among other prospective applications, serve as a convenient tool for the fast, nondestructive, noncontact and spatially resolved characterization of silicon and other semiconductors.

The use of THz imaging for the characterization of semiconductor layers was first demonstrated by *Mittleman* et al. [2, 38]. They used a specialized THz imaging setup with a magnetic field and two detectors, making use of the rotation of the polarization plane of THz radiation due to the "terahertz Hall effect" in order to determine the carrier concentration and Hall mobility of a thin n-doped GaAs layer. This technique can determine the carrier concentration and Hall mobility of a thin layer to a good accuracy, but it needs a high layer mobility and can be expected to have problems with Si, where the mobility is roughly a factor 10 lower than in n-GaAs and can drop by another factor 10 for strongly doped layers.

In this Chapter, we present the spatially resolved characterization of semiconductors, particularly Si, with a standard THz imaging setup both for homogeneous wafers as well as thin layers. An advantage over other electrical and optical methods based on the Hall effect is that the technique described here directly determines the drift mobility, which is more relevant in electronic transport than the Hall mobility. Since homogeneous Si wafers have already been extensively studied with THz radiation [18, 32, 33, 35, 36, 37] though not yet spatially resolved, we will present only one example of the characterization of bulk Si with our imaging setup and then focus on the characterization of thin Si layers.

4.1 THz Radiation and Plasmons in Semiconductors

In this section we will develop the theoretical tools for connecting the electrical properties of a semiconductor sample via the frequency-dependent optical refractive index to the amplitude and phase shift of a transmitted THz pulse. This will happen in two steps. In the first step we will review plasma oscillations in a semiconductor. Since this issue is found in many textbooks on general and solid-state physics, we will do it in a very short and concise manner mainly assembling the background for step 2. In the second step we will address the transmission of EM radiation through dielectric samples with respect to particular configurations relevant to the characterization of semiconductors with THz radiation.

To start from scratch, in the Drude model the equation of motion for a free electron in an electric field E is

$$m^*\ddot{s} + \gamma\dot{s} = -eE, \tag{6}$$

where m^* is the effective mass of a charge carrier, s its deflection, e the electric charge, and γ is introduced as an arbitrary damping constant, but we realize from the stationary case ($\ddot{s} = 0$) that $\gamma = e/\mu = m^*/\tau$, where μ is the drift mobility and τ the momentum relaxation time.

The electric field is related to the polarization as

$$-Nes = P = \varepsilon_0 \left(\varepsilon - \varepsilon_\infty\right) E, \tag{7}$$

where N is the charge carrier concentration, ε_0 is the vacuum permittivity, ε (or ε_∞) the dielectric function (at "infinite" frequency, respectively), resulting in

$$m^*\ddot{s} + \frac{m^*}{\tau}\dot{s} = \frac{Ne^2}{\varepsilon_0 \left(\varepsilon - \varepsilon_\infty\right)}s. \tag{8}$$

For oscillations depending on time as $e^{-i\omega t}$ (in agreement with the time-dependent Schrödinger equation), we can switch over to the Fourier transform and rearrange (8), arriving at the well-known formula describing the frequency dependence of plasma oscillations

$$\varepsilon\left(\omega\right) = \varepsilon_\infty - \frac{\varepsilon_\infty\omega_{\mathrm{p}}^2}{\omega^2 - i\omega\tau^{-1}} = \tilde{n}^2 = \left(n + ik\right)^2, \tag{9}$$

where ω_{p} is the plasma frequency with $\omega_p^2 = Ne^2/m^*\varepsilon_0\varepsilon_\infty$ and ω is the frequency of the incoming radiation. In this first step the two most important quantities for semiconductor characterization, N and μ, have been related to the complex refractive index $\tilde{n}$ and its components n and k.

We want to emphasize at this point that this method does not determine the Hall mobility, which is accessed in various other methods, but rather the drift mobility that is more important for characterizing the charge carrier transport porperties of semiconductors [39]. The method has worked well with our samples, but special cases may need refinements, for example, with respect to the effective mass [32] (which can increase at very high doping concentrations [40]) and high-frequency contributions to the mobility [39].

In the following, we want to relate the refractive index to the transmission spectrum of a THz pulse transmitted through a semiconductor sample. There are two different approaches to calculating the transmission of EM radiation through a sample. In an approach typically found in textbooks of quantum mechanics [41], combinations of plane waves $A_n e^{i(\pm qx - \omega t)}$ (with the wave vector q and the dispersion relation $q = \frac{\omega}{c}$) are sent through a potential barrier, the coefficients A_n are determined using the boundary conditions of continuity of the wave function and its derivative. This derivation is straightforward and simple, but the equations are lengthy. The result automatically includes multiple reflections in the sample.

In another approach often preferred in textbooks of optics [42] continuity of the electric and magnetic fields are considered at the interface between two media. The results from two interfaces are then combined to form the total transmittance through a sample. This approach needs more reasoning, but is mathematically undemanding as long as reflections within the sample are neglected. Multiple reflections can be added, but then the derivation becomes intricate.

The results of these two approaches are, of course, the same. For a sample consisting of one homogeneous layer of thickness d,

$$\tilde{t}_{\text{one}}(\omega) = \frac{\tilde{E}_{\text{sample}}}{\tilde{E}_{\text{ref}}} = \frac{4\tilde{n}}{(\tilde{n}+1)^2} \frac{e^{-i(\tilde{n}-1)d\omega/c}}{1 - \frac{(\tilde{n}-1)^2}{(\tilde{n}+1)^2}e^{-2i\tilde{n}d\omega/c}} \tag{10}$$

for the amplitude relation between a wave transmitting the sample and a reference wave without the sample present. The first term on the right-hand side of (10) is an interface term describing transmission losses from reflections at the sample surfaces. The numerator of the second term is a bulk term. It describes the time delay and attenuation in the bulk of the sample. The denominator takes account of multiple reflections bouncing back and forth within the sample. It can be set to one or replaced by a finite progression if the measurement is not sensitive to multiple reflections or only to a limited number of them.

For samples that have a strongly doped layer implanted on the substrate, a double-layer model must be used. It can be calculated in the same ways as outlined above, resulting in

$$\tilde{t}_{\text{two}}(\omega) = \frac{8\tilde{n}_s\tilde{n}_l}{(1+\tilde{n}_s)(\tilde{n}_s+\tilde{n}_l)(\tilde{n}_l+1)} \frac{e^{i(\tilde{n}_s-1)d_s\omega/c} e^{i(\tilde{n}_l-1)d_l\omega/c}}{1 - \alpha_s - \alpha_l - \alpha_{sl}} \tag{11}$$

with

$$\alpha_s = \frac{(\tilde{n}_s-1)(\tilde{n}_s-\tilde{n}_l)}{(\tilde{n}_s+1)(\tilde{n}_s+\tilde{n}_l)}e^{2i\tilde{n}_s d_s\omega/c}$$

$$\alpha_l = \frac{(\tilde{n}_l-\tilde{n}_s)(\tilde{n}_l-1)}{(\tilde{n}_l+\tilde{n}_s)(\tilde{n}_l+1)}e^{2i\tilde{n}_l d_l\omega/c}$$

$$\alpha_{sl} = \frac{(\tilde{n}_s-1)(\tilde{n}_l-1)}{(\tilde{n}_s+1)(\tilde{n}_l+1)}e^{2i(\tilde{n}_s d_s+\tilde{n}_l d_l)\omega/c},$$

where $\tilde{n}_s$, $\tilde{n}_l$, d_s and d_l are the complex refractive index and thickness of the substrate (index s) or the implanted layer (index l), respectively.

Equation (11) may be used for explaining THz spectra through semiconductor samples with implanted layers by fitting four to six parameters

N_s, μ_s, N_l, μ_l, and potentially d_s and d_l. Fitting spectra with many parameters is not particularly costly any longer in terms of computational time, but the resulting precision may suffer. Often, the properties of the substrate will be known or can be measured separately and then a substrate spectrum should serve as a reference for determining the layer properties. A spectrum through the layer relative to the substrate spectrum is calculated by dividing (11) over (10) where the doped layer is in, not on, the substrate, which means that the thickness of the reference must be chosen as $d = d_s + d_l$. This approach becomes particularly simple in pulsed THz spectroscopy when multiple reflections through the substrate can be neglected because the reflected pulses, arriving much later than the main pulse, are cut off. In this case

$$\tilde{t}_{\text{layer}}(\omega) = \frac{\tilde{t}_{\text{two}}}{\tilde{t}_{\text{one}}} = \frac{2\tilde{n}_l(\tilde{n}_s + 1)}{(\tilde{n}_l + \tilde{n}_s)(\tilde{n}_l + 1)} \frac{e^{i(\tilde{n}_l - \tilde{n}_s)d_l\omega/c}}{1 - \frac{(\tilde{n}_l - \tilde{n}_s)(\tilde{n}_l - 1)}{(\tilde{n}_l + \tilde{n}_s)(\tilde{n}_l + 1)}e^{2i\tilde{n}_l d_l\omega/c}}. \tag{12}$$

An important point with respect to the precision of this formula is the question of the layer thickness, since it cannot be determined easily and may not always be known precisely. If the layer thickness is not important, the layer may be considered infinitely thin. Such a thin-layer model has the additional advantage that it results in much simpler equations from which it is easier to assess the influence of certain parameters such as, for example, the substrate refractive index. Thus it is important to develop a thin-layer model and to determine its range of validity.

For deriving a thin-layer model, care must be taken to fix the side conditions properly. While the thickness $d_l \to 0$, we need the total number of charge carriers to be constant. This means that the bulk carrier concentration N becomes infinite and must be replaced by a finite sheet carrier concentration $N_{2D} = Nd_l$. Since $N \to \infty$, the plasma frequency and the refractive index also become infinite, a closer look at (9) showing us that $\tilde{n}_l^2 \propto d_l^{-1}$. A thin-layer model can now be developed either from scratch using the above-mentioned theoretical approach on a delta-function potential, or by inserting our findings into (12). In this latter case, we find that the e-function in the numerator can be reduced to 1, while in the denominator e^x has to be worked out to $1 + x$ because the 0-th order in the denominator cancels out. The result is

$$\tilde{t}_{\text{thinlayer}}(\omega) = \frac{1}{1 + \frac{\tilde{\alpha}(\omega)}{\tilde{n}_s + 1}}, \tag{13}$$

with $\tilde{\alpha}(\omega) = -i\tilde{n}_l^2 d_l\omega/c$. Here, $\tilde{n}_l^2$ can be inserted from (9) with the constant addend ε_∞ neglected since N is large, resulting in

$$\tilde{\alpha}(\omega) = \frac{i\varepsilon_\infty \omega_p^2 d_l\omega/c}{\omega^2 + i\omega\tau^{-1}} = \frac{N_{2D}e^2\tau}{\varepsilon_0 m^* c} \frac{i}{\omega\tau + i} = \frac{N_{2D}\mu e}{\varepsilon_0 c} \frac{1}{1 - i\omega\tau}. \tag{14}$$

The relaxation time τ in a strongly doped layer is typically low, for silicon 10 fs to 20 fs. For not too high frequencies $f = \omega/2\pi \ll 10$ THz the absolute value of the transmitted amplitude is frequency independent to within 1% between 200 GHz and 1 THz. This means that we can derive the absolute value of the amplitude transmittance through a thin layer simply by ignoring the imaginary part

$$t_{\text{thinlayer,abs}} = \frac{1}{1 + \frac{N_{2D}\mu e}{\varepsilon_0 c (n_s + 1)}} \tag{15}$$

with $\tilde{n}_s = n_s$ assuming that the substrate is transparent. Since the imaginary part is small, the delay time is also small (in comparison with the period of a THz wave, 1 ps). Here, also assuming $\omega\tau \ll 1$, it works out to

$$\Delta t = \frac{1}{\omega} \arctan\left(\frac{\Im(t)}{\Re(t)}\right) = -\frac{\tau}{1 + \frac{\varepsilon_0 c (n_s + 1)}{N_{2D}\mu e}}, \tag{16}$$

which, for high doping, reduces further to $\Delta t = -\tau$. Unfortunately, though this result is small, the contributions from its components (surface, bulk, reflections) are not small, as is seen from (12). Consequently, the thin-layer approximation of the delay time reacts strongly to high-order contributions, and though it can be helpful in some special cases, it is not generally useful. In the case of our samples (see next section), (16) reflects the dependence on the doping level correctly and agrees with the measured delay times within approximately 5 fs; and while this is an excellent agreement as a time difference in a THz experiment, the relative error is still large because the observed delay times are also < 10 fs. It may appear surprising at first sight that (15) is precise while (16) is not, but it reflects the simple physical fact that, when adding up waveforms that are shifted against each other on the time-axis, the position of the resulting waveform depends strongly on the positions of the components, while its amplitude may be calculated by adding up the amplitudes of the components without too much consideration to their positions.

Since (15) does not depend on frequency, the pulse shape is conserved, and we can determine the transmission coefficient just by measuring the pulse amplitude at one position of the time-delay line, opening up the way to speeding up the measurements by a factor 10 to 100. The downside of this is that (15) does not depend on N_{2D} and μ independently any longer but rather on the product $N_{2D} \cdot \mu$. Sheet carrier concentration and mobility cannot both be determined in this way. Additional information should come from the delay time, but as the problems around (16) show, this requires both very precise measurement and also modeling. Alternatively, one may determine τ, and thus μ, independently from N_{2D}, by extending the frequency beyond 3 THz where the simplified model of (15) does not hold any longer. Or finally, one may be able to make some assumptions about the quality of the sample, and assume a relation between N and μ as tabulated by *Jacoboni* et al. [43], *Sze* [44], *Zeghbroeck* [45] or others. In order to use these relations for determining μ, one needs the bulk carrier concentration N that has been assumed infinite within the thin-layer model. However, since μ depends only weakly on N, it may often be sufficient to determine N from N_{2D} by reasonably guessing the layer thickness.

4.2 Applications to Si Wafers

The experimental verification of this method of characterizing semiconductors has been carried out with various n-type and p-type specimens cut from $380\,\mu m$ thick Si wafers. Three of the n-type samples with $4.5\,\Omega\cdot cm$ resistivity were doped on half of one surface by implanting B ions at a dosage of $(5 \times 10^{13})\,cm^{-2}$, $(5 \times 10^{14})\,cm^{-2}$ or $(5 \times 10^{15})\,cm^{-2}$, respectively, and subsequent annealing, resulting in strongly doped p-type layers. The layer thickness was determined for one of the samples with a spreading-resistance profile measurement as $0.55\,\mu m$. The THz imaging measurements were carried out in air.

Figure 18a shows a THz image of the sample with $(5 \times 10^{15})\,cm^{-2}$, implanted ions, where the maximum THz signal of the pulse at any sample position has been transformed into a brightness level. The sample covers the left-hand 80% of the image area, while the right-hand side has been left free, and the THz beam here serves as reference for calculating the pulse delay time.

The left-hand part of the image shows two regions, the lower of which corresponds to the unimplanted part of the sample where the THz amplitude is roughly 50% of the reference amplitude. The upper, dark region displays low transmittance through the strongly doped layer. In Fig. 18b, the THz waveforms in the three regions are plotted. The graph displays the different amplitudes and confirms the time delay caused by the substrate. It amounts to 2.99 ps that corresponds to a bulk refractive index of 3.36, almost 2% lower than the value of 3.418 for undoped Si [33] and thus in excellent agreement with the value of 3.362 according to Drude theory for a $4.5\,\Omega\cdot cm$ sample. Figure 18c shows the maximum signal of the THz waveform depending on the sample position along the line a–b in Fig. 18a. The transition from a high- to a low-intensity level is clearly observed. However, there are an intensity maximum and a minimum close to the border between the two regions. Another strong minimum can be observed at the edge of the sample in Fig. 18a. This effect is most likely caused by refraction and interference. With THz waves just as with visible light, structures with sharp boundaries can cause intensity patterns [25], although in experiments with pulsed radiation the patterns do not extend far away from the boundary.

The substrates and the implanted layers were characterized in two separate steps. For investigating the substrates, we Fourier transformed the measured waveforms and calculated the relative transmission coefficient using an undoped sample as reference. The latter is a better reference than no sample at all because inserting a "thick" sample at the focus of a THz imaging setup changes the beam alignment at the position of the next focus, the detector

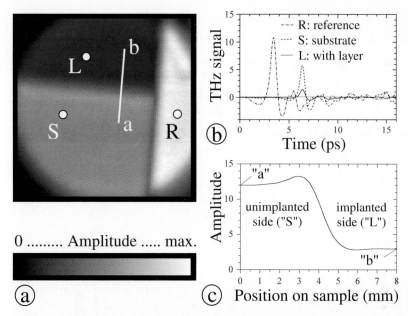

Fig. 18. (a) THz image of a Si sample with a $(5 \times 10^{15})\mathrm{cm}^{-2}$ implanted layer: Reference region (R, no sample), substrate region (S, not implanted) and layer region (L, implanted and strongly absorbing). The image size is 20 mm × 20 mm. The dark areas in the edges are shadows from the sample holder. (b) THz time-domain waveforms taken in the regions R, S and L. (c) Pulse maxima along the line a–b of (a) (from a separate measurement). There are refraction effects resulting in a maximum and a minimum at the border between the implanted and unimplanted regions

antenna, in such a way that it can significantly influence the received spectra. This problem is avoided by using another sample, preferably an undoped one, with the same thickness as the sample under test.

The spectra obtained in this way were then compared to spectra calculated according to (10), where $\tilde{n}$ was calculated from (9) using N and μ as fit parameters. We used a best square fit for the region from 0.20 THz to 0.95 THz excluding the vicinity of the two strong water-vapor absorption lines at 557 GHz and 752 GHz. A corresponding spectrum and its best fit are shown in Fig. 19. Our results for carrier concentration and mobility agree typically within 30% with the data supplied by the manufacturers. *Van Exter* et al. [32] report similar agreements between their measurements and manufacturer data.

From our own measurements and from the literature [18, 32] it can be concluded that the determination of N and μ in Si substrates (wafers with a typical thickness of 0.3 mm to 0.4 mm) is feasible for p- and n-type Si within $N = (4 \times 10^{14})\mathrm{cm}^{-3}$ to $(2 \times 10^{16})\mathrm{cm}^{-3}$ with an estimated accuracy of 5%

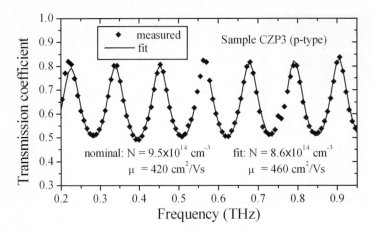

Fig. 19. Transmission spectrum of a *p*-type Si sample, and its best fit

Fig. 20. THz images of three *n*-type Si samples the lower parts of which have been ion-implanted with B. Implantation dosages vary from left to right: $(5 \times 10^{13})\,cm^{-2}$, $(5 \times 10^{14})\,cm^{-2}$ and $(5 \times 10^{15})\,cm^{-2}$. The image size is 8 mm × 8 mm each

in ω_p and τ [18, 37]. Samples considerably outside this range are too transparent or too opaque to be measured in the same way unless the frequency scale is extended considerably. In many cases, however, this problem may be overcome by choosing a convenient sample thickness.

The result of the quantitative thin-layer evaluation is shown in Fig. 20 for the three samples with $4.5\,\Omega \cdot cm$ *n*-doped substrates and *p*-doped implanted layers on half of a surface each. The color in this image corresponds to the carrier concentration in the implanted layer. The fit was based on (15) with the relation between N and μ as tabulated by *Jacoboni et al.* [43]. The sheet carrier concentrations determined with this technique amount to $5.1 \times 10^{13}\,\Omega \cdot cm^{-2}$ ($5.0 \times 10^{14}\,\Omega \cdot cm^{-2}$, $4.7 \times 10^{15}\,\Omega \cdot cm^{-2}$) in very good agreement with the known implantation dosages of $5 \times 10^{13}\,\Omega \cdot cm^{-2}$ ($5 \times 10^{14}\,\Omega \cdot cm^{-2}$, $5 \times 10^{15}\,\Omega \cdot cm^{-2}$).

The weakly doped sample can be considered at the limit of what can be measured without additional averaging, an error of 1% in the amplitude measurement here results in 15% error in the product $N\mu$. This error goes down to 1% for higher doping rates. Although transmittance drops when the doping increases, this technique should be suitable for layers even with the highest doping rates possible in Si.

The spatial resolution of this technique is currently determined by the diameter of the focus that can be adjusted to ≈ 1 mm, but it may be greatly improved using near-field techniques [46, 47].

5 Imaging of Supercurrent Distributions

A supercurrent distribution in a YBCO thin film can be visualized by observing THz radiation from the film irradiated with fs optical pulses. An image is obtained by scanning the film and detecting the emitted THz radiation, since the radiation amplitude is proportional to the supercurrent density in the irradiated area. Furthermore, the distributions of the supercurrent flow direction in the film can be monitored by detecting the polarization of the emitted THz radiation using polarizers.

5.1 Supercurrent Distribution in a High-T_c Bridge

When a supercurrent in an $YBa_2Cu_3O_{7-\delta}$ (YBCO) film is irradiated with fs optical pulses, the supercurrent is modulated very rapidly and emits THz radiation into the free space [48, 49, 50, 51, 52, 53, 54] (see Chapter by *Kida* et al.). Since the radiation amplitude is proportional to the local supercurrent density at the optically excited area, the supercurrent distribution is obtained by scanning the YBCO film while irradiating it with the focused fs optical pulses and detecting the emitted THz radiation with a PC antenna or a hot-electron bolometer [55, 56, 57, 58, 59, 60, 61, 62, 63, 64]. The irradiating laser power was chosen so as to avoid perturbation of the supercurrent distribution in the film [62]. In order to discuss the supercurrent distribution quantitatively, the local supercurrent density was estimated by comparing a supercurrent distribution with a bias current [62, 63]. The spatial resolution of the observed supercurrent density was limited by the irradiating beam diameter. The advantages of this system are discussed by comparing it with other methods to observe supercurrent distributions.

Figure 21 shows an optical setup to observe the supercurrent distribution in YBCO films. A YBCO film was cooled below the transition temperature on a cold finger in a closed-cycle helium cryostat and was exposed to a bias current or a magnetic field to carry a supercurrent. The film was irradiated with the fs optical pulses focused by a lens to emit the THz radiation, which was collimated and focused by a pair of paraboloidal mirrors onto the detector, a PC antenna or an InSb hot-electron bolometer. Although bolometers are easier to align, PC antennas, particularly bow-tie antennas, were employed in most experiments, since they have a higher sensitivity and can detect the

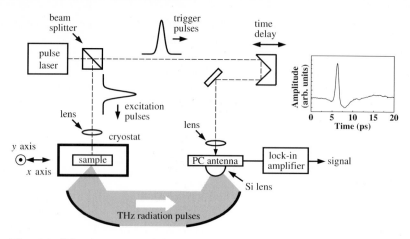

Fig. 21. Schematic view of the experimental setup that employs a PC antenna as detector. In some measurements, an InSb hot-electron bolometer is used instead of the PC antenna. The *inset* shows a typical waveform of the THz radiation that is obtained by scanning the time delay while monitoring the photocurrent induced in the PC antenna. During the two-dimensional scan of the YBCO film, the time delay was set to 6.2 ps, giving rise to maximum photocurrent

polarity of the radiation, while bolometers can detect only the radiation intensity. The THz-radiation waveforms were obtained by monitoring the photocurrent induced in the PC-antenna detector while changing the time delay. After fixing the time delay at the position giving rise to maximum photocurrent (see inset in Fig. 21), the YBCO was scanned two-dimensionally (x and y directions) and the THz radiation was detected. The scanning sample holder was connected to a cold base of the cryostat through a metal mesh and also to an exterior x–y stage controlled by a computer through a thermally insulating polyethylene rod.

The samples were made from commercially available YBCO films with a thickness of about 100 nm grown on 0.5 mm thick MgO substrates. They were patterned into a bow-tie structure as illustrated in Fig. 22a to emit THz radiation efficiently [48,49,50,62], using conventional photolithographic techniques. A hemispherical MgO lens with a diameter of 3 mm was attached to the back side of the sample substrates with vacuum grease (Fig. 22b) [48,62] to increase the collection efficiency of the THz radiation by about ten times, though it limits the observable area to about 200 μm in diameter because of the focusing effect of the lens. Thus the supercurrent distributions were observed in the central strip of the bow-tie pattern.

The intensity of the fs laser irradiation was set to a level that does not perturb the supercurrent distributions in the YBCO films. Magnetic flux penetrating a superconductor accompanies a supercurrent around the flux [65]. The intense irradiation of the fs optical pulses causes changes in the distrib-

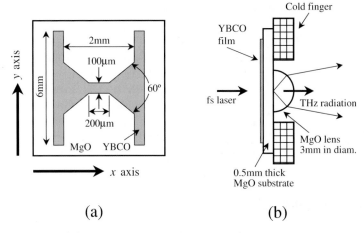

(a)   (b)

Fig. 22. (a) Schematic structure of the fabricated YBCO sample and orientation of the coordinates. **(b)** Schematic configuration of the sample mount (cross section) [62]

ution of the magnetic flux and in that of supercurrent [55, 62]. The threshold of the intensity level causing a perturbation was investigated by observing the THz amplitude with various intensities of the irradiation. It was found to be about 10 mW and the irradiation laser power was set below this threshold [62].

A transfer coefficient was estimated to calculate the local supercurrent density from the THz-radiation amplitude. The coefficient was obtained by comparing the bias current with the distribution of the THz-radiation amplitude on a YBCO strip. Note that the integration of the distribution on the full width of the YBCO strip is proportional to the bias current. Using the transfer coefficient, the supercurrent distributions were quantitatively estimated also in the flux-penetrated films. Since the radiation efficiency depends on the samples, the transfer coefficient has to be determined for each YBCO film. In addition, since the THz radiation efficiency also depends on temperature [48, 52, 53], the transfer coefficient has to be determined at each temperature when the temperature-dependent distribution is observed [62, 63, 64]. The noise level of the supercurrent images was estimated using the transfer coefficient, showing that the system can observe a minimum current density of less than $0.02 \, MA/cm^2$ using optical pulses with a power of 5 mW and a diameter of 25 μm [62].

The spatial resolution of the supercurrent distribution was limited by the laser spot size. In the experiments described below, the full width at half maximum (FWHM) of the spot was typically 25 μm to 30 μm, which was measured with a beam profiler [60, 62]. However, this could be reduced to the size of the laser wavelength of about 800 nm by focusing the beam with a large converging angle (diffraction limit) [66].

The supercurrent distributions have hitherto been obtained by measuring magnetic-field distributions with such methods as scanning Hall probe [67], scanning SQUID microscope [68] and magneto-optical (MO) sampling [69, 70, 71]. Among these, MO sampling has been the most powerful because it can provide a noncontact and a nondestructive method to obtain a supercurrent distribution up to a critical value with a high spatial resolution. However, the MO material layer must be fabricated directly on the superconductor to have a resolution as high as about $1\,\mu m$, which results in a contact measurement [72]. Compared to these methods, the THz imaging technique is advantageous because it combines noncontact and nondestructive measurements with a high spatial resolution that could be about 800 nm.

Prior to the description of the experimental results, the critical-state model and its application to thin films are briefly introduced. Usually, the critical-state model treats a superconductor with a long shape in a parallel magnetic field under quasistatic changes of applied magnetic fields and bias currents [65]. The model describes distributions of supercurrent and magnetic flux in a cross section perpendicular to the magnetic field around the central part and neglects the shielding supercurrent near the end surfaces perpendicular to the magnetic field. As the magnetic field increases, the shielding current density j at the side surface increases to prevent the flux penetration. When the shielding current density j reaches the critical value j_c, the magnetic flux starts to penetrate into the superconductor. As the flux penetrates into the inner part of the superconductor, the distribution of the shielding current density j changes. The shielding current density j is equal to j_c in the whole region of the flux penetration, while $j = 0$ in the region without the flux. However, in the case of a superconductor film perpendicular to a magnetic field, the supercurrent at the surfaces perpendicular to the magnetic field is no longer negligible. When the magnetic flux has partly penetrated the film and the supercurrent-distributed region with $j = j_c$ appears near the edges of the superconductor films, the supercurrent flows over the entire region of the film to shield the central flux-free region [73, 74].

For a demonstration of this imaging system, the supercurrent distribution was observed in a current-biased YBCO strip cooled without a magnetic field [58, 59, 60, 61, 62]. The supercurrent makes a magnetic field that is perpendicular to the film at the surface. The supercurrent distributions agree well with calculations based on the critical-state model.

Figure 24 shows an example of the supercurrent distribution, where the YBCO film was cooled to 16.5 K and current-biased at 30 mA [62]. Figure 24 shows the variation of the supercurrent distribution across the center strip of the YBCO sample with increasing and then decreasing bias current between 0 and 200 mA at 16 K [60]. When the bias current was supplied, the supercurrent distribution had peaks near the edges of the strip as mentioned above. As the bias current increased, they became larger and progressed to the inner part of the strip. As the bias current decreased, the decrease in the super-

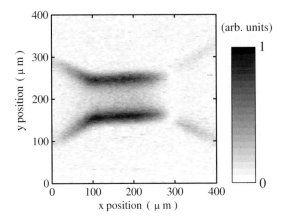

y position (μ m)

x position (μ m)

(arb. units)

Fig. 23. Two-dimensional plot of the THz-radiation amplitude near the center of the bow-tie antenna pattern of the YBCO film [62]

current density started near the edge of the strip. Finally, the bias current was set at 0 mA but in the inner part the supercurrent remained flowing in the same direction as the bias current. On the other hand, near the edges the supercurrent was observed to flow in the opposite direction, resulting in the total supercurrent of 0 mA. This behavior is explained by the pinning effect of the magnetic flux in superconductors. Once the magnetic flux was induced by the bias current and penetrated into the strip, the flux was trapped by the pinning effect and accompanied by a shielding supercurrent. The magnetic flux and the supercurrent left in the strip caused an irreversible change of the supercurrent distribution as shown in Fig. 24.

The observed results in Fig. 24 were compared with a theoretical calculation based on the critical-state model. Since the critical current of this sample was 220 mA, the critical current density was obtained as $j_c = 2.2 \times 10^6 \, \text{A/cm}^2$ and the supercurrent distributions were calculated as shown in Figs. 25a and b. They were convoluted with a Gaussian function with a FWHM of 25 μm, the diameter of the irradiating laser spot. The results are shown in Figs. 25c and d, which show good agreement with the observed distributions in Fig. 24.

5.2 Vector Imaging of a Supercurrent Flow in a High-T_c Thin Film

When a PC antenna is employed as a detector, the measured supercurrent distributions contain information on directions of the supercurrent flow. If the distributions are obtained for two orthogonal directions, the supercurrent can be reconstructed with its direction (i.e., a vector image). For this purpose, a sophisticated method utilizing wire-grid polarizers has been proposed [75].

When a PC antenna is employed in the supercurrent imaging system, the obtained image shows the distribution of the supercurrent component projected onto the sensitive direction of the PC antenna. For example, the

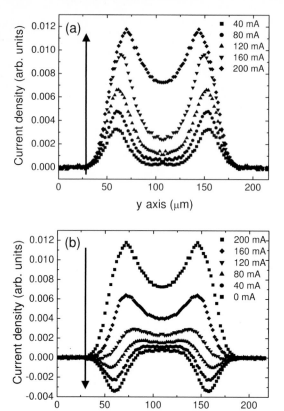

Fig. 24. Cross-sectional distribution of the supercurrent in a current-biased YBCO thin film strip at 16 K. The change of the supercurrent distribution with increasing the bias current from 0 mA to 200 mA is shown in (a) and that with decreasing the bias current down to 0 mA is shown in (b) [60]

sensitive direction of a PC bow-tie antenna is parallel to the bow-tie axis (Fig. 26). If the radiation is incident on the PC antenna with its electric field E at an angle of θ to the sensitive direction, the induced signal is proportional to $\cos\theta$. Since the YBCO film emits THz radiation polarized along the supercurrent flow, the detection with the PC antenna produces images of the supercurrent component along the bow-tie axis. Detecting the supercurrent flow along two orthogonal directions at various places of the film, we obtain a distribution of supercurrent flow vectors.

Two orthogonal directions of sensitivity can be achieved with any of the three methods depicted in Fig. 27. First, they can be achieved by rotating a PC antenna itself (Fig. 27a). However, this method requires the realignment of the optical gate pulses after the rotation. Since a PC gap of the PC antennas is typically several μm, it is very difficult to keep the focusing condition of the optical pulses completely equivalent to that before the rotation. Second, it can be achieved by using a cross-shaped PC antenna (Fig. 27b). However, this method limits the design of the PC antenna. Designs with high radiation

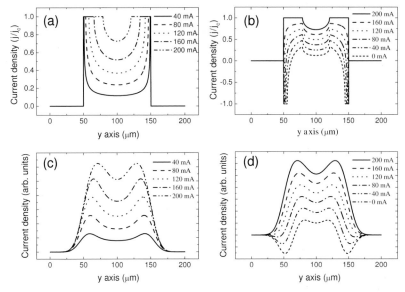

Fig. 25. The change of the transport-current distribution calculated on the basis of the critical-state model for the thin film with (**a**) increasing and (**b**) decreasing bias current. The figures (**c**) and (**d**) show calculated results by taking the convolution between the distribution of (**a**) and (**b**) and the laser-spot profile [60]

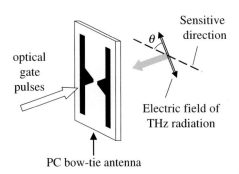

Fig. 26. The sensitive direction of the PC bow-tie antenna is parallel to the bow-tie axis. If radiation is incident with its electric field at an angle of θ to the sensitive direction, the induced signal is proportional to $\cos\theta$

efficiency such as a bow-tie cannot be used. Third, it can be achieved by using linear polarizers such as wire-grids (Fig. 27c). The disadvantages caused by the above methods do not occur, since the radiation path length does not change even if the polarizer rotates, and the antenna design is not restricted.

In the experiment, the configuration in Fig. 27c was employed, where a PC bow-tie antenna was set to detect radiations with horizontal electric fields. The polarizer was set to transmit radiations with electric fields inclined at ±45°, which is referred to as "wire-grid A" in Fig. 27c. Another polarizer was used to guarantee that radiations with a horizontal electric field were incident

(a)

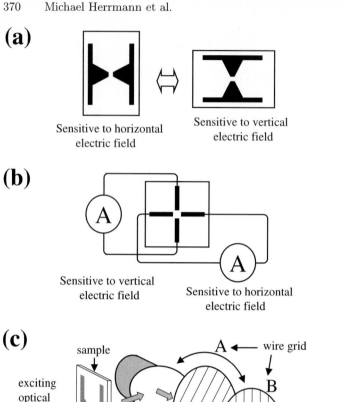

Sensitive to horizontal
electric field

Sensitive to vertical
electric field

(b)

A

Sensitive to vertical
electric field

A

Sensitive to horizontal
electric field

(c)

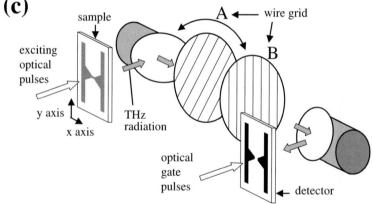

sample   A ← wire grid

exciting
optical
pulses

B

y axis   THz
radiation

x axis

optical
gate
pulses

detector

Fig. 27. Methods of preparing two sensing directions of THz radiation perpendicular to each other. (**a**) Rotation of the PC antenna, (**b**) cross-shaped antenna and (**c**) polarizers

on the PC-antenna detector, which is referred to as "wire-grid B" in Fig. 27c. The polarizers were grids of 10 μm diameter tungsten wires with 25 μm spacing. Except for the polarizers, the experimental system was the same as that shown in Fig. 21 [75]. A YBCO film was patterned into a bow-tie structure and cooled below its transition temperature without a magnetic field. Then an external magnetic field of 15 mT was applied for 5 s and removed and the magnetic flux was left penetrated in the YBCO film near the edges. The

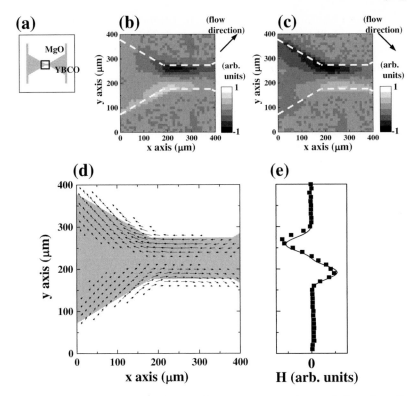

Fig. 28. Supercurrent distribution flowing in two orthogonal directions (**b,c**), and a vector map image (**d**) obtained near the center of a vortex-penetrated YBCO thin film. The square near the center of (**a**) denotes the region of (**b**), (**c**) and (**d**). The images of (**b**) and (**c**) show the distributions of the supercurrent component pointing to upper right (↗) and lower right (↘), respectively. The vector map image obtained from (**b**) and (**c**) is shown in (**d**). The observed supercurrent distribution at $x = 300\,\mu\mathrm{m}$ (■) and the calculation based on the critical-state model with $j_\mathrm{c} = 10\,\mathrm{MA/cm}^2$ (*solid line*) is shown in (**e**)

system was used to observe the supercurrent distribution accompanying the magnetic flux.

Figure 28 shows a set of the supercurrent vector images observed near the center of the YBCO bow-tie structure. Figures 28b and c show the recorded images, the distributions of the supercurrent component pointing to upper right (↗) and lower right (↘), respectively, where the broken lines denote the edges of the bow-tie pattern. In the left half of Fig. 28b, the positive current or the current pointing upright is seen along the lower edge, which is explained by the critical-state model. Since the magnetic flux near the edges was accompanied by the shielding current, the supercurrent flowed along the edges. It emitted THz radiation with polarization of $+30°$ near the lower edge, almost parallel to the sensitive direction of $+45°$ (↗) and was detected with high efficiency, resulting in the positive current. On the other hand, the distribution of the THz radiation with polarization of $-30°$ near the upper edge, almost perpendicular to the sensitive direction and was detected with low efficiency, resulting in the small current. In the left half of Fig. 28c, the large negative current or the current pointing upper left along the upper edge is explained similarly. Figure 28d shows the vector image calculated from data shown in Figs. 28b and c, where the supercurrent flows along the edges in directions parallel to the film edges. Figure 28e shows the cross section of the current distribution at $x = 300\,\mu m$. The square symbols represent the experimental data, which is well fit by the calculation based on the critical-state model (solid line), where the critical supercurrent density $j_c = 10\,MA/cm^2$ and the calculated curve has been convoluted by the Gaussian function with the width of the laser spot size.

Figure 28d shows that the supercurrent appears only around the edges and not in the center of the triangular-shaped part. However, the supercurrent distribution spreading in the whole film was observed in the case of the flux-trapped state that was prepared by applying a magnetic field and then cooling the YBCO film to the superconducting state [76]. The difference can be explained as follows. In the case of Fig. 28 the flux enters the film though the edges when the magnetic field is applied and remains only around the edges after removing the field. On the other hand, in the case of cooling in the magnetic field, the flux penetrates the film already before the superconductive transition and can remain in the center part of the film. These behaviors agree with the observation and theory reported so far [69, 70, 71, 72, 73, 74].

6 Real-Time THz Imaging

6.1 Real-Time THz Images

In this section we present several images obtained with Tochigi Nikon's real-time THz imaging system. It takes only 0.2 s to capture a single image. Figure 29 shows real-time THz images from a leaf illuminated with THz radiation. Though the leaf transmits THz radiation overall, this is not as true of the veins. THz radiation is readily absorbed by water, and this property can be utilized to investigate the water content of leaves, for example.

Figure 30 shows a THz image of a hobby knife blade hidden inside an envelope. The paper easily passes THz radiation, making it possible to check quickly for objects inside.

Using the real-time THz imaging system, we can see not only static objects but moving ones, too. A sample used as a moving object was an insectivorous plant (Venus Flytrap). Figure 31 shows THz transmission images of the insectivorous plant. Visible light images of the sample are shown in the lower-left part and the lower-right part of Fig. 31. Venus Flytrap has foldable leaves, which are usually open. In the real world, when insects walk on the leaves,

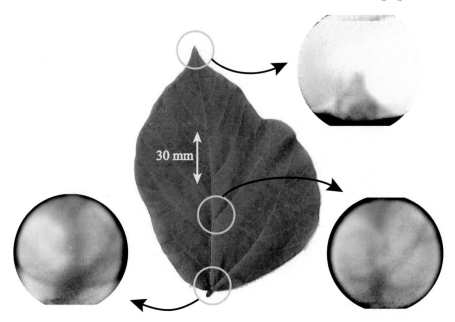

Fig. 29. THz transmission images of a leaf

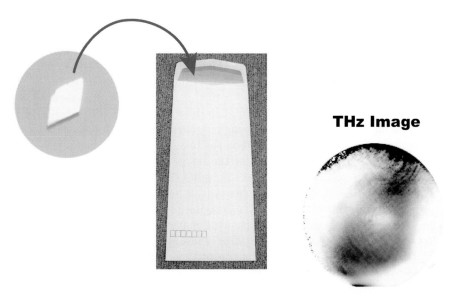

Fig. 30. THz transmission image of a hobby knife blade hidden inside an envelope

the plant catches them by closing its leaves. Previously, it was impossible to make streaming videos in the THz frequency region. It took about 0.1 s to capture each image, for a total of ten THz images per second.

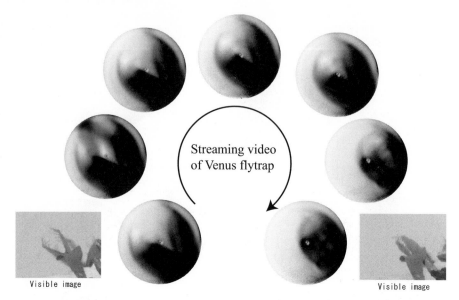

Fig. 31. THz transmission images of a moving object

6.2 Time-Domain THz Imaging

The time-domain THz imaging method, in a system as indicated in Fig. 32a, was used to demonstrate that radiation emitted from the THz emitter is propagated in a spherical wavefront. Physics textbooks claim that light is a wave, spreading out in a spherical wavefront from the source, but it is amazingly difficult to find evidence for this claim from the time domain directly. The photographs shown in Fig. 32b show the cross section of this THz wavefront as it passes through the imaging plate. The time-dependent spatial distribution of the THz electric field on the imaging plate can be observed by scanning the time-delay line. At first the image is bright only in the center (left photo), but as time passes (to the right) the white area moves toward the periphery while the center darkens. This is in agreement with the passage of a shperical wavefront through the imaging plate as depicted in Fig. 32c. The time-resolved images of the THz field distribution were captured at the extremely short interval of 0.33 ps. This series of photographs clearly shows that time-domain THz imaging can be utilized with this setup to explore ultrafast phenomena in semiconductors and other materials.

6.3 THz Spectroscopic Imaging

In addition to simply acquiring images, the use of spectroscopic imaging also makes it possible to capture which chemical compounds are present in which spatial distribution. Spectroscopic images can be obtained with a

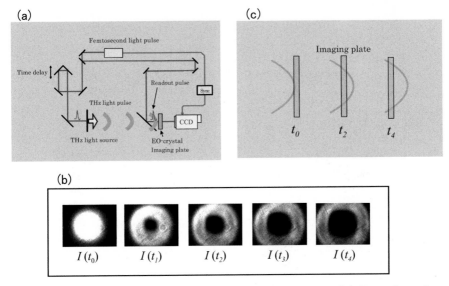

Fig. 32. (a) Schematic setup for time-domain THz imaging. (b) Time-dependent spatial distribution of the THz electric field on the imaging plate. (c) Passage of the wavefront through the imaging plate

time-domain THz imaging mode. The movement of mirrors on the delay line changes the arrival time of the readout beam at the detector in order to record the temporal change of the image in terms of the THz electric field. The spectrum is obtained through a Fourier transform of the waveform, which represents the temporal progression of the recorded electric field for each pixel. Figure 33 shows the typical THz waveform of a pixel and the spectral distribution obtained through a Fourier transform. As can be seen from this figure, the THz waveform looks like a half-cycle of the EM wave and the THz spectrum extends from 0.05 THz to 1 THz. This process is performed for every pixel, and the spectroscopic image is created by isolating a specified frequency component. Figure 34 demonstrates the THz spectroscopic images for the test pattern shown in the upper part of Fig. 34. Each image is extracted from a specific frequency, and demonstrates that the spatial resolution of the spectroscopic image is improved as the frequency increases. The spatial resolution of the THz spectroscopic images depends on the wavelength of the THz radiation. Near-field measurements are an effective way to improve the spatial resolution of THz imaging [46].

7 Summary

As little as two decades ago the word "terahertz" was, to many scientists and engineers, little more than a theoretical structure permitted by the rules of

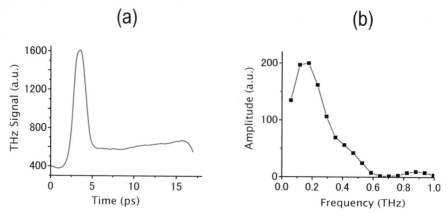

Fig. 33. (a) Measured electric field of a freely propagating terahertz pulse. (b) Amplitude spectrum of (a)

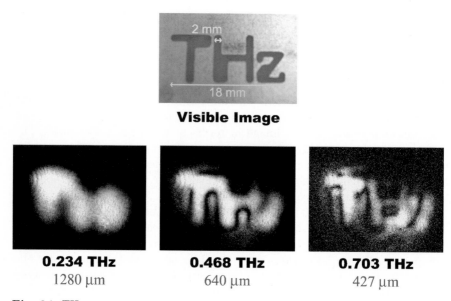

Fig. 34. THz spectroscopic images of a test pattern with the letters 'THz'

adding SI prefixes to units, and the "terahertz gap" was the pole of inaccessibility at the far end of infrared radiation. The fs laser and substantial advances in the field of electro-optics have closed the terahertz gap and permitted access by mainstream scientists to the terahertz world. One decade ago, THz imaging was developed and has turned into a road connecting the terahertz world to the real world of scientific and commercial applications. This Chapter has shown a map of this road as well as a detailed view of some of the scenic spots. Other places had to be left aside disregarding their relevance. Even more places are being opened up while the road is being extended into new areas. The authors wish to mention particularly the developments in the fields of medical diagnosis and biological analysis with THz radiation as well as near-field THz imaging. With commercial THz imaging setups now available, the progress into the next decade should continue faster than ever.

References

[1] B. B. Hu, M. C. Nuss: Imaging with terahertz waves, Opt. Lett. **20**, 1716 (1995)

[2] D. M. Mitteleman, R. H. Jacobsen, M. C. Nuss: T-ray imaging, IEEE J. Sel. Top. Quantum Electron. **2**, 679 (1996)

[3] M. Herrmann, M. Tani, M. Watanabe, K. Sakai: Terahertz imaging of objects in powders, IEEE Proc.-Optoelectron. **149**, 116 (2002)

[4] Q. Chen, X. C. Zhang: Polarization modulation in optoelectronic generation and detection of terahertz beams, Appl. Phys. Lett. **74**, 3435 (1999)

[5] K. Kawase, Y. Ogawa, Y. Watanabe: Non-destructive terahertz imaging of illicit drugsusing spectral fingerprints, Opt. Express **11**, 2549 (2003)

[6] D. Abbott, B. Ferguson, D. Gray, S. Wang, X. C. Zhang: T-ray Computed Tomography, Opt. Lett. **27**, 1312 (2002)

[7] D. M. Mittleman, M. Gupta, R. Neelamani, R. G. Baraniuk, J. V. Rudd, M. Koch: Recent advances in terahertz imaging, Appl. Phys. B **68**, 1085 (1999)

[8] D. A. Crawley, C. Longbottom, B. E. Cole, C. M. Ciesla, D. Arnone, V. P. Wallace, M. Pepper: Terahertz pulse imaging: A pilot study of potential applications in dentistry, Caries Res. **37**, 352 (2003)

[9] T. Löffler, T. Bauer, K. Siebert, H. G. Roskos: Terahertz dark-field imaging of biomedical tissue, Opt. Express **9**, 616 (2001)

[10] P. Knoblauch, C. Schildknecht, T. Kleine-Ostmann, M. Koch, S. Hoffman, M. Hofmann, E. Rehberg, M. Sperling, K. Donhuijsen, G. Hein, K. Pierz: Medical THz imaging: an investigation of histo-pathological samples, Phys. Med. Biol. **47**, 3875 (2003)

[11] R. M. Woodward, V. P. Wallace, R. J. Pye, B. E. Cole, D. D. Arnone, E. Linfield, M. Pepper: Terahertz pulse imaging of ex vivo basal cell carcinoma, J. Invest. Derm. **73**, 120 (2003)

[12] Q. Wu, T. D. Hewitt, X. C. Zhang: Two-dimensional electro-optic imaging of THz beams, Appl. Phys. Lett. **69**, 1026 (1996)

[13] F. Miyamaru, T. Yonera, M. Tani, M. Hangyo: Terahertz two-dimensional electrooptic sampling using high-speed complementary metal-oxide semiconductor camera, Jpn. J. Appl. Phys. **43**, L489 (2004)

[14] M. Herrmann, M. Tani, K. Sakai, M. Watanabe: Towards multi-channel time-domain terahertz imaging with photoconductive antennas, in *Microwave Photonics* (International Topical Meeting, Technical Digest P4-9, Awaji, Japan 2002) pp. 317–320

[15] M. Herrmann, M. Tani, M. Watanabe, K. Sakai: An electric read-out circuit for time-domain terahertz imaging with photoconductive antennas, Phys. Med. Biol. **47**, 3711 (2002)

[16] D. Arnone, C. Ciesla, M. Pepper: Terahertz imaging comes into view, Phys. World **April 35** (2000)

[17] M. C. Nuss, J. Orenstein: Terahertz time-domain spectroscopy, in G. Grüner (Ed.): *Millimeter and Submillimeter Wave Spectroscopy of Solids*, Top. Appl. Phys. **74** (Springer 1998)

[18] M. van Exter, D. Grischkowsky: Optical and electronic properties of doped silicon from 0.1 to 2 THz, Appl. Phys. Lett. **56**, 1694 (1990)

[19] N. Katzenellenbogen, D. Grischkowsky: Electrical characterization to 4 THz of n- and p-type GaAs using THz time-domain spectroscopy, Appl. Phys. Lett. **61**, 840 (1992)

[20] T. I. Jeon, D. Grischkowsky: Characterization of optically dense, doped semiconductors by reflection THz time domain spectroscopy, Appl. Phys. Lett. **72**, 3032 (1998)

[21] M. Herrmann, M. Tani, K. Sakai, R. Fukasawa: Terahertz imaging of silicon wafers, J. Appl. Phys. **91**, 1247 (2002)

[22] C. M. Ciesla, D. D. Arnone, A. Corchia, D. Crawley, C. Longbottom, E. H. Linfield, M. Pepper: Biomedical applications of terahertz pulse imaging, in *Commercial and Biomedical Applications of Ultrafast Lasers II*, vol. 3934 (Proc. SPIE 2000) p. 73

[23] R. H. Jacobsen, D. M. Mittleman, M. C. Nuss: Chemical recognition of gases and mixtures with terahertz waves, Opt. Lett. **24**, 2011 (1996)

[24] M. Usami, T. Iwamoto, R. Fukasawa, M. Tani, M. Watanabe, K. Sakai: Development of a THz spectroscopic imaging system, Phys. Med. Biol. **47**, 3749 (2002)

[25] M. Herrmann, M. Tani, K. Sakai: Display modes in time-resolved terahertz imaging, Jpn. J. Appl. Phys. **39**, 6254 (2000)

[26] J. T. Darrow, X. C. Zhang, D. H. Auston, J. D. Morse: Saturation properties of large-aperture photoconducting antennas, IEEE J. Quantum Electron. **28**, 1607 (1992)

[27] J. B. Knight, C. G. Fandrich, C. N. Lau, H. M. Jaeger, S. R. Nagel: Density relaxation in a vibrated granular material, Phys. Rev. B **51**, 3957–3963 (1995)

[28] S. Kawato, T. Hattori, T. Takemori, H. Nakatsukaand: Short-range interference effect in the diffusion of light in random media, Phys. Rev. B **58**, 6180–6193 (1998)

[29] J. Pearce, Z. Jian, D. M. Mittleman: Propagation of terahertz pulses in random media, Phil. Trans. R. Soc. Lond. A **362**, 301–314 (2004)

[30] J. Pearce, D. M. Mittleman: Scale model experimentation: Using terahertz pulses to study light scattering, Phys. Med. Biol. **47**, 3823–3830 (2002)

[31] *The First International Conference on Biomedical Imaging & Sensing Applications of THz Technology* (Leeds 2001)

[32] M. van Exter, D. Grischkowsky: Carrier dynamics of electrons and holes in moderately doped silicon, Phys. Rev. B **41**, 12140 (1989)

[33] D. Grischowsky, S. Keiding, M. van Exter, C. Fattinger: Far-infrared time-domain spectroscopy with terahertz beams of dielectrics and semiconductors, J. Opt. Soc. Am. B **7**, 2006 (1990)

[34] R. Fukasawa, K. Sakai, S. Perkowitz: Far-infrared reflectance study of coupled longitudinal-optical phonon-hole plasmon modes and transprt properties in heavily doped p-type GaAs, Jpn. J. Appl. Phys. **36**, 5543 (1997)

[35] O. Morikawa, M. Tonouchi, M. Hangyo: Sub-THz spectroscopic system using a multimode laser diode and photoconductive antenna, Appl. Phys. Lett. **75**, 3772 (1999)

[36] O. Morikawa, M. Tonouchi, M. Hangyo: A cross-correlation spectroscopy in sub-terahertz region using a incoherent light source, Appl. Phys. Lett. **76**, 1519 (2000)

[37] L. Duvillaret, F. Garet, J. L. Coutaz: Influence of noise on the characterization of meterials by terahertz time-domain spectroscopy, J. Opt. Soc. Am. B **17**, 452 (2000)

[38] D. M. Mittleman, J. Cunningham, M. C. Nuss, M. Geva: Noncontact semiconductor wafer characterization with the terahertz hall effect, Appl. Phys. Lett. **71**, 16 (1997)

[39] K. Seeger: *Semiconductor Physics*, 7th ed. (Springer, Berlin, Heidelberg) Sect. 4.2 and 4.14

[40] M. Auslender, S. Hava: IR optical constants of doped n-silicon, in E. D. Palik (Ed.): *Handbook of Optical Constants and Solids III* (Academic 1997) Chap. 6, pp. 155–186

[41] L. D. Landau, E. M. Lifshitz: Quantum mechanics – non-relativistic theory, in *Course of Theoretical Physics*, vol. 3, 2nd ed. (Pergamon, Oxford 1965) Chap. 25

[42] M. Born, E. Wolf: *Principles of Optics*, 7th ed. (Pergamon, Oxford 1999) Sect. 1.6.4 and 7.6.1

[43] C. Jacoboni, C. Canali, G. Ottaviani, A. A. Quaranta: A review of some charge transport properties of silicon, Solid State Electron. **20**, 77 (1977)

[44] S. M. Sze: *Physics of Semiconductor Devices*, 2nd ed. (Wiley, New York 1981) Sect. 1.5.1, p. 29

[45] B. Van Zeghbroeck: *Principles of Semiconductor Devices*
URL `ece-www.colorado.edu/~bart/book/transpor.htm#silicon`

[46] S. Hunsche, M. Koch, I. Brener, M. C. Nuss: THz near-field imaging, Opt. Commun. **150**, 22 (1998)

[47] Q. Chen, Z. Jiang, G. X. Xu, X. C. Zhang: Near field THz imaging with dynamic aperture, Opt. Lett. **25**, 1122 (2000)

[48] M. Tonouchi, M. Tani, Z. Wang, K. Sakai, M. Hangyo, N. Wad, Y. Murakami: Enhanced THz radiation from YBCO thin film bow-tie antennas with hyperhemispherical MgO lens, IEEE Trans. Appl. Supercond. **7**, 2913–2916 (1997)

[49] M. Hangyo, S. Tomozawa, Y. Murakami, M. Tonouchi, M. Tani, K. Sakai: THz emission from high-T_c superconductors by optical pulse excitation, IEEE Trans. Appl. Supercond. **7**, 3730–3733 (1997)

[50] M. Tani, M. Tonouchi, Z. Wang, K.Sakai, M. Hangyo, S. Tomozawa, Y. Murakami: Enhanced sub-picosecond electromagnetic radiation from $YBa_2Cu_3O_{7-\delta}$ thin-film bow-tie antennas excited with femtoseconds laser pulses, Jpn. J. Appl. Phys. **35**, L1184–L1187 (1996) part 2

[51] M. Tonouchi, M. Tani, Z. Wang, K. Sakai, S. Tomozawa, M. Hangyo, Y. Murakami, S. Nakashima: Ultrashort electromagnetic pulse radiation from YBCO thin films excited by femtosecond optical pulse, Jpn. J. Appl. Phys. **35**, 2624–2632 (1996) part 1

[52] M. Tani, M. Tonouchi, M. Hangyo, Z. Wang, N. Onodera, K. Sakai: Emission properties of $YBa_2Cu_3O_{7-\delta}$-film photoswitches as terahertz radiation sources, Jpn. J. Appl. Phys. **36**, 1984–1989 (1997) part 1

[53] M. Tonouchi, M. Tani, Z. Wang, K. Sakai, N. Wada, M. Hangyo: Terahertz emission study of femtosecond time-transient nonequilibrium state in optically excited $YBa_2Cu_3O_{7-\delta}$ thin films, Jpn. J. Appl. Phys. **35**, L1578–L1581 (1996) part 2

[54] M. Hangyo, S. Tomozawa, Y. Murakami, M. Tonouchi, M. Tani, Z. Wang, K. Sakai, S. Nakashima: Terahertz radiation from superconducting $YBa_2Cu_3O_{7-\delta}$ thin films excited by femtosecond optical pulses, Appl. Phys. Lett. **69**, 2122–2124 (1996)

[55] T. Fukui, A. Moto, H. Murakami, M. Tonouchi: Distribution of optically-generated vortices in YBCO thin film strips, Physica C **357–360**, 454–456 (2001)

[56] M. Tonouchi, M. Yamashita, M. Hangyo: Vortex penetration in YBCO thin film strips observed by THz radiation imaging, Physica B **284–288**, 853–854 (2000)

[57] T. Kondo, S. Shikii, M. Yamashita, T. Kiwa, O. Morikawa, M. Tonouchi, M. Hangyo, M. Tani, K. Sakai: A novel two-dimentsional mapping system for supercurrent distribution using femtosecond laser puses, in Koshizuka, S. Tanjima (Eds.): *Advances in Superconductivity XI* (Springer, Tokyo 1999) pp. 1285–1288

[58] M. Tonouchi, A. Moto: Vortex penetration effects on supercurrent distributions in YBCO thin-film strips, Physica C **367**, 33–36 (2002)

[59] M. Tonouchi, A. Moto, M. Yamashita, M. Hangyo: Terahertz radiation imaging of votices penetrated into YBCO thin films, IEEE Trans. Appl. Supercond. **11**, 3230–3233 (2001)

[60] Y. Yamashita, M. Tonouhi, M. Hangyo: Supercurrent distriuion in YBCO strip lines under bias current and magnetic fields observed by THz radiation imaging, Physica C **355**, 217–224 (2001)

[61] S. Shikii, T. Kondo, M. Yamashita, M. Tonouchi, M. Hangyo, M. Tani, K. Sakai: Observation of supercurrent distribution in $YBa_2Cu_3O_{7-\delta}$ thin films using THz radiation excited with femtosecond laser pulses, Appl. Phys. Lett. **74**, 7317–1319 (1999)

[62] M. Tonouchi, M. Yamashita, M. Hangyo: Terahertz radiation imaging of supercurrent distribution in vortex-penetrated $YBa_2Cu_3O_{7-\delta}$ thin film strips, J. Appl. Phys. **87**, 7366–7375 (2000)

[63] A. Moto, H. Murakami, M. Tonouchi: Temperature dependence of supercurrent distribution in YBCO thin film strips observed by terahertz radiation imaging, Physica C **357–360**, 1603–1606 (2001)

[64] M. Yamashita, M. Tonouchi, M. Hangyo: Visualization of supercurrent distribution by THz radiation mapping, Physica B **284–288**, 2067–2068 (2000)

[65] M. Tinkham: *Introduction to Superconductivity*, 2nd ed. (McGraw-Hill, Singapore 1996)

[66] A. Yariv: *Quantum Electronics*, 3rd ed. (Wiley, Singapore 1988)

[67] W. Xing, B. Heinrich, H. Zhou, A. A. Fife, A. R. Cragg: Magnetic flux mapping, magnetization, and current distributions of $YBa_2Cu_3O_7$ thin films by scanning Hall probe measurements, J. Appl. Phys. **76**, 4244–4255 (1994)

[68] J. R. Kirtley, M. B. Kethen, K. G. Stawiasz, J. Z. Sun, W. J. Gallagher, S. H. Blanton, S. J. Wind: High-resolution scanning SQUID microscope, Appl. Phys. Lett. **66**, 1138–1140 (1995)

[69] T. Schuster, M. V. Indenbom, M. R. Koblischka, H. Kuhn, H. Kronmüller: Observation of current discontinuity lines in type-ii superconductors, Phys. Rev. B **49**, 3443–3452 (1994)

[70] T. H. Johansen, M. Baziljevich, H. Bratsberg, Y. Galperin, P. E. Lindelof, Y. Shen, P. Vase: Direct observation of the current distribution in thin superconducting strips using magneto-optic imaging, Phys. Rev. B **54**, 16264–16269 (1996)

[71] Z. W. Lin, J. W. Cochrane, N. E. Lumpkin, G. J. Russell: Magneto-optical observations of magnetic flux distribution in a high-temperature superconductor x-array, Physica C **312**, 247–254 (1999)

[72] C. Jooss, R. Warthmann, A. Forkl, H. Kronmüller: High-resolution magneto-optical imaging of critical currents in $YBa_2Cu_3O_{7-\delta}$ thin films, Physica C **299**, 215–230 (1998)

[73] E. H. Brandt, M. Indenbom: Type-II-superconductor strip with current in a perpendicular magnetic field, Phys. Rev. B **48**, 12893–12906 (1993)

[74] E. Zeldov, J. R. Clem, M. McElfresh, M. Darwin: Magnetization and transport currents in thin superconducting films, Phys. Rev. B **49**, 9802–9822 (1994)

[75] O. Morikawa, M. Yamashita, H. Saijo, M. Morimoto, M. Tonouchi, M. Hangyo: Vector imaging of supercurrent flow in $YBa_2Cu_3O_{7-\delta}$ thin films using terahertz radiation, Appl. Phys. Lett. **75**, 3387–3389 (1999)

[76] O. Morikawa, M. Yamashita, M. Tonouchi, M. Hangyo: Vector map imaging of supercurrent distribution in high-T_c superconductive thin films, Physica B **284–288**, 2069–2070 (2000)

Index

absorption coefficient, 206, 207
acceleration, 100, 117, 119, 122
acetone, 247
amino acid, 234
amplifier, 335
anisotropic, 217
anticrossing, 140, 226
astronomical observation, 194
attenuated total reflection (ATR), 242
autocorrelation-type interferometer, 21, 129

bacteriorhodopsin, 229, 241, 242
bandwidth, 165, 180, 187
barium borate (BBO), 37
biotin, 231
birefringence, 31
bismuth titanate $Bi_4Ti_3O_{12}$ (BIT), 213
Bloch oscillation, 125
Bloch oscillator, 126, 132
bolometer, 102
Boltzmann transport, 134
bovine serum albumin, 237
bow-tie PC antenna, 23
Brewster angle, 248
broadband self-complementary antenna, 164
built-in potential, 147

cadmium telluride (CdTe), 37, 81, 84
calf thymus DNA, 237
carrier accumulation, 151
carrier concentration, 355, 362
carrier density, 206, 207, 209, 211
carrier lifetime, 41, 59, 170
cavity-mode, 140
cavity-polariton, 139
CCD camera, 336

Cherenkov radiation, 4
coherent LO phonon, 15, 77–78
coherent-control experiments, 39
collagen, 236, 237
colossal magnetoresistance (CMR), 271–274, 291, 292, 303
complex (electrical) conductivity, 205, 207–209, 211
complex dielectric constant, 205, 206, 208, 212, 215–217
complex refractive index, 205, 206, 212, 213, 248, 249, 357
conductivity, 134, 271, 298, 303, 304, 307–309, 316–318
continuous-wave (CW), 157
copper oxide, 271
critical-state model, 366
cross-correlation, 261
cryostat, 210
cytochrome c, 238–241

D-galactose, 234
D-glucose, 233, 234
DC conductivity, 208, 259
deceleration, 117, 119
delay stage, 334
δ-function, 41
dephasing, 144
depletion field, 101
depletion layer, 13
dielectric constant, 205, 206, 208, 212, 213, 215–218, 221, 223
dielectric parameter, 221
difference frequency mixing (DFM), 64, 67
difference-frequency generation, 36
dispersion curve, 140
dispersion relation, 215, 218, 223, 225

Topics in Applied Physics